Surveys in Differential Geometry

Vol. 1: Lectures given in 1990
edited by S.-T. Yau and H. Blaine Lawson

Vol. 2: Lectures given in 1993
edited by C.C. Hsiung and S.-T. Yau

Vol. 3: Lectures given in 1996
edited by C.C. Hsiung and S.-T. Yau

Vol. 4: Integrable systems
edited by Chuu Lian Terng and Karen Uhlenbeck

Vol. 5: Differential geometry inspired by string theory
edited by S.-T. Yau

Vol. 6: Essays on Einstein manifolds
edited by Claude LeBrun and McKenzie Wang

Vol. 7: Papers dedicated to Atiyah, Bott, Hirzebruch, and Singer
edited by S.-T. Yau

Vol. 8: Papers in honor of Calabi, Lawson, Siu, and Uhlenbeck
edited by S.-T. Yau

Vol. 9: Eigenvalues of Laplacians and other geometric operators
edited by A. Grigor'yan and S-T. Yau

Vol. 10: Essays in geometry in memory of S.-S. Chern
edited by S.-T. Yau

Vol. 11: Metric and comparison geometry
edited by Jeffrey Cheeger and Karsten Grove

Vol. 12: Geometric flows
edited by Huai-Dong Cao and S.-T. Yau

Vol. 13: Geometry, analysis, and algebraic geometry
edited by Huai-Dong Cao and S.-T.Yau

Vol. 14: Geometry of Riemann surfaces and their moduli spaces
edited by Lizhen Ji, Scott A. Wolpert, and S.-T. Yau

Vol. 15: Perspectives in mathematics and physics: Essays dedicated to Isadore Singer's 85th birthday
edited by Tomasz Mrowka and S.-T. Yau

Volume I

Surveys in Differential Geometry

Proceedings of the conference on geometry and topology
held at Harvard University, April 27–29, 1990

edited by

H. Blaine Lawson, Jr.
S.-T. Yau

International Press
www.intlpress.com

Surveys in Differential Geometry, Vol. 1
Proceedings of the conference on geometry and topology held at Harvard University, April 27-29, 1990

Editors:
H. Blaine Lawson, Jr., State University of New York, Stony Brook
S.-T. Yau, Harvard University

Published by International Press, Somerville, Massachusetts, U.S.A.

ISBN 978-1-57146-231-2

Paperback re-issue 2012. Originally published in 1991 under ISBN 0-8218-0168-6 (paperback). The preface and papers herein are reproduced in facsimile from the original publication.

Printed in the United States of America.

Contents

PREFACE

As a service to the mathematics community, the editors of the Journal of Differential Geometry decided to organize a conference, to be held every three years, at which invited speakers would survey the field of differential geometry and related subjects. In choosing the speakers, we look at geometry in a very broad way and hope geometers will go beyond the classical perception of other fields. (Among the speakers at our first conference were people working in algebraic geometry, mathematical physics and other fields.) We decided that the speakers should be prominent specialists in their respective areas who are able to give a broad overview of recent trends and make predictions and suggestions for future research. We believe that these conferences will be useful gatherings for experts and nonexperts alike, and in particular for those who are isolated from the mainstream activity in geometry.

Our first conference, held at Harvard University from April 27–29, 1990, was made possible by the generous funding of the Journal of Differential Geometry. We are also grateful to Harvard University and the faculty of the Mathematics Department who were strongly supportive of this endeavor. Dean Michael Spence delivered the opening remarks commending our efforts to bring together geometry and other fields.

Central to the success of this conference were the speakers who prepared not only two hours' worth of lectures, but also the manuscripts published here. Their lectures were enthusiastically received by several hundred participants. We would like to express our gratitude to them for their efforts.

It is our hope that these conferences will become a tradition. We look forward to the next conference, which will take place at Harvard University in 1993.

S. T. Yau
H. Blaine Lawson, Jr.

SURVEYS IN DIFFERENTIAL GEOMETRY
1 (1991) 1–18

STABLE BUNDLES REVISITED

RAOUL BOTT

The topological classification of complex vector bundles over a Riemann surface is of course very simple: they are classified by one integer $c_1(E)$, corresponding to the first Chern class of E. On the other hand, a classification in the complex-analytic—or algebro-geometric—category leads to "continuous moduli" and subtle phenomena which have links with number theory, gauge theory, and conformal field theory. I will try to report briefly on some of these developments here.

The simplest instance of our problem occurs when $\dim E = 1$, that is, when we are dealing with complex analytic line bundles over M—a compact Riemann surface of the genus g. Because line bundles form a group under the tensor product $L, L' \rightsquigarrow L \otimes L'$, the set of isomorphism classes of line bundles $J(X)$ will in this case inherit an abelian group structure, and the first fundamental theorem of the subject, going back to Riemann, Abel, and Jacoby, asserts that $J(M)$ can be given the structure of a complex analytic abelian group. More precisely the first Chern class gives rise to a homomorphism $J(M) \to \mathbb{Z} \to 0$ whose kernel $J_0(M)$ is a complex analytic, and indeed *algebraic*, torus of $\dim_{\mathbb{C}} = g$:

$$0 \to J_0(M) \to J(M) \to \mathbb{Z} \to 0.$$

A proof of this exact sequence on the complex analytic level is actually quite easy, granted the basics of sheaf theory and Hodge theory. Indeed, the exact sequence

$$0 \to \mathbb{Z} \xrightarrow{e^{2\pi i}} \mathbb{C} \to \mathbb{C}^* \to 0$$

induces an analogous sequence on the corresponding sheafs of germs of functions, and hence leads to an exact sequence in cohomology:

$$H^0(M; \underline{\mathbb{C}}^*) \to H^1(M; \underline{\mathbb{Z}}) \to H^1(M; \underline{\mathbb{C}}) \to H^1(M; \underline{\mathbb{C}}^*) \xrightarrow{\delta} H^2(M; \underline{\mathbb{Z}}).$$

But $H^1(X; \underline{\mathbb{C}}^*)$ is seen to be precisely $J(M)$, by passing from a line bundle to its transition functions, while by the Dolbaux resolution, $H^1(M; \underline{\mathbb{C}}) \simeq H^{0,1}(M)$, that is, the vector space of type $(0, 1)$ in $H^1(M; \mathbb{R})$.

Received May 29, 1990.

Finally, the δ in this sequence corresponds to c_1, so that the exactness of the cohomology sequence together with the obvious computation that $H^0(M;\mathbb{C}) = H^0(M;\mathbb{C}^*) = \mathbb{C}$, leads to the description of $J_0(M)$ as the quotient

$$J_0(M) = H^{0,1}(M)/H^1(M;\mathbb{Z})$$

of a g-dimensional complex vector space by a lattice, i.e., a torus. The *algebraic* structure on $J_0(M)$ is much less trivial.

In any case, these classical results provide us with as beautiful an answer as we could hope for: this Jacobian $J(M)$ varies functorially with M, and its own function theory is intimately related to that of M. In particular, there is a natural map

$$M \xrightarrow{\iota} J_1(M)$$

which maps a point $p \in M$ to the isomorphism class of the line bundle L_p on M determined by p as a divisor on M. (In terms of a local coordinate z_p centered at p in M, L_p can also be described by the data: let $U = \{U_0, U_1\}$ be the cover of M consisting of a small disc, U_0, about p, and let $U_1 = M - p$. Let $g_{U_0,U_1} = \frac{1}{z_p}$, so that the assignment $f_0 = z_p$, $f_1 \equiv 1$ defines a holomorphic section of L_p with precisely one zero at p. Thus $c_1(L_p) = 1$, and so $p \to L_p$ takes values in the component $J_1(M)$ of bundles with $c_1 = 1$.) Iterating this map $(g-1)$ times leads to the diagram:

$$\begin{array}{ccccc}
\underbrace{M \times \cdots \times M}_{(g-1)} & \to & \underbrace{J_1 \times \cdots \times J_1}_{(g-1)} & \xrightarrow{m} & J_{g-1}(M) \\
\downarrow & & & \nearrow_{\iota^{g-1}} & \\
M^{(g-1)} & & & &
\end{array}$$

and so to a canonical map of the $M^{(g-1)}$, the $(g-1)$st *symmetric power* of M into $J_{g-1}(M)$. The image of this arrow is now of codimension 1 in $J_{g-1}(M)$ and so defines a divisor (the θ-divisor), and hence a line bundle $\mathscr{L}_\theta$, over $J_{g-1}(M)$.

In the terminology of our century the classical θ-functions of Jacoby of level k now appear as the space of holomorphic sections of the bundle $\mathscr{L}_\theta^k$ over J_{g-1}:

$$\theta\text{-functions of level } k \cong H^0(J_{g-1}(M);\mathscr{L}_\theta^k).$$

From the classical formulas one can then also compute the number of such θ functions to be k^g:

$$\dim H^0(J_{g-1}(M);\mathscr{L}_\theta^k) = k^g.$$

To summarize, for $n = 1$, the classical theory teaches us that the moduli space of line bundles is

A) a complex variety each of whose components is a torus of $\dim 2g$ over $\mathbb{R}$ so that its Poincaré series is given by

$$P_t\{J_k(M)\} = (1+t)^{2g},$$

and

B) the component $J_{g-1}(M)$ carries a natural line bundle $\mathscr{L}_\theta$, with

$$\dim H^0\{J_{g-1}(M); \mathscr{L}_\theta^k\} = k^g.$$

The primary difficulty of extending these results to higher-dimensional vector bundles is that the problem immediately becomes infinite dimensional. This comes about because bundles E of $\dim > 1$ will in general have many nontrivial automorphisms, and to properly deal with this situation the Mumford notion of stability is indispensable. The condition is as follows:

(1) E is stable if and only if for any subbundle $F \subset E$,

$$\frac{c_1(F)}{\dim F} < \frac{c_1(E)}{\dim E}.$$

Let me illustrate the power of this condition, say for a bundle E of $\dim 2$. First, remark that E always admits a line subbundle $L \subset E$ and so fits into an exact sequence

$$0 \to L \to E \to Q \to 0$$

with L and Q line bundles.

Indeed, any rational section of E will determine such an L, and of course the Chern classes of the three constituents are related by

$$c_1(E) = c_1(L) + c_1(Q).$$

Now, if E is stable, we will have $c_1(L) < c_1(E)/2$ whence $c_1(Q) > c_1(E)/2$, and it follows that the sequence cannot split if E is stable!

Similarly a stable E can have no nontrivial automorphism. Indeed, consider an automorphism

$$\varphi \colon E \to E.$$

The characteristic polynomial of φ is clearly constant because M is compact. If the eigenvalues of φ differ, φ will split E into two line bundles, which is ruled out by stability. Hence the eigenvalues are equal to λ, say, so that $\psi = \varphi - \lambda 1$ will have to be a nilpotent endomorphism: $\psi^2 = 0$. Now if ψ is not identically 0, and has constant rank, then we have an exact sequence

$$0 \to L \to E \overset{\psi}{L} \to 0$$

where $L = \ker \psi$. But by stability $c_1(L) < c_1(E)/2$, which rules out this alternative. Hence $\psi \equiv \lambda \cdot 1$. q.e.d.

The nonconstant rank case is eliminated similarly.

In any case, armed with this concept, the structure of vector bundles now becomes tractable. First of all, Mumford shows that the isomorphism classes of stable bundles do admit a natural structure as a smooth algebraic variety $J^n(M)$, which varies functorially with M. The tangent space to $J^n(M)$ at E is given by

$$T_E(J^n(M) \simeq H^1(M; \operatorname{End} E).$$

By Riemann-Roch and stability ($\dim H^0 = 1$!) we see that $J^n(M)$ has dimension $n^2(g-1)+1$. Just as for $J = J^1$, these spaces fall into components $J_k^n(M)$, $k = c_1(E)$ according to the topological type of E, but these will not be isomorphic in general. In fact,

$$J_k^n \text{ is compact } \Leftrightarrow n \text{ and } k \text{ are relatively prime.}$$

In short, for $n > 1$, the "true" analogues of the Jacobian $J(M)$ are the $J_r^n(M)$, with $(n, r) = 1$. When $(n, r) \neq 1$, one usually compactifies $J_r^n(M)$ by semistable bundles (replace $<$ by $\leq$) modulo a certain equivalence relation, to get a complete, but in general *singular*, variety, and in the sequel $J_r^n(M)$ will always denote this compactified version.

Actually this stability notion also helps to explain the structure of *all* bundles over M. For instance, suppose E has dimension two but is *not* semistable. Then one finds that E contains a *unique* line bundle $L_E \subset E$ of maximal c_1, so that we can attach to E a unique exact sequence

$$0 \to L_1 \to E \to L_2 \to 0.$$

If we set $\mu_i = c_1(L_i)$, then the bundles of this type are therefore naturally parametrized by the set

$$\mathscr{C}_\mu \colon J_{\mu_1}(M) \times J_{\mu_2}(M) \times H^1(M; L_2^* \otimes L_1), \qquad u_1 + u_2 = c_1(E); \mu_1 \geq \mu_2,$$

where the last factor measures the extension determined by E. But note that this is no longer a true parametrization of *isomorphism classes*. For, as is easily seen, two bundles whose extensions are multiples of each other, $\eta_1 = \lambda \eta_2$, $\lambda \neq 0$ in $H^1(M; L_2^* \otimes L_1)$, will lead to isomorphic bundles.

Still this seems to be the best one can do while staying in a *Hausdorff* framework, and it is therefore plausible to consider the disjoint union

$$\mathscr{M}_r^{(2)} = J_r^2(M) \coprod_\mu \mathscr{C}_\mu, \qquad \mu = (u_1, u_2) \text{ with } \mu_1 + \mu_2 = r,\ \mu_1 \geq \mu_2,$$

as some sort of covering of the set of all complex 2-dimensional bundles with $c_1 = r$ over M. In the algebraic category this data leads to what I believe is called the "stack of bundles" over M.

In any case, this stratification plays an essential role in the approaches that Harder [5], Harder & Narasimhan [6], and Atiyah & Bott [1] took to the problem of computing $H^*(J_r^n)$. The Harder-Narasimhan method is number-theoretical and has as its denouement the beautiful counting formula which is a function field analogue (A. Weil, Tamagawa) of a corresponding counting formula going back to Minkowski and C. L. Siegel in the number field case:

$$\sum_E \frac{1}{|\operatorname{Aut} E|} = \frac{1}{q-1} q^{(n^2-1)(g-1)} \zeta_M(2) \dots \zeta_M(n).$$

Here on the left one is counting the isomorphism classes of vector bundles (of fixed determinant!) over a curve of genus g, defined over a finite field F_q, each vector bundle contributing the reciprocal of the number of automorphisms it has. On the right the answer is given in terms of the ζ-function of M:

$$\zeta_{M(s)} = \frac{\prod_1^{2q}(1-\omega_i q^{-s})}{(1-q^{-s})(1-q\cdot q^{-s})},$$

the $\omega_i(M)$ being intrinsically defined algebraic integers with $|\omega_i| = q^{1/2}$, and in terms of these $\{\omega_i\}$ the number of rational points on the Jacobian $J(M)$ of M is given by:

$$|J(M)| = \prod_1^{2g}(1-\omega_i).$$

Using these formulas it is an easy but beautiful computation to find the contribution of each stratum in $\mathscr{M}_r^{(2)}$ to the left-hand side of the Minkowski-Siegel-Weil formula. For $\mathscr{M}_1^{(2)}$ this then leads to the relation:

$$\frac{|J_1^2|}{q-1} + \frac{|J|^2}{(q-1)^2} \cdot \sum_{r=0}^{\infty} \frac{1}{q^{2r+2-g}} = \frac{1}{(q-1)} \cdot q^{3q-3} \zeta_M(2), \tag{2}$$

and this relation in turn enabled Harder to compute $|J_1^2|$ by subtraction, and hence, via the Weil conjectures, to obtain a hold on the Poincaré polynomial of J_1^2! Actually these computations were done *before* the solution of the Weyl conjectures by Deligne and Grothendieck, so that at that time they merely used this example to check these Weyl conjectures against the computations Newstead [10] had made for $P_t(J_1^2)$, using a topological method which will be explained below.

First, however, one parting comment on the "covering" $\mathcal{M}_r^{(2)}$. It is unfortunately not true that the vector spaces $H^1(X; L_2^* \otimes L_1)$ in C_u are constant in dimension: For large genus this vector space jumps as L_1 and L_2 vary over J_{u_1} and J_{u_2}. A proper algebro-geometric model of the "set of all bundles" should therefore, I think, be built out of the virtual bundles

$$V_u \xrightarrow{\pi_u} J_u \times J_u$$

with fiber $H^1(X; L_2^* \otimes L_1) - H^0(X; L_2^* \otimes L_1)$.

These bundles have a constant (even if negative!) dimension, and the disjoint union

$$\widetilde{\mathcal{M}}_1^{(2)} = J_1^{(2)} \coprod V_1 \coprod V_2 \coprod \cdots$$

now precisely corresponds to the left-hand side of (2), if we agree to count the number of points of a virtual vector space, V, by $q^{\dim V}$.

Indeed, with this understanding, the expression

$$|J_1^2| + \sum_{i=1}^{\infty} |V_i|$$

precisely goes over into the left-hand side of (2), of course multiplied by $|J|$, because we have not fixed the determinant in the present discussion.

Now in topology it often happens that we are given a decomposition of a space Y, by strata of the form:

$$Y = Y_0 \coprod W_1 \coprod W_2 \cdots \coprod W_k \cdots$$

with the W_i vector bundles over Y_i, so that Y is finally put together by successively attaching the boundary sphere-bundle of W_i to what has already been built. Indeed, any smooth function f on a manifold Y with reasonable critical behavior induces such a decomposition. Here Y_0 is the absolute minimum of f, and W_i is the "negative bundle" of f over the critical set Y_i of "index i". Thus these W_i are spanned by the direction of steepest descent for f along Y_i.

What is novel in the algebraic framework about our $\mathcal{M}_1^{(2)}$ is therefore that the bundles V_μ are virtual, and actually get more and more negative as $\mu \to \infty$. In topology such a state of affairs is usually remedied by "suspending" to convert the virtual bundles into honest ones. In the present situation that would indicate that one has to suspend more and more as the strata are added—in short, one seems to be building something "down from ∞", a phenomenon often present in the heuristics of modern physics.

It is time to turn to the Yang-Mills version of this story. I first learned about the $J_r^n(M)$ when I visited India in 1976 and was taught the fundamental results of Seshadri, Narasimhan, and Ramanan personally by Ramanan, and became intrigued by the problem of computing the cohomology of the $J_r^n(M)$. Later that year I went to Oxford where Atiyah and Hitchin were in the midst of Yang-Mills theory on the four sphere. It was therefore natural for Atiyah and me to test this new mathematical toy on the problem I had brought along from India.

In short we started to explore the Yang-Mills theory of unitary bundles over a Riemann surface of genus g.

At first, the problem looks quite different. One now deals with a principal $U(n)$ bundle P over M, and considers the space $\mathscr{A}$, of connections over P. Technically such a connection A is then an equivariant 1-form A on P with values in the Lie algebra u_n of U_n whose curvature

$$F(A) = dA + A^2$$

is seen to descend to M and appears that as a 2-form with values in u_n:

$$F(A) \in \Omega^2(M; u_n).$$

The *size* of the curvature is now measured with the aid of Riemann structure on M via the formula

$$\mathrm{YM}(A) = \int_M \mathrm{tr}\{F(A) \wedge *F(A)\}.$$

A crucial property of YM is that it is *invariant under the group of automorphism* $\mathscr{G}(P)$, of P, which cover the identity on M, and it is *this* group which eventually furnishes us with the proper geometric and topological analogue of the right-hand side in the Minkowski-Siegel-Weil formula.

In fact, Atiyah and I found that the "proper" description of the homotopy type of this "stack of all bundles" is precisely that of the topological classifying space $B\mathscr{G}$ of the group $\mathscr{G}(P)$.

But first a word about the relation of our two subjects. What do complex structures on E have to do with connections on P? The link is the following one. A principal U_n bundle P has associated to it a canonical C^∞ Hermitian vector bundle E over M. On the other hand, a Riemann structure on M induces a complex structure on M. Finally a holomorphic structure on E compatible with the one on M together with the Hermitian structure defines a unique "connection of type 1-1" on P. Conversely a connection of type $(1, 1)$ on P defines a complex structure on $E(P)$, and under this correspondence the group of gauge transformations $\mathcal{G}$ mediates between equivalent complex structures on $E(P)$. Grosso Modo then, the space of complex structures on E should be the quotient space $\mathcal{A}/\mathcal{G}$. On the other hand, $\mathcal{G}$ does not act freely on $\mathcal{A}$, and so one has to take this quotient in the "equivariant sense". But now the contractibility of $\mathcal{A}$ implies that homotopy theoretically one is dealing precisely with $B\mathcal{G}(P)$.

In this topological framework of our basic question, the critical points of Yang-Mills are now easily determined: Certainly the absolute minimum of YM on $\mathcal{A}(P_0)$, where P_0 is the trivial U_n-bundle, occurs when $F(A) = 0$. The isomorphism classes of these are therefore precisely the *flat* U_n *bundles* over M, and these are parametrized by their holonomy—that is, by the conjugacy class of the homomorphism of $\pi_1(M)$ to U_n they induce. Thus

$$\operatorname{Min} \mathrm{YM}|_{c_1=0} = \operatorname{Hom}(\pi_1(M); U_n)^{U_n}.$$

But the cornerstone of the whole analytic theory of stable bundles was precisely the result of Seshadri [13] that the semistable bundles of trivial topological type correspond to this same space; that is,

$$J_0^n = \operatorname{Hom}(\pi_1(M); U_n)^{U_n}.$$

Armed with this knowledge it is then easy to see that the higher critical points of Yang-Mills, say in the two-dimensional case, correspond to splittings of the structure group from $U(2)$ to $U(1) \times U(1)$. Thus the critical sets in the Morse stratification for the Yang-Mills function $\mathrm{YM}: \mathcal{A} \to \mathbb{R}$ fall into components

$$Y_1^{(2)} = J_1^2 \coprod_{\mu} J_{\mu_1}^1 \times J_{\mu_2}^1, \qquad \mu_1 \geq \mu_2,\ u_1 + u_2 = 1,$$

which are clearly in 1-1 correspondence with the stratification of $\mathcal{M}_r^{(2)}$ discussed earlier.

Next, the "negative-bundles" W_μ over these critical sets are now found to be $H^1(X; L_1^* \otimes L_2)$, with L_1 and L_2 as before. Note that $L_1^* \otimes L_2$ is just the dual bundle to the extension bundle we found earlier. It is therefore strictly negative, and hence does *not* jump when L_1 and L_2 vary in J_{μ_1} and J_{μ_2}. To compute its dimension we again use $R - R$ which now yields

$$-\dim W_\mu = c_1(L_1^* \otimes L_2) - (g-1),$$

or, equivalently,

$$\dim W_\mu = c_1(L_1 \otimes L_2^*) + (g-1).$$

This compares to

$$\dim V_\mu = -c_1(L_1 \otimes L_2^*) + (g-1),$$

so that

$$\dim W_\mu + \dim V_\mu = 2(g-1),$$

and explains the dual relation of the two bundles we have been discussing.

In *truly* favorable situations for the Morse theory, the critical sets with their negative bundles determine the cohomology of the total space in the sense that for any coefficient field k the Poincaré polynomials $P_t(Y_i; k) = P_t(Y_i)$ of the critical sets Y_i determine $P_t(Y)$ according to the simple law

$$P_t(Y_i) + \sum_{i \geq l} t^{\dim W_i} P_t(Y_i) = P_t(Y).$$

(In general, these two sides are of course only related by the Morse inequalities.)

In the equivariant theory, there would be a corresponding formula with $P_t(Y_i)$ replaced by $GP_t(Y_i)$, the Poincaré series of the equivariant cohomology of Y_i. Recall now that the equivariant cohomology for a group G "counts" a point by the ordinary cohomology of its classifying space BG, and in general "counts" an orbit G/H, by the ordinary cohomology of BH. In favorable situations the equivariant contribution of Y_i, that is, $GP_t(Y_i)$ is therefore simply equal to $P_t(Y_j) \times P_t(BH)$ where H is the stabilizer of Y_j.

Now although $\mathscr{G}(P)$ and $\mathscr{A}(P)$ are infinite dimensional, the stabilizers of the action of $\mathscr{G}$ on $\mathscr{A}$ are easily determined:

The stable bundles J_1^2 are stabilized only by U_1, the center of $\mathscr{G}(P)$, whereas the $J_{\mu_1} \times J_{\mu_2}$ are all centralized by $U_1 \times U_1$.

Thus continuing in our assumption of the "most favorable state of affairs", these considerations lead to the formula:

$$\frac{P_t(J_1^2)}{1-t^2} + \sum \frac{(1+t)^{4g}}{(1-t^2)(1-t^2)} t^{2(2r+g)}$$

as the equivariant count of the strata in $\mathscr{A}$. (Recall here that $P_t(BU_1) = 1/(1-t^2)$.) On the other hand, the total equivariant cohomology of $\mathscr{A}$ is counted by $P_t(B\mathscr{G})$, which in the case under consideration is seen to be:

$$P_t(B\mathscr{G}) = \frac{\{(1+t)(1+t^3)\}^{2g}}{(1-t^2)^2(1-t^4)}.$$

In short, *if YM is a perfect equivariant Morse function*, on $\mathscr{A}$, then we should have the equality:

$$\frac{P_t(J_1^2)}{1-t^2} + \sum \frac{(1+t)^{4g}}{(1-t^2)^2} t^{2(2r+g)} = \frac{((1+t)(1+t^3))^2}{(1-t^2)^2(1-t^4)}. \tag{3}$$

When Michael Atiyah and I reached this point in our tentative arguments, we were unaware of the Harder's computation, and it was only later that we realized how precise the analogy was, and how precisely the right-hand side of C. L. Siegel computes $P_t(B\mathscr{G})$ in the U_n case, if one lets $q \to t^{-2}$ and the $\omega_1 \to -1^{-1}$. (See [1, p. 596].)

In any case, after we checked that (3) agreed with the computations Newstead had made in this case for $g = 1, \cdots, 6$, we pushed on to prove that YM was indeed equivariantly perfect for the group U_n, so that one obtains an inductively defined procedure for counting $P_t(J_r^n)$. Over the rationals this result is then really quite equivalent to the Harder-Narasimhan procedure, but our method produces better results over the integers. In particular, the Yang-Mills approach shows that J_r^n, $(n, r) = 1$, has no torsion.

By the way, Newstead's computations are based on the Narasimhan-Seshadri theorem, which can be used to give the following explicit model for J_1^n:

Let $F_g = \{X_i, Y_i : i = 1 \dots g\}$ be the free group generated by X_i and Y_i, and let

$$\mathscr{E}_g = [X_i, Y_i] \dots [X_g, Y_g]$$

be the product of the indicated commutators, so that $\pi_1(M) \simeq F_g/[\mathscr{E}_g]$. With this understood, the Narasimhan-Seshadri result yields the identity:

$$J_r^n = \mathrm{Hom}(F_g; SU(n))^{SU(n)},$$
$$\varphi(\mathscr{E}_g) = \xi^r \qquad \xi = e^{2\pi i/n}.$$

Put differently, consider the map:

$$\underbrace{SU_n \times \cdots \times SU_n}_{2g} \xrightarrow{\mu} SU_n$$

sending $\{X_1, Y_1; \dots; X_g Y_g\}$ to $\Pi([X_i, Y_i])$.

Then

$$J_r^n = (\mu^{-1}\xi^r)/SU_n,$$

where SU_n acts on all factors by conjugation.

This gives a no-nonsense C^∞ description of J_r^n which, however, is difficult to translate into direct cohomological information, except for

$n = 2$, where $SU_2 = S^3$, and μ is seen to have only two critical values. This fact, used with considerable ingenuity, then produces Newstead's original tables for $g = 1, \cdots, 6$.

It would be interesting to derive our basic induction formula directly from this picture, but in view of the infinite summation that underlies the other approaches I expect this to be unlikely.

Finally, let me remark that a truly Morse-theoretic approach to this problem is really only possible now, due to the work of K. Uhlenbeck and her student, Georgios D. Daskalopoulos. We based our consideration essentially on the stratification $\mathcal{M}_r^{(n)}$ which in the general case was studied in detail by Shatz [14].

Let me now turn to the question of extending the classical θ-functions to the nonabelian case. Interestingly enough the recent impetus for this development originates in consideration of modern physics, rather than number theory or algebraic geometry proper, and I think it is fair to say that a definitive algebro-geometric treatment of this whole subject is still lacking. On the more historical side, there can be no question that the basic reference to all we have been, and will be, talking about is Andre Weil [17].

The first question to be settled is of course the definition of the θ-divisor in the general case. The one-dimensional construction J outlined earlier has no known generalization, but there is one alternate definition of this classical θ which does extend. Indeed, consider a stable bundle E of Chern class k. Then by Riemann-Roch:

$$\dim H^0(M; E) - \dim H^1(M; E) = k - n(g-1).$$

Hence on the component $J^n_{n(g-1)}(M)$ this number vanishes, and in the classical case ($n = 1$) it is well known that the θ-divisor can also be identified with the set of those $L \subset J_{g-1}(M)$ for which

$$\dim H^0(M; L) = \dim H^1(M; L) \neq 0.$$

On the other hand, this definition has the plausible extension to all dimensions:

$$E \in \theta \Rightarrow \dim H^0(M; E) = \dim H^1(M; E) \neq 0.$$

Unfortunately, however, $J^n_{n(g-1)}(M)$ is a singular space, and so it has to be checked that the "jumping" condition defining θ satisfies the nontrivial technical requirement of being a Cartier divisor. A theorem to this effect is, I believe, to be published by Seshadri and some of his collaborators. Of course the physicists, never much concerned with such technical hurdles, not only assume that this generalized θ-divisor exists, but even furnish us with a remarkable formula for the number of these generalized θ-functions over M. This is a formula due to E. Verlinde for $\dim H^0(J^n_{n(g-1)}(M); \mathcal{L}^k)$, where $\mathcal{L}$ is the line bundle associated to the θ as defined above.

Let me close this account by describing this remarkable "Verlinde formula" [16], but in a representation theoretic transcription, developed by Andras Szenes and me. We of course hope that ultimately this formulation, or one close to it, might lead to a more direct proof than the one that is at present available, say, via the difficult and very technical papers of T. Tsuchiya, K. Ueno, and T. Yamada [15].

In any case here is our version of the Verlinde formula, at least in the case $n = 2$. First of all one separates the nonabelian case from the abelian one, by fixing the determinant of E. Thus Verlinde is computing $\dim H^0(SJ^{(2)}_{n(g-1)}(M); \mathcal{L}^k)$, where SJ^n describes the fiber of the determinant map $J^n \to J$:

$$1 \to SJ^n \to J^n \overset{\det}{\to} J \to 1.$$

This step reduces the considerations from U_2 to SU_2 and we formulate Verlinde's answer purely in terms of the representative, or character, ring of $G = SU_2$. Recall then that as a ring over $\mathbb{C}$ this character ring $R(G)$ is freely generated by the standard representation of SU_2, say V, so that $R(G)$ can be thought of quite simply as the polynomial ring in one variable V:

$$R(G) = \mathbb{C}[V].$$

On the other hand, as the representative ring of a compact Lie group, $R(G)$ has a natural linear functional defined on it, which I write Φ, and which on any finite-dimensional G-module W simply counts the dimension of the G-invariant subspace W^G in W:

$$\Phi(W) = \dim W^G = \int_G \chi(W)\, dg.$$

Essentially, then, Φ is given by the Haar measure on G normalized to 1. The ring $R(G)$ has a natural involution, sending W to its dual G-module W^*, and our Φ is *positive* on $R(G)$ in the sense that

$$\Phi(WW^*) \geq 0.$$

Finally this Φ is seen to define a positive definite inner product on $R(G)$, by the rule

$$\langle W, W' \rangle = \Phi(W^* \otimes W'),$$

and, as is well known, the classes of the irreducible representations in $R(G)$ then form an *orthonormal basis* for $R(G)$.

For example, in the case under discussion, $G = SU_2$, we have $V^* = V$, so that * is the identity. On the other hand, $\Phi(V^k)$ is not so easy to compute, but is clearly determined recursively in terms of Clebsh-Jordan formulas. Indeed, if we write $V_0, V_1, \ldots$ for the irreducibles in ascending order: V_i of dimension $i+1$, then they are recursively given by

$$V_0 = 1; \qquad V_k \cdot V = V_{k+1} + V_{k-1},$$

so that the first four terms are:

$$\begin{array}{ll} V_0 = 1 & V_3 = V^3 - 2V \\ V_1 = V & V_4 = V^4 - 3V^2 + 1 \\ V_2 = V^2 - 1 & \text{etc.} \end{array}$$

Consequently, by Shur's lemma, $\Phi(V_k) = 0$ for $k > 0$, and so $\Phi(V) = 0$, $\Phi(V^2) = 1$, $\Phi(V^3) = 0$, and $\Phi(V^4) = 2$, etc.

In any case, consider now the ideal $I_k \subset R(G)$ generated by the irreducible V_k, and the corresponding quotient ring

$$R_k(G) = R(G)/I_k, \qquad I_k = \{V_k\}.$$

A first lemma is now that the *additive* subspace spanned by the irreducibles $V_0, \cdots, V_{k-1}$ spans a complement to I_k in $R(G)$:

$$R(G) = I_k \oplus \{V_0, \cdots, V_{k-1}\}.$$

Additively we therefore have

$$R_k(G) \cong \{V_0, \cdots, V_{k-1}\},$$

which enables us to push Φ (forward!) to a linear function Φ_* on $R_k(G)$. (Put more mundanely: if $x \in R_k(G)$, then $\Phi_*(x)$ is defined as the coefficient of V_0 when x is written in the additive base furnished by the natural projections $\underline{V}_i$ of V_i; $i = 0, \cdots, k-1$ to $R_k(G)$.)

With all this understood consider now the element

$$\underline{k} = \underline{V}_0 \otimes \underline{V}_0^* + \underline{V}_1 \otimes \underline{V}_1^* + \cdots + \underline{V}_{k-1} \otimes \underline{V}_{k-1}^*$$

in $R_k(G)$. In terms of this $\underline{k}$ Verlinde's formula is equivalent to

$$\dim H^0(SJ^n_{n(g-1)}(M; \mathscr{L}^k)) = \Phi_*(\underline{k}^g). \tag{4}$$

Remark. (1) I like this formulation, because it gives such a natural extension of the classical formula: $\dim H^0(J; \mathscr{L}^k) = k^g$. However, to bring it more into line with the physics literature, or say the concepts of topological field theory, the following quite equivalent way of writing the right-hand side of (4) is useful.

Equipped with Φ_* the algebra $R_k(G)$ becomes a finite-dimensional commutative H^*-algebra and as such is isomorphic to the algebra of functions on a finite measure space X_k, of cardinality k. Let us take this set to be the integers $0 \le i \le k-1$, and let $w(i)$ be the measure of the ith point, so that the Φ_*-induced norm on $R_k(G)$ corresponds to the integral:

$$f\cdot : \to \sum |f(i)|^2 \cdot \omega_i .$$

Now under the inner product induced on $R_k(G)$ by Φ_* the elements $\{\underline{V}_i\}$, $0 \le i \le k-1$, are seen to be an orthonormal basis for $R_k(G)$ so that the element

$$\underline{k} = \sum_{i=0}^{k} \underline{V}_i \otimes \underline{V}_i^*$$

is canonically defined in $R_k(G)$, and thus is *independent* of the orthonormal basis chosen. But the δ-functions

$$\delta_i(j) = \delta_{ij},$$

clearly, have norm $= \omega_i$, so that

$$e_i = \frac{\delta_i}{\sqrt{\omega_i}}$$

form an orthonormal system in terms of which

$$\underline{k} = \sum \frac{\delta_i}{\omega_i},$$

and on this basis, the idempotent property of the δ_i implies that

$$\underline{k}^g = \sum \frac{\delta_i}{\omega_i^g}.$$

Hence

$$\Phi_*(\underline{k}^g) = \sum_{i=0}^{k-1} \omega_i^{(1-g)}.$$

The right-hand side now takes on an even better form if we pass to the square roots of the ω_i:

(5) $$\Phi_*(\underline{k}^g) = \sum_{i=0}^{k-1} (\omega_i)^{1/2 \cdot \chi(M)}.$$

In this form one is now also closer to the Verlinde formula in [16] and, as Szenes first remarked, the ω_i are now identified with the volumes of certain adjoint orbits in SU_2—or equivalently to the value of the Weyl measure on the maximal torus T of SU_2 on certain points of finite order in T.

(2) The element $\underline{k} = \sum V_i \otimes V_i^*$ is of course canonically defined in $R(G)$ for any finite group. (The sum here is over *all* the irreducibles of $R(G)$.) In this situation one has

$$\Phi(\underline{k}^g) = \sum w_i^{1/2\chi(M)},$$

where i ranges over the conjugacy classes of G, and the weight $\omega(x)$ of a given conjugacy class $x \in G$ is 1 over the number of elements in its stabilizer:

$$\omega(x) = 1/|s_x|.$$

Thus in the Verlinde formula the $R_k(G)$ plays an analogous role to $R(G)$ for a finite group. There are also connections to "quantum group" constructions, which, however, are not sufficiently clear to me to report on here.

In low genuses one can actually check this formula by independent procedures. For instance, when $g = 2$, Ramanan had shown long ago that

$$SJ_0^2 = \mathbb{C}P_3.$$

(This is the only case where the "even" SJ_0^2 has no singularities.) More generally for $SU(n)$ the Verlinde formula also checks the recent computations of Beauville, Narasimhan and Ramanan [3] for $k = 1$. Finally, Szenes has conjecturally extended (6) to the other components SJ_k^n as well, and there one can then check matters against Riemann-Roch computations. For instance, for $SJ_1^{(2)}$ he finds:

$$\dim H^0(SJ_1^{(2)}; \mathscr{L}^k) = \Phi_*((2\underline{k})^g \cdot V_{2k}). \tag{6}$$

On the right-hand side one is therefore dealing with the ring $R_{2k}(G)$, and on the left-hand side $\mathscr{L}$ is now defined as the generator of the Picard group of $SJ_2^{(2)}$. For low genus one can then compute the left-hand side using Riemann-Roch either from Ramanan's explicit description of $SJ_1^{(2)}(M)$ for hyperelliptic M, or from the cohomological data in [1].

Unfortunately the Verlindes' direct arguments in [16] are largely unintelligible to us. It is possible, though, to trace a path (a very long one) from the computation of [15] to the Verlinde formula, via the notion of parabolic bundles, due to Mehta and Seshadri [9], and an informal account of such a voyage can be found in the Oxford Notes of Atiyah-Hitchin-Segal [11]. The essential new ingredient which the physics-inspired literature adds to this subject is a sort of multiplicative Meyer-Vietoris theorem, which holds when two punctured surfaces M_1 and M_2 are sewed together. Here let me just point the way with the following comments.

Note that as a consequence of the orthogonality of the V_i, $i = 0, \cdots, k-1$, in $R_k(G)$ one has the recurrence relation

$$\sum_{\lambda=0}^{k-1} \Phi(k^{g_1} \cdot V_\lambda)\Phi(k^{g_2} V_\lambda^*) = \Phi(k^{g_1+g_2}). \tag{7}$$

Indeed, by the orthogonality we can compute the left-hand side of (9) using the idempotent basis

$$\text{Left-hand side} = \sum \Phi(k^{g_1} e_\lambda)\Phi(k^{g}\bar{e}_\lambda).$$

But $k^{g_1} e_\lambda = \delta_\lambda / \omega_\lambda^{g_1+1/2}$ whence $\Phi(k^{g_1} e_\lambda) = 1/\omega^{g_1-1/2}$ so that the λth term of (7) contributes $\omega_\lambda^{1/2-g_1} \cdot \omega_\lambda^{1/2-g_2}$. But then (5) becomes clear.

The recursion satisfied by Verlinde's answer is the dimensional consequence of the Künneth property, of the functor $M \to H^0(SJ_0^n(M); \mathscr{L}^k)$ expressed by the "fusion rules". Indeed, the choice of $p \in M$ and a representation, V_λ, $0 \le \lambda \le k$, determine a "parabolic structure" on M in the sense of Seshadri, and a corresponding space of semistable bundles, which I will notate $SJ_0^n(M; V_\lambda)$. This space again inherits a 0-divisor with corresponding line bundle $\mathscr{L}$, so that the spaces $H^0(SJ_0^n(M; V_\lambda); \mathscr{L}^k)$ become well defined and provide the constituents of the Künneth decomposition:

$$\begin{aligned} \sum_{\lambda=1}^{k-1} H^0(SJ_0^n(M_1; V_\lambda); \mathscr{L}^k) \otimes H^0(SJ_0^n(M_2; V_\lambda^*); \mathscr{L}^k) \\ \simeq H^0(SJ_0^n(M_1 \vee M_2), \mathscr{L}^k). \end{aligned} \tag{8}$$

How might such a formula fit into the more standard algebraic geometry picture? Presumably the answer is the following:

The assignment $M \to H^0(SJ_0^n(M); \mathscr{L}^k)$ defines a vector bundle over the moduli space of curves of a fixed genus g which can be endowed with a projectively *flat* connection. See, for instance, the very interesting papers of Axelrod, Della Pietra and Witten [2] on the one hand and Hitchin [7] and Ramadas [12] on the other. The holonomy of this bundle around paths linking the compactification divisor of the moduli space near the stable curve $M_1 \vee M_2$ should then hopefully induce the decomposition

(8), and this procedure might eventually lead to a better mathematical understanding of Verlinde's arguments. However at the moment we still seem far away from a direct and conceptual proof of their formula.

Finally, let me remark that in view of Riemann-Roch, the dimensions of the $H^0(J_r^n(M); \mathcal{L}^k)$ are given in terms of certain characteristic numbers of the varieties $J_r^n(M)$, and these in turn would be completely understood if we knew the homological position of $J_r^n(M) \subset B\mathcal{G}$, i.e., if we knew the homomorphism which fundamental cycle of $J_r^n(M)$ determines on $H^*(B\mathcal{G}_n)$, whose generators have direct geometric meaning. Computations of this sort would also be useful in the better understanding and computations of the Donaldson invariants for four-manifolds.

Added in proof. A recent Oxford preprint by M. Thaddeus entitled "Conformal field theory and the cohomology of the moduli space of stable bundles" carries out this program in the rank-2 case.

Bibliography

[1] M. F. Atiyah & R. Bott, *The Yang-Mills equations over Riemann surfaces*, Philos. Trans. Roy. Soc. London Ser. A **308** (1982) 523–615.

[2] S. Axelrod, S. Della Pietra & E. Witten, *Geometric quantization of Chern-Simons gauge theory*, J. Differential Geometry **33** (1991), 787–902.

[3] A. Beauville, M. S. Narasimhan & S. Ramanan, *Spectral curves and the generalized theta divisor*, J. Reine Angew. Math. **398** (1989) 169–179.

[4] G. D. Daskalopoulos, *The topology of the space of stable bundles on a compact Riemann surface*. I, to appear.

[5] G. Harder, *Eine Bemerkung zu einer Arbeit von P. E. Newstead*, J. Reine Angew. Math. **242** (1970) 16–25.

[6] G. Harder & M. S. Narasimhan, *On the cohomology groups of moduli spaces of vector bundles over curves*, Math. Ann. **212** (1975) 215–248.

[7] N. J. Hitchin, *Flat connections and geometric quantization*, Preprint, 1990.

[8] D. Mumford, *Geometric invariant theory*, Ergebnisse Math., Vol. 34, Springer, Berlin 1965.

[9] V. B. Mehta & C. S. Seshadri, *Moduli of vector bundles on curves with parabolic structures*, Math. Ann. **248** (1980) 205–239.

[10] P. E. Newstead, *Topological properties of some spaces of stable bundles*, Topology **6** (1967) 241–262.

[11] Oxford Seminar on Jones-Witten Theory, Michaelmas Term, 1988.

[12] T. R. Ramadas, *Chern-Simons gauge theory and projectively flat vector bundles on* $\mathcal{M}_g$, MIT Preprint, 1989.

[13] C. S. Seshadri, *Space of unitary vector bundles on a compact Riemann surface*, Ann. of Math. **85** (1967) 303–336.

[14] S. S. Shatz, *The decomposition and specialization of algebraic families of vector bundles*, Compositio Math. **35** (1977) 163–187.

[15] T. Tsuchiya, K. Ueno & T. Yamada, *Conf. on theory on universal family of stable curves with gauge symmetries*, Advanced Studies in Pure Math. **19** (1989) 459–565.

[16] H. Verlinde & E. Verlinde, *Conformal field theory and geometric quantization*, Preprint, Institute for Advanced Study, Princeton, 1989.
[17] A. Weil, *Generalisation des fonctions abéliennes*, J. Math. Pures Appl. **47** (1938) 17.
[18] E. Witten, *Quantum field theory and the Jones polynomial*, Commun. Math. Phys. **121** (1989) 351–399.
[19] ——, *Topological quantum field theory*, Comm. Math. Phys. **117** (1988) 353–386.

HARVARD UNIVERSITY

SURVEYS IN DIFFERENTIAL GEOMETRY
1 (1991) 19–111

LECTURES ON TRANSFORMATION GROUPS: GEOMETRY AND DYNAMICS

G. D'AMBRA & M. GROMOV

Table of Contents

0. Introduction

In these lectures we study groups of diffeomorphisms of smooth manifolds such that the action of the group, say G acting on V, preserves some geometric structure φ given on V. For example, φ may be a Riemannian or pseudo-Riemannian metric on V (i.e., a nonsingular quadratic differential form), and then φ-preserving diffeomorphisms are called *isometries* of (V, φ). In the sequel, we shall use the same "isometric" terminology for general structures φ on V.

0.1. Basic notation and conventions. Throughout these lectures, we will consider C^∞-manifolds. The geometric objects which we are interested in are assumed to be C^∞-smooth unless otherwise stated.

We denote the action of an element $g \in G$ on V by $g: V \to V$ for all $g \in G$, the action of g on each $v \in V$ by $gv \in V$, the tangent bundle of V by $T(V)$, and by $\mathrm{Is}(V, \varphi)$ the group of (global) isometries of (V, φ), i.e., of those diffeomorphisms of V which preserve φ.

We use $\mathrm{Is}^{\mathrm{loc}}(V, \varphi)$ to denote the pseudogroup of local isometries of (V, φ). (These are isometries $(U_1, \varphi|U_1) \to (U_2, \varphi|U_2)$ for all open subsets U_1 and U_2 in V.)

The local isotropy subgroup consisting of the germs of diffeomorphisms at $v \in V$, which fix φ and v, is denoted by $\mathrm{Is}^{\mathrm{loc}}(v) = \mathrm{Is}^{\mathrm{loc}}(V, v, \varphi)$.

Received May 8, 1990.

We shall often speak of an action of G on (V, φ) meaning an action on V preserving φ. For the most part our G-actions are faithful and thus correspond to subgroups $G \subset \mathrm{Is}(V, \varphi)$.

0.2. We begin this subsection by indicating the basic problems we are concerned with.

0.2.A. *The existence problem for invariant structures.* Assume we are given V and an action of G on V, and we want to find an invariant structure φ on V of prescribed type. For example, we may look for an invariant pseudo-Riemannian metric on V of prescribed signature (p, q) for $p + q = \dim V$.

If the group G is *compact*, then the existence of such a φ is a purely topological problem. For example, if G is finite and the action is free, then G-invariant structures 1-1 correspond to structures on the quotient space V/G. In the general case (G is compact, infinite, the action is nonfree) the situation is more complicated, but still one may relate G-invariant structures on V with appropriate structures on V/G. The key fact is the existence of the quotient V/G which is a *Hausdorff* topological space for compact G.

The properties of compact actions are very beautiful and useful for a geometric and topological study. However, in these lectures we are interested in the actions with nontrivial dynamics, where the word "dynamics" refers to the asymptotic behavior of orbits gv, $v \in V$, $g \in G$, for $g \to \infty$. Of course, this definition leaves no room for any nontrivial dynamics of compact group actions.

0.2.B. There is no simple recipe to decide whether there exists some kind of invariant φ, but there are dynamical constructions which are interesting in this regard.

As an example we point out the following construction which often provides an invariant subbundle T^+ of the tangent bundle $T(V)$ (compare 2.1). Let $G = \mathbf{Z}$ or $G = \mathbf{R}$. Define $T^+ \subset T(V)$ by

$$T^+ = \left\{ \tau \in T(V) | \lim_{g \to +\infty} \|Dg(\tau)\| = 0 \right\},$$

where $\| \ \|$ denotes some Riemannian metric given on V. Notice that this T^+ is not in general a subbundle of $T(V)$ since the dimension of the "fiber" over v, $T_v^+ \subset T^+$, may be nonconstant in $v \in V$. Clearly, T^+ does not depend on the choice of the Riemannian metric in case V is compact. This shows T^+ is indeed *invariant* in the compact case.

A geometrically interesting case where T^+ *is* a subbundle occurs, when V is the unit tangent bundle $UT(W)$ of a complete Riemannian manifold W of negative sectional curvature, and $G = \mathbf{R}$ acts by the geodesic flow.

One knows in this case (see [3]) that $\dim T^+ = n-1$ for $n = \dim V$, and moreover there exists a foliation of V into $(n-1)$-dimensional submanifolds (called the *stable* manifolds for this flow (see [3] and also see 2.2(b)) such that T^+ is the tangent bundle of this foliation. If W is simply connected, the projections of stable manifolds to V, under the projection $V = UT(W) \to W$, are *horospheres* (i.e., spheres of infinite radius) in W. If $n = 2$ and horospheres are 1-dimensional, then they are called horocycles.

We shall see in 2.6 that the above geodesic flow admits an invariant pseudo-Riemannian metric which is continuous but in general not smooth. In fact, such nonsmoothness is typical for the invariant structures obtained in differential dynamics by limit arguments. For example, an action may easily admit an invariant measure without having any smooth invariant measure (see, e.g., [13, p. 43]).

0.2.C. *The isometry group problem.* Now, let us *start* with a geometric structure φ on V and ask what is the isometry group G of (V, φ). Note that if we want to have a sufficiently large (or even just nontrivial) isometry group G, then we must start with a very special structure φ, as for most "sufficiently rigid" φ the full isometry group, called $\mathrm{Is}(V, \varphi)$, is trivial. For example, everybody knows that $\mathrm{Is}(V, \varphi) = \mathrm{Id}$ for *generic* pseudo-Riemannian metrics φ on V, for $\dim V \geq 2$. This is also true for generic subbundles $T \subset T(V)$ (viewed as geometric structures on V) such that $2 \leq \dim T \leq \dim T(V) - 2$, where "dim" refers to the dimension of the fibers of T. In fact, the same applies to general *rigid structures* which generalize pseudo-Riemannian and connection type structures (see 0.4 and §5 for the definition of a rigid structure).

0.3. In view of the above discussion, one does not expect rigid geometry to be accompanied by rich dynamics. In fact, a cohabitation of a big enough G with a rigid φ makes both G and φ extremely special. However, these special situations are very often encountered in mathematical practice. In fact, by looking at available examples (especially at those which arise in homogeneous surroundings (see 0.5 below)) one may come to the conclusion that genericity is exceptional while nongenericity is predominant.

0.4. Special or nonspecial, we want to study an action of G on (V, φ) where G is a *noncompact* Lie group, and (V, φ) is a compact manifold with a *rigid* geometric structure φ. The precise definition of the term rigid is given in 5.10 and 5.11. Here the reader may restrict to φ being one of the following:

(a) A pseudo-Riemannian metric.

(b) The conformal structure associated to a pseudo-Riemannian metric.

Notice that the conformal structure is rigid for dimension $n \geq 3$ but is not in our sense for $n = 1$ and $n = 2$. In fact, the local isometry pseudogroup (i.e., the group of local conformal transformations) is infinite dimensional for $n = 1, 2$, which is incompatible with rigidity according to 5.16.E.

(c) A subbundle or a system of subbundles in $T(V)$ with a certain non-degeneracy condition (see 2.6) which should rule out, for example, a single *integrable* subbundle $T \in T(V)$. (The nonrigidity of an integrable φ is manifested by the fact that the local isometry group of (V, φ) is infinite dimensional.) Since we have not yet given the definition of rigidity, we shall make two comments. First, one can think of rigidity of a geometric structure as, essentially, finite dimensionality of the local isometry pseudogroup of (V, φ), called $\mathrm{Is}^{\mathrm{loc}}(V, \varphi)$.

Secondly, we should notice that if one starts with a rigid structure φ and then add another structure φ' which does not have to be rigid, then the structure represented by the pair (φ, φ') is rigid.

(d) An affine connection φ on V is a rigid structure in our sense. In fact, every rigid structure (see 5.11) can be viewed as a kind of higher order connection (see 5.16.C).

0.5. There are two radically different aspects in the study of G and (V, φ).

0.5.A. *Dynamical aspect.* Here one wants to understand the dynamical properties of an action by taking into account the fact that the action preserves some geometric structure. A special (and probably most interesting) case is that of the action of a Lie group G (or of a subgroup $H \subset G$) on some homogeneous space G/Γ, where $\Gamma \subset G$ is a discrete subgroup (see 6.7). Such "homogeneous" actions often come along with natural invariant structures. For example, the standard conformal structure φ_0 on S^2 underlies the theory of *Kleinian groups* which are discrete subgroups in $\mathrm{PO}(3, 1) = \mathrm{Is}(S^2, \varphi_0)$. (Some properties of these are briefly discussed in 1.7.)

The most striking example of an interaction of homogeneous local geometry and ergodic theory is the recent theorem of Margulis concerning the action of $H = \mathrm{O}(2, 1) \subset G = \mathrm{SL}(3, \mathbf{R})$ on G/Γ for $\Gamma = \mathrm{SL}(3, \mathbf{Z})$. Namely, Margulis has proven that every compact minimal H-invariant subset is a smooth submanifold in G/Γ and hence consists of a single compact $\mathrm{O}(2, 1)$-orbit (see [51] for the proof and spectacular applications to the arithmetic of quadratic forms. See [61 $\frac{1}{2}$] for further developments.)

We want to emphasize once again that the dynamical depth and the beauty of the above examples is related to the presence of invariant structures and is unparalleled by what one sees in the systems of generic type, which preserve no smooth rigid structure.

0.5.B. *Geometric aspect.* Here we ask ourselves what is the *geometry* of (V, φ) provided that the isometry group $G = \mathrm{Is}(V, \varphi)$ is "sufficiently large". The most general "largeness" condition is noncompactness of G. One may impose stronger conditions by requiring, for example, that G has sufficiently fast rate of growth, or by insisting that G contains a given group G (i.e., a free group on two generators or such group as $\mathrm{SL}(2, \mathbf{R})$). Besides conditions imposed on G one may also require that the *action* of G on V is dynamically speaking "large" or, better to say, "ample". Two such ampleness conditions which are especially useful are *ergodicity* and *topological transitivity*. First, let us recall the definitions.

(i) *Ergodicity*. An action of a group G on a space V with a measure μ is called *ergodic* if $A \subset V$ is G-invariant implies

$$\mu(A) = 0 \quad \text{or} \quad \mu(V \backslash A) = 0.$$

Any transitive action is clearly ergodic. More generally, any essentially transitive action (i.e., transitive on the complement of a null set) is ergodic.

(ii) *Topological transitivity*. We say that an action of G on V is *topologically transitive* if there exists a dense orbit $G(v) \subset V$, $v \in V$. The following example shows how such a condition may effect an invariant structure.

0.6. Example. Let V be a compact connected surface, and φ a C^2-smooth pseudo-Riemannian metric on V. Notice that such a V of type $(1, 1)$ is homeomorphic to the torus or to the Klein bottle. Indeed, the existence of a Lorentz metric on V gives a vector field on a double covering of V, and so the Euler-Poincaré characteristic $\chi(V) = 0$ if V is a closed manifold (surface in our case).

Assume the action of G on V is topologically transitive. Then the Gauss curvature K_φ (being invariant as φ is invariant) is constant on each orbit $G(v) \subset V$, $v \in V$, and by continuity it is constant on all of V, as V equals the closure of a dense orbit. Since $\chi(V) = 0$, the Gauss-Bonnet theorem shows that the constant is zero, $K_\varphi = 0$ on V, and so V is locally flat. (The notion of Gauss curvature and the Gauss-Bonnet theorem automatically extend to the case of an indefinite metric. In fact, the reader who remembers the proof of the Gauss-Bonnet theorem will see that the positivity of the metric is never used there. See [5] for a proof of

Gauss-Bonnet formula for Lorentzian manifolds).

Now, it is not difficult to show (see, e.g., [5]) that $V = \mathbf{R}^{1,1}/\mathbf{Z}^2$, and G is a subgroup of affine transformations acting on the torus T^2 (compare 6.6.B(ii)). Thus we have got a good picture of both V and G in this case, and from our (geometric) point of view this is the end of the story. (But, of course, one may insist on further study of the dynamics of G.)

0.7. Remarks. The key step in the above argument is the passage from topological transitivity to local homeogeneity by means of the Gauss curvature which displays very well how the tensorial nature of the structure influences the dynamics. We shall see in §5 that the same idea can be applied to all geometric structures which have (essentially) tensorial nature. Unfortunately, this requires a somewhat unpleasant but unavoidable formal language of higher order jets and their infinitesimal transformations (see 5.2), and the conclusion is weaker. Namely, we have the following theorem (see 5.14.C).

0.7.A. *If the isometry group $G = \mathrm{Is}(V, \varphi)$ is topologically transitive on V, then there exists an open dense subset $U \subset V$ such that the structure φ is locally homogeneous on U, i.e., every two points in U have φ-isometric neighborhoods.*

Notice that there are simple (but not very natural) examples where $U \neq V$, and it would be nice to ensure the equality $U = V$ by a reasonable condition on (V, φ).

Also notice that in the case of an invariant Riemannian metric the passage from topological transitivity to local homogeneity is trivial, since the full isometry group G of every compact Riemannian manifold is compact, and therefore a dense orbit is necessarily equal to all of V.

0.8. As we have already remarked, generic manifolds have no isometries at all, therefore the presence of an isometry on V makes the manifold V quite special. Now, if we insist on the assumption that the isometry group $G = \mathrm{Is}(V, \varphi)$ is noncompact, then this makes the manifold V even more special. These considerations seem to indicate that there are good reasons to conjecture that it should be possible to classify all compact rigid manifolds having noncompact isometry groups. More precisely, we have the following.

Vague general conjecture. All triples (G, V, φ), where V is compact (or has finite volume) and G is "sufficiently large" (e.g., G is noncompact), are almost classifiable.

We are still far from proving (or even starting) this conjecture, but there are many concrete results which confirm it (see, e.g., 0.9.A below). On the other hand, we shall give in §6 a list of known (G, V, φ), which gives the idea of what kind of classification one may expect.

0.9. The following is a theorem supporting the conjecture.

0.9.A. **Theorem** (Obata [59], Lelong-Ferrand [44]). *Let V be a compact connected Riemannian manifold of dimension n. If the group of conformal transformations of V is noncompact, then V is conformally equivalent to the Euclidean sphere S^n. (Note that the group of conformal transformations of S^n equals* $\mathrm{PO}(n+1, 1) = \mathrm{O}(n+1, 1)/\{\pm 1\}$.

This statement confirms the fact that the existence of an action of a noncompact group on a manifold which preserves some geometric structure is a rather unusual phenomenon, and this completely agrees with the philosophy on which the conjecture was based.

0.10. It should be noted that the noncompactness of the group of conformal transformations of S^n is a nontrivial phenomenon which contradicts everybody's geometric intuition. It is not clear at all why there exists a single conformal transformation of S^n, which is not a rigid rotation. Similarly, one cannot see by a plain eye not equipped with the mathematical machinery any nontrivial conformal transformation of $\mathbf{R}^n$ (which, as we know, maps round spheres to round spheres) where "trivial" refers to the similarity transformation.

Even geometrically minded artists, designers of symmetric patterns, could not overcome this limitation of human imagination. If we look at the incredible variety of ornaments designed through the centuries all over the world, we see all kinds of translational and rotational symmetries but never a conformal symmetry. Yet, in recent times conformal symmetries were displayed in many beautiful drawings by Escher. However, the idea of those was communicated to the artist by a mathematician, namely Coxeter.

The most important transformation group in the world is the Lorentz group $\mathrm{O}(3, 1)$ of the special relativity. The group is noncompact; this appears to be one of the major obstacles for intuitively understanding the special relativity. Notice that the special relativity has replaced one infinity by another, namely it has banished the infinite (or unbounded) speed of motion but introduced arbitrarily large Lorentz transformations. But it is not easy (at least for a mathematician) to reconcile any kind of infinity with the intuitive vision of the physical universe.

0.10.A. **Remark.** The group of conformal transformations of S^n may be hard to see intuitively, but it can be easily introduced from the linear algebraic point of view.

In fact, this group is identified with the projectivized group $O(n+1, 1)$ of linear transformations of $\mathbf{R}^{n+2}$ which preserve the quadratic form

$$q(x_0, x_1, \cdots, x_{n+1}) = \sum_{i=1}^{n+1} (x_i)^2 - (x_0)^2$$

(see 6.5.D). The sphere S^n appears here as the set of those lines in $\mathbf{R}^{n+1}$ passing through the origin on which $q = 0$. (The reader can see 1.6 and 6.3 for a general discussion on algebraic actions.)

Notice that algebraically the group $O(n+1, 1)$ is very close to the group $O(n+2)$ of the linear transformations of $\mathbf{R}^{n+2}$, which preserve the *definite* quadratic form $\sum_{i=1}^{n+2}(x_i)^2$. Yet geometrically $O(n+1, 1)$ and $O(n+2)$ are radically different as one is compact and the other noncompact.

0.11. **Remark.** Now, we want to exhibit a rigid group action of non-algebraic nature which is infinitely more complicated than what we have seen earlier.

0.11.A. *Very important example.* Our group here is $SL(2, \mathbf{R})$ which acts on the homogeneous space $V = SL(2, \mathbf{R})/\Gamma$. If Γ is a connected subgroup, this action is of the same level of complexity as the action of $O(n+1, 1)$ on S^n.

But now we take a *discrete* subgroup $\Gamma \subset SL(2, \mathbf{R})$ such that $V = SL(2, \mathbf{R})/\Gamma$ is compact. (The existence of such Γ is not at all a trivial matter: see 1.9.D(ii).) This action is far from anything algebraic, and one cannot gain much intuition here by appealing to algebra. Yet, the geometry remains useful here as this action preserves a rigid geometric structure (see 5.11, 5.12 for the definition).

In fact, there exists on this V an invariant pseudo-Riemannian metric of type $(1, 2)$ coming from the Killing form on the Lie Algebra $sl(2, \mathbf{R})$. (For more details on this $SL(2, \mathbf{R})/\Gamma$-example the reader can see 1.9.C.)

0.11.B. **Remark.** The fundamental group $\pi_1(V)$ of $V = SL(2, \mathbf{R})/\Gamma$ is quite large. In fact, it is at least as large as Γ since the quotient map $SL(2, \mathbf{R}) \to V$ is a covering map. ($\pi_1(SL(2, \mathbf{R})/\Gamma)$ is *strictly* larger than Γ as $\pi_1(SL(2, \mathbf{R})) \neq \{0\}$.)

The significance of $\pi_1(V)$ being large will be clarified later on (see 1.12).

0.12. It is hard to reconcile our intuition on isometries of compact manifolds with having such a huge noncompact group as $SL(2, \mathbf{R})$ for the group of isometries. But the intuition regains some ground if we are willing to sacrify the fundamental group of V. Namely, we can prove the following.

0.12.A. **Theorem** [14]. *Let (V, φ) be a compact simply connected real analytic Lorentz manifold (i.e., φ is a pseudo-Riemannian metric of type*

$(n-1, 1))$. *Then the isometry group* $\operatorname{Is}(V, \varphi)$ *is compact.*

0.12.B. **Remarks.** (i) If (V, φ) is a compact Riemannian manifold, i.e., φ is a *definite* quadratic differential form in V, then the group $\operatorname{Is}(V, \varphi)$ is compact without any extra assumption on V.

(ii) Our Theorem 0.12.A is vaguely similar to the Obata–Lelong-Ferrand theorem (see 0.9.A) on conformal transformations.

(iii) Our compactness theorem does not directly generalize to pseudo-Riemannian manifolds (V, φ), where φ has type (p, q) for $\min(p, q) \geq 2$ (see §4 in [14]). However, one has the following weaker compactness result which can be used to prove the compactness of $\operatorname{Is}(V, \varphi)$ in the Lorentz case (see [14]).

0.12.C. **Theorem** [29] (Compare 3.2.B.(i)). *If* V *is a compact simply connected real analytic pseudo-Riemannian manifold then the orbits of the full isometry group* $\operatorname{Is}(V)$ *are compact.*

This result is a specialized (to the case of a pseudo-Riemannian metric φ) version of a more general statement in [29] which, under the same assumption on V as in 0.12.B, ensures the compactness of the $\operatorname{Is}(V, \varphi)$-orbits when φ is a C^{an} smooth rigid structure of algebraic type (see 5.5, 5.11, 5.12, for the definitions) provided that $\operatorname{Is}(V, \varphi)$ preserves a smooth volume element on V.

Notice that by imposing certain conditions on V and φ one necessarily gets that compactness of orbits (see 3.7.A in [29] and also 1.11.B in these lectures) strongly restricts the range of all possible (G, V, φ). In fact, rigid actions with compact orbits should be regarded as classifiable in our sense (compare with the twisted rotation example in 1.11.D).

0.12.D. *Warning.* Amazingly, there exists examples of actions with compact but not uniformly compact orbits (see [64]), but these do not appear in our framework.

0.13. While the above Theorems 0.9.A and 0.12.A indicate that noncompactness of $\operatorname{Is}(V, \varphi)$ makes V rather special, the following Theorem 0.13.A states that if $\operatorname{Is}(V, \varphi)$ is a *very large* noncompact group, then V is very special.

0.13.A. **Splitting theorem** (See 0.8.B and 5.4. in [29]). *Let* (V, φ) *be a connected Lorentz manifold of finite volume (e.g., compact) such that the isometry group* $\operatorname{Is}(V, \varphi)$ *contains* $\mathrm{SL}(2, \mathbf{R})$ *as a subgroup. Then the action of* $\mathrm{SL}(2, \mathbf{R})$ *on* V *is everywhere locally free (i.e., the isotropy subgroup is discrete at all* $v \in V$*). The metric* φ *is nonsingular on the* (3-*dimensional) orbits and the normal subbundle to the orbits is integrable with totally geodesic leaves. Furthermore, some infinite covering* $\tilde{V}$ *of* V

is split by the lifts to $\tilde{V}$ *of the two* (*into the orbits and the normal leaves*) *foliations* (*see* 4.9 *for a more detailed discussion*).

0.13.B$_1$. **Remark.** Our theorem refines an earlier result of Zimmer who proved that the above action of $SL(2, \mathbf{R})$ on V is almost everywhere locally free and that the full isometry group $Is(V)$ containing $SL(2, \mathbf{R})$ differs from $SL(2, \mathbf{R})$ roughly by a compact group (see [72], [69] for the precise statements and more general results).

0.13.B$_2$. **Historical remark and references.** Until recently, the transformation groups of (V, φ) were pursued from a geometric point of view, as can be seen by looking in the monographs by Lichnerowicz [45], Kobayashi [41], Koszul [42].

A dynamical approach was developed in a series of papers by Zimmer (see, e.g., [68]–[76]) and an intermediate approach was attempted in [29].

This paper grew out of the lectures given by the second author in Cagliari in October 1988 and made into notes by the first author. The problem section (§7) was written later by the second author following the suggestion by Professor S.-T. Yau. The second author wants to thank the Department of Mathematics at the University of Cagliari for organizing the lectures, while the first author acknowledges the hospitality of IHES where the preparation of the manuscript was essentially concluded.

1. Dynamics of A-actions

1.1. As we mentioned earlier the structure of an action of a *compact* group G on a smooth manifold V is quite easy and transparent from the dynamical point of view. Namely, each orbit of G is closed, the quotient space V/G is Hausdorff, and each orbit has a "nice" invariant neighborhood by the slice theorem (see [60]) if the action is smooth. In fact, the whole subject of compact (in particular finite) group actions belongs to the algebraic and geometric topology rather than to the dynamics.

1.2. Now, if we turn to the really interesting case where G is noncompact, we shall see a more complicated and tantalizing picture. For example, a noncompact isometry group G may easily have a *dense* orbit $G(v) \subset V$, such that $\dim G(v) < \dim V$. Certainly, such an orbit cannot be closed in V! The standard example of that is an *irrational rotation* of the torus $T^2 = \mathbf{R}^2/\mathbf{Z}^2$ which is a flow (i.e. an action of $\mathbf{R}$) on T^2 induced by parallel translations of $\mathbf{R}^2$. These are defined with a fixed real number α, by

$$(x_1, x_2) \to (x_1 + t, x_2 + \alpha t),$$

where $x_1, x_2 \in \mathbf{R}^2$, and $t \in \mathbf{R}$ is the flow parameter.

If α is a *rational* number, then all orbits of the induced action of $\mathbf{R}$ on T^2 are compact. In fact, such a rational action of $\mathbf{R}$ factors through that of the circle $\mathbf{R}/\mathbf{Z}$ on T^2, where $\mathbf{Z}$ is the intersection of the line $\mathbf{R} = \{t, \alpha t\} \subset \mathbf{R}^2$ with the integral lattice $\mathbf{Z}^2 \subset \mathbf{R}^2$. On the contrary, if α is irrational, each orbit $\mathbf{R}(v)$ in T^2 is noncompact. Moreover, by the famous theorem of Kronecker each orbit is dense (moreover, uniformly distributed) in T^2.

1.3. Another important feature of noncompact group actions is the *recurrency* phenomenon.

1.3.A. **Definition.** A point $v \in V$ is called *recurrent* for an action of G on V if $g(v)$ comes back infinitely often to an arbitrary small neighborhood $U \subset V$ of v. That is, the set $\{g \in G | g(v) \in U\} \subset G$ is noncompact for every neighborhood U of v.

Notice that recurrency does not exclude closed orbits. In fact if the orbit $G(v)$ is compact, then v is obviously recurrent, insofar as G is noncompact. But the real case of interest is that where some orbit $G(v)$ is noncompact while v is recurrent. In this case the closure of $G(v)$ in V is significantly bigger than $G(v)$ itself. For example, each point $v \in V = T^2$ is recurrent for the above action of $\mathbf{R}$ on T^2, in the rational as well as in the irrational case. This fact can easily be derived from the famous

1.3.B. **Poincaré recurrence theorem** (see [71], [13]). *If an action of G preserves a finite measure on V, then almost all points $v \in V$ are recurrent. ("Almost all" here signifies, according to the usual convention, that nonrecurrent points constitute a set of measure zero.)*

1.3.B_1. **Remark.** The Poincaré theorem ensures a somewhat stronger recurrency than that given by Definition 1.3.A. Namely, the theorem guarantees a certain kind of "relative density" of the subset $\{g \in G | g(v) \subset U\}$ in G.

1.4. Instead of dealing with individual orbits one may look at the quotient space V/G and see many features of the orbit structure of the action reflected in the topology of V/G. The simplest actions from our point of view are those where the orbit space V/G is Hausdorff.

An especially simple class of group actions which have this property is described in the following.

1.4.A. **Definition.** An action of G on a manifold V is called *proper* if for each pair x, y of points in V there exist neighborhoods U_x of x and U_y of y in V such that the subset $\{g \in G | gU_x \cap U_y \neq \emptyset\}$ is relatively compact in G.

1.4.B. **Examples.** (i) A well-known and easy theorem from the theory of Lie transformation groups states that the standard (transitive) action of G on G/H, where H is a Lie subgroup of the group G, is proper iff H is compact (see, e.g., [66]).

(ii) It follows from (i) above that every closed (e.g., discrete) subgroup H of a connected Lie group G admits a smooth proper action on some Euclidean space. The proof of this fact follows from a well-known result about Lie groups, which states that if G is a connected Lie group, there is a (maximal) compact subgroup H (unique up to conjugation) such that G/H is diffeomorphic to a Euclidean space (see, e.g., [35]).

A typical example of this situation is the hyperbolic space H^n (which is topologically $\mathbf{R}^n$) viewed as the homogeneous space $SO(n, 1)/SO(n-1)$.

1.5. After proper actions come stratified actions which can be defined as follows:

1.5.A. **Definition.** We call an action of G on V *stratified* if V can be decomposed into the union of *locally closed* subsets called strata, $V = V_0 \cup V_1 \cup \cdots \cup V_n$, where V_0 is open, V_1 is open in $V_1 \cup V_2 \cup \cdots \cup V_n$, V_2 is open in $V_2 \cup V_3 \cup V_4 \cup \cdots$, and so on, such that V_i/G is a Hausdorff space for all i. (We remind the reader that a subset in (a topological space) V is called locally closed if it is contained and closed in some open subset $U \supset S$ in V.)

1.5.B. **Example.** Let G be a Lie group, and $H \subset G$ a closed subgroup. Consider the one-point compactification $(G/H)^* = G/H \cup \{\infty\}$. Then the action of G on $(G/H)^*$ is stratified with two strata, $\{\infty\}$ and G/H. For example, if $G = \mathbf{R}^n$ and $H = \{0\}$, then $(G/H)^* = S^n$, and a remarkable additional property of the resulting action of $\mathbf{R}^n$ on S^n is the existence of an invariant structure. Namely, the conformal structure of $\mathbf{R}^n$ extends to a smooth conformal structure on $S^n \supset \mathbf{R}^n$ (via the stereographic projection).

1.5.C. **Exercise.** Take $G/H = \mathbf{R}^n$ with an action of G equal to $GL(n, \mathbf{R})$, $SL(n, \mathbf{R})$, or the group of all orthogonal isometries. Then the corresponding action on $S^n = \mathbf{R}^n \cup \{\infty\}$ is obviously stratified. The question for the reader is to decide in which case there is an invariant A-structure (see 1.8, 5.5) on S^n.

1.5.D. **Remarks.** The dynamical complexity of a stratified action is comparable to that of transitive actions. For example, one can easily prove the following:

(i) If a point $v \in G$ is recurrent for a stratified action of G on V, then v is recurrent for the (transitive) action of G on the orbit $G(v) = G/G_v$.

(ii) Another important (and easy to prove) property states that *every finite invariant measure μ on V decomposes into finite invariant measures on the orbits*. In fact, the existence of μ on V implies that the orbit G/G_v admits a finite invariant measure for almost all $v \in V$.

1.6. The importance of stratified actions stems from the fact that every *algebraic* action is *stratified* (see, e.g., [71] and §2 in [29]), where *algebraic* means that the manifold V is given a structure of *real algebraic manifold* such that the group G acts on V as an *algebraic group*. We do not need and shall not give the detailed abstract definition in these lectures. On the other hand, we want to point out that all basic features of algebraic actions can be seen in the following example (compare discussion in 6.3, 6.4, 6.5).

1.6.A. *Linear actions.* A subgroup $G \subset \mathrm{GL}(N, \mathbf{R})$ is called *algebraic* if it equals to zero set of a polynomial map $f\colon \mathrm{GL}(N, \mathbf{R}) \to \mathbf{R}^k$. Notice that $\mathrm{GL}(N, \mathbf{R})$ is an open subset in the Euclidean space $\mathbf{R}^{N^2}$ of $(N \times N)$-matrices, and polynomials on $\mathrm{GL}(N, \mathbf{R})$ by definition are the functions which are polynomial in the Euclidean coordinates. Also notice that by taking $\|f\|^2$ one may restrict oneself to a single polynomial $\mathrm{GL}(N, \mathbf{R}) \to \mathbf{R}$.

1.6.B. *Basic example.* Let φ be a *tensor* on $\mathbf{R}^N$, e.g., a multilinear (say quadratic) form. Then the subgroup G of $\mathrm{GL}(N, \mathbf{R})$ consisting of transformations of $\mathbf{R}^N$ preserving φ is *algebraic* as a simple (and well-known) argument shows. Let us indicate an important construction leading to many concrete examples. Start with the standard action of $\mathrm{GL}(n, \mathbf{R})$ on $\mathbf{R}^n$, and consider the induced action of $\mathrm{GL}(n, \mathbf{R})$ on $\mathbf{R}^M$ for $M = n^k$ on the kth tensor power $\mathbf{R}^M = \bigotimes_k \mathbf{R}^n$. This defines an embedding $\mathrm{GL}(n, \mathbf{R}) \to \mathrm{GL}(M, \mathbf{R})$ whose image is an *algebraic* subgroup in $\mathrm{GL}(M, \mathbf{R})$. Moreover, for any algebraic subgroup $G \subset \mathrm{GL}(n, \mathbf{R})$ (notice that the case $G = \mathrm{GL}(n, \mathbf{R})$ is already interesting) and for every G-invariant subspace $\mathbf{R}^N \subset \mathbf{R}^M$ (e.g., the subspaces of symmetric or antisymmetric tensors) the homomorphism $G \to \mathrm{GL}(N, \mathbf{R})$ corresponding to the restriction to $\mathbf{R}^N$ has algebraic image in $\mathrm{GL}(N, \mathbf{R})$.

1.6.C. **Remarks.** (i) Note that the geometry (as opposed to the dynamics) of an algebraic action can be quite intricate. To see an example of this, we suggest that the readers first look at the action of $(\mathbf{R}^\times)^2$ on the projective space $P^{n-1} = P(\mathbf{R}^n)$, given by

$$(t_1, t_2)(x_1, \cdots, x_n) = (t_1^{k_1} t_2^{m_1} x_1, t_1^{k_2} t_2^{m_2} x_2, \cdots, t_1^{k_n} t_2^{m_n} x_n)$$

for given integers $k_1, m_1, k_2, m_2, \cdots, k_n, m_n$, and then try to understand the geometry of the closure of a given orbit in P^{n-1}. (Compare with the discussion in 6.5.A.)

(ii) Notice that the dynamical simplicity of algebraic actions does not extend to *subalgebraic* actions. This term refers to an action of a subgroup $G_0 \subset G$, where G acts algebraically on V. If G_0 is not algebraic, the action of G_0 may have nontrivial dynamics. One gets especially interesting actions by looking at actions of discrete subgroups $G_0 \subset G$. The classical example is that of a discrete subgroup $G_0 \subset \mathrm{PGL}(2, \mathbf{C}) = \mathrm{PSO}(3, 1)$ acting on $\mathbf{C}P^1 \cong S^2$. These are Kleinian groups already mentioned in 0.5.A.

1.7. Little digression. Kleinian groups are very beautiful, and there is a wide variety of productive ways to think about them. The basic feature of Kleinian groups is the *existence* of the *limit set* $\Omega \subset S^2$ with the following properties:

(i) Ω is a closed G_0-invariant subset in S^2 such that the closure of the orbit $G_0(\omega)$ equals Ω for every $\omega \in \Omega$.

(ii) The action of G_0 on $S^2 \setminus \Omega$ is proper (see 1.4.A).

(iii) For every point $s \in S^2$ the set of the accumulation points of the orbit $G_0(s) \subset S^2$ equals Ω. It is well known (and easy to prove) that the limit set exists and is unique for all *nonelementary* Kleinian groups, where G_0 is called elementary if some orbit of G_0 is finite. What is much harder and really exciting is understanding the geometry of limit sets. One knows, for instance, that the limit set of a nonelementary Kleinian group is either the whole sphere S^2 or a round circle or an amazingly complicated fractal subset.

Nowadays one can see fractal limit sets displayed on beautiful multicoloured posters (see, e.g., [48]).

1.8. Now we turn to a class of actions which generalize algebraic and subalgebraic actions and play the central role in these lectures.

1.8.A. *Basic nondefinition.* We say that an action of a Lie group G on V is *A-rigid* if the action preserves some rigid structure φ of algebraic type given on V or, for brevity, an *A-structure.* The precise definitions of the terms "rigid" and "algebraic type" are given in §5. Here we only recall that most structures encountered in differential geometry are of algebraic type and that the rigidity for these structures is essentially equivalent to the finite-dimensionality of the pseudogroup $\mathrm{Is}^{\mathrm{loc}}(V, \varphi)$ of local isometries of the structure φ. For example, every tensor field on V is an A-structure, and most of tensorial structures are rigid (see 5.11).

In the language of §0 (see 0.2), A-actions correspond to subgroups $G \subset \mathrm{Is}(V, \varphi)$ for a rigid A-structure φ. Thus, for example, A-actions include isometries of pseudo-Riemannian manifolds, groups of conformal transformations, and connection preserving transformations.

1.9. One of the reasons for a nontrivial dynamics of an A-rigid action may be the fact that the group G in question is *strictly smaller* than the full isometry group $\mathrm{Is} = \mathrm{Is}(V, \varphi) \supset G$. For example, let us look again at the Kleinian group acting on S^2 (see 1.7). The relevant *rigid* structure φ here is the *flat complex projective structure* for which $\mathrm{Is}(S^2, \varphi) = \mathrm{PSO}(3, 1)$. (The conformal structure on S^2 does not formally fit into our discussion since it is nonrigid).

As we have mentioned before, the action of a Kleinian group $G \subset \mathrm{Is}$ on S^2 may have a remarkably rich dynamics. This is due to two properties:

(a) Is is noncompact;

(b) G is a proper subgroup in Is.

Moreover G, being discrete infinite, is not an algebraic subgroup in $\mathrm{Is} = \mathrm{PSO}(3, 1)$.

1.9.A. **Remark.** If we start with a structured manifold (V, φ) with compact isometry group $\mathrm{Is}(V, \varphi)$, then a subgroup $G \subset \mathrm{Is}(V, \varphi)$ may have a dynamically interesting action on V only if G is *nonclosed* in Is. What happens in this case can be seen in the example of the irrational translation of the torus (see 1.2).

1.9.B. Now we want to look at A-rigid actions where $G = \mathrm{Is}(V, \varphi)$, but the dynamics is at least as rich as that of any subalgebraic (e.g., Kleinian) action. We start by looking at our "very important example" of 0.11.A.

1.9.C. The group G in this example is $\mathrm{SL}(2, \mathbf{R})$, and V is a compact G-homogeneous space. Namely, $V = \mathrm{SL}(2, \mathbf{R})/\Gamma$, where Γ is a discrete cocompact subgroup ("cocompact" signifies the compactness of $\mathrm{SL}(2, \mathbf{R})/\Gamma$).

We already mentioned in 0.11.A that the action of $\mathrm{SL}(2, \mathbf{R})$ on this V is A-rigid as there exists an invariant pseudo-Riemannian metric of type $(1, 2)$ defined on V. To see this, let us start with the Killing form φ_0 on the Lie algebra $\mathrm{sl}(2, \mathbf{R})$ identified with the tangent space $T_e(\mathrm{SL}(2, \mathbf{R}))$ of $\mathrm{SL}(2, \mathbf{R})$ at the identity. We get a pseudo-Riemannian metric $\tilde{\varphi}$ on $\mathrm{SL}(2, \mathbf{R})$ by left translations of $T_e(\mathrm{SL}(2, \mathbf{R}))$ to the tangent spaces $T_g(\mathrm{SL}(2, \mathbf{R}))$, $g \in \mathrm{SL}(2, \mathbf{R})$.

Since φ_0 is Adj-invariant, the form $\tilde{\varphi}$ is invariant under the right translations as well as under left translations. It follows that $\tilde{\varphi}$ descends to an *invariant* metric φ on the quotient $V = \mathrm{SL}(2, \mathbf{R})/\Gamma$, and thus $\mathrm{Is}(V, \varphi) \supset \mathrm{SL}(2, \mathbf{R})$. Furthermore, one can show that the full isometry group $\mathrm{Is}(V, \varphi)$ equals $\mathrm{SL}(2, \mathbf{R})$ or $\mathrm{SL}(2, \mathbf{R})/\{\pm 1\}$.

1.9.D. *Digression.* (i) We have considered so far *cocompact* discrete subgroups $\Gamma \subset G = \mathrm{SL}(2, \mathbf{R})$, but noncocompact subgroups may also be interesting from our point of view. For example, one may distinguish (discrete) *lattices* $\Gamma \subset G$, where a subgroup Γ is called a lattice if G/Γ admits a finite G-invariant measure. Notice that every discrete cocompact subgroup Γ in a unimodular Lie group G (e.g., in $\mathrm{SL}(2, \mathbf{R})$) is necessarily a lattice as the Haar measure on G (being bi-invariant on the unimodular group) descends to an invariant measure on G/Γ.

(ii) The existence of (cocompact and noncocompact) discrete lattices in a Lie group is a highly nontrivial matter. Of course some cases are quite easy. For example, if $G = \mathbf{R}^n$, one can immediately see a discrete lattice in there, namely $\mathbf{Z}^n \subset \mathbf{R}^n$, and then one can prove that any lattice is obtainable from $\mathbf{Z}^n$ by an automorphism of $\mathbf{R}^n$. The situation is by far more subtle and complicated for semisimple Lie groups G, such as $\mathrm{SL}(2, \mathbf{R})$. The simplest example here is $\Gamma = \mathrm{SL}(n, \mathbf{Z})$, the group of matrices with integral entries and determinant one.

There is a nontrivial (though not very difficult) theorem saying that $\mathrm{SL}(n, \mathbf{R})/\mathrm{SL}(n, \mathbf{Z})$ has *finite measure*, and so $\mathrm{SL}(n, \mathbf{Z})$ is indeed a *lattice* in $\mathrm{SL}(n, \mathbf{R})$. This lattice is not cocompact, and getting cocompact lattices requires more effort. In general, there exist three methods of constructing discrete lattices in semisimple Lie groups G.

1. *Arithmetic method.* This amounts to getting an appropriate monomorphism $G \to \mathrm{SL}(n, \mathbf{R})$ for large n and obtaining $\Gamma \subset G$ as the intersection $G \cap \mathrm{SL}(n, \mathbf{Z})$. By this method Borel [9] has proven that every semisimple G contains a cocompact as well as a noncocompact lattice.

2. *Geometric method.* This is especially clear for $\mathrm{SL}(2, \mathbf{R})$ viewed as the isometry group on the hyperbolic plane $\mathbf{H}^2$. One gets our Γ by taking, for example, a regular n-gon, $n > 5$, in $\mathbf{H}^2$ with $90°$-angles and then generating Γ by the n-reflection of $\mathbf{H}^2$ in the sides of this n-gon.

3. *Differential equations.* Monodromy groups of some totally integrable system of differential equations are lattices in their Zariski closures. (See Mostow [56] for an extensive discussion of all these matters.)

1.10. To appreciate the action of $\mathrm{SL}(2, \mathbf{R})$ on $V = \mathrm{SL}(2, \mathbf{R})/\Gamma$ it is useful to look on how one-parameter subgroups of $\mathrm{SL}(2, \mathbf{R})$ act on this V. A somewhat unexpected fact is that every such subgroup $G_0 \subset \mathrm{SL}(2, \mathbf{R})$ equals the *full* isometry group of some rigid A-structure on V. In fact let X_0 denote the vector field on V generating G_0, and consider the "sum" of X_0, viewed as a geometric structure, with the Killing metric φ on V. What we get (by the definition of the "sum") is the pair $\varphi' = (\varphi, X_0)$ and the isometry group of (V, φ') is the intersection $\mathrm{Is}(V, \varphi) \cap \mathrm{Is}(V, X_0)$.

In other words, $\mathrm{Is}(V, \varphi')$ consists of those isometries of (V, φ) which also preserve X_0. Since X_0 is a vector field, to "preserve" X_0 means to commute with X_0. Therefore, $\mathrm{Is}(V, \varphi')$ consists of the centralizer C_0 of G_0 in $\mathrm{SL}(2, \mathbf{R})$. As it is well known (and obvious), every subgroup C in $\mathrm{SL}(2, \mathbf{R})$ with 1-dimensional center is 1-dimensional, and so C_0 is at most a finite extension of G_0.

The action of noncompact one-parameter subgroups G_0 on $V = \mathrm{SL}(2, \mathbf{R})/\Gamma$ displays a wide spectrum of beautiful dynamical properties such as topological transitivity, ergodicity, etc. Notice that there are only two essentially different actions corresponding to two conjugacy classes of noncompact one-dimensional subgroups in $\mathrm{SL}(2, \mathbf{R})$. Namely, we have the following.

(a) The action of the one-parameter subgroup

$$g_t = \begin{pmatrix} e^t & 0 \\ 0 & e^{-t} \end{pmatrix} \quad \text{on } V = \mathrm{SL}(2, \mathbf{R})/\Gamma.$$

If the group Γ in question has no torsion, then this action can be identified with the geodesic flow on a compact surface W of constant negative curvature. Namely, one notices first that the (left) quotient $\mathrm{SL}(2, \mathbf{R})/S^1$ admits a left invariant metric of constant curvature -1 and equals the hyperbolic plane $\mathbf{H}^2$. Then the unit tangent bundle $UT(\mathbf{H}^2)$ can be identified with $\mathrm{PSL}(2, \mathbf{R}) = \mathrm{SL}(2, \mathbf{R})/\{\pm 1\}$, such that the action of g_t corresponds to the geodesic flow on $UT(\mathbf{H}^2)$. Finally, we take the quotient $\mathbf{H}^2/\Gamma$ for our W which is a *smooth* surface as Γ has no torsion and so acts *freely* on $\mathbf{H}^2$.

(b) The second action is that of the (unipotent) subgroup $O_t = \left(\begin{smallmatrix} 1 & 0 \\ t & 1 \end{smallmatrix}\right)$ on V. This corresponds to the horocycle flow on the unit tangent bundle of the above W. (This flow moves each tangent vector along the oriented horocycle normal to and normally oriented by this vector; see [4].) Notice that, in the above $\mathrm{SL}(2, \mathbf{R})/\Gamma$-example, rich dynamics goes along with a large fundamental group $\Pi = \pi_1(V)$. In fact, Π is the central extension

$$0 \to \mathbf{Z} \to \Pi \to \pi \to 0$$

where $\mathbf{Z}$ equals the Galois group of the universal covering $\widetilde{\mathrm{SL}(2, \mathbf{R})}$ of $\mathrm{SL}(2, \mathbf{R})$.

The following Theorem 1.11.A shows, in contrast, that for simply connected manifolds the dynamics of the full isometry group is almost as simple as that of an algebraic group.

1.11.A. **Theorem** [29]. *Let (V, φ) be a compact manifold with a rigid real analytic structure φ of algebraic type (i.e., a pseudo-Riemannian metric* *or a conformal structure). If V is simply connected, then the action of $G = \mathrm{Is}(V, \varphi)$ is stratified. Moreover, the isotropy subgroup $G_v \subset G$ of each point $v \in V$ has at most finitely many connected components.*

An easy corollary of 1.11.A is the following:

1.11.B. (Compare 0.12.C.). *Let V be as in 1.11.A and assume, in addition, $G = \mathrm{Is}(V, \varphi)$ preserves a smooth measure on V. Then all G-orbits are compact.*

1.11.C. *Unbounded volume property* (see 6.4.B$_5$). The above $\mathrm{Is}(V, \varphi)$ actions for $\pi_1(V) = 0$ may look rather nonalgebraic in spite of the above stratification Theorem 1.11.A.

For example, the volumes of the graphs $\Gamma_g \subset V \times V$ of the transformation $g: V \to V$ may be unbounded as $g \in G$ goes to infinity (compare 6.4.B$_5$).

1.11.D. *Twisted rotation example.* Let us indicate a specific A-rigid action on $V = S^3$ where $\mathrm{Vol}(\Gamma_g) \to \infty$ for $g \to \infty$. Let S^3 be fibered over S^2 in the usual way, $p: S^3 \to S^2$, and let $S^1 \times \mathbf{R}$ act on S^3 in the following way. S^1 acts by the usual rotations having the Hopf fibers for the orbits, while $\mathbf{R}$ rotates the fibers with variable speed. That is, for a given function, say, a on S^1, let $Y = aX$ be the vector field on S^3 where X is the generating field for the S^1 action. Then Y integrates to an action of $\mathbf{R}$ which rotates the circle $p^{-1}(s) \subset S^3$, $s \in S^2$, with the speed $a(s)$. If a is nonconstant, then by an easy argument $\mathrm{Vol}(\Gamma_g)$ is unbounded for $g \to \infty$. On the other hand, this action is A-rigid analytic for all real analytic a (see 6.6).

1.11.E. **Remark.** Another way to see that the above twisted action is not algebraic is by looking at the corresponding diagonal action of our $G = S^1 \times \mathbf{R}$ on the products $S^3 \times S^3 \times \cdots \times S^3$ and by observing that there exist no nonempty open invariant subsets where this action is proper (compare 6.4.B$_4$).

1.12. The role of the simply connectedness condition in Theorem 1.11.A is based on the fact that every local isometry of a *rigid simply connected real analytic* manifold V extends to a global isometry of V. If the manifold V is not simply connected, then local isometries do not always extend to global isometries. For example, take $V = \mathrm{SL}(2, \mathbf{R})/\Gamma$ where the group $\mathrm{SL}(2, \mathbf{R})$ has the Lorentz metric $\tilde{\varphi}$ determined by the Killing form on $\mathrm{sl}(2, \mathbf{R})$. Then we know (see 1.9.C) that $\mathrm{Is}(V, \varphi) = \mathrm{SL}(2, \mathbf{R})$.

Yet, there are many local isometries of V which are not extendible to all of V. Namely, every isometry of $(\tilde{V}, \tilde{\varphi}) = (\mathrm{SL}(2, \mathbf{R}), \tilde{\varphi})$ which covers (V, φ) defines a local isometry of (V, φ). Now, as $\tilde{\varphi}$ is bi-invariant, both the left and the right translations are $\tilde{\varphi}$-isometric. Thus $\mathrm{Is}(\mathrm{SL}(2, \mathbf{R}), \tilde{\varphi}) \supset \mathrm{SL}(2, \mathbf{R}) \times \mathrm{SL}(2, \mathbf{R})$ and so $(\mathrm{SL}(2, \mathbf{R})/\Gamma, \varphi)$ has (at least) twice as much local isometries compared to global isometries.

Nonextendible local isometries are the major source of dynamical complexity of the action of $G = \mathrm{Is}(V, \varphi)$ on non-simply-connected V. To see this, let us show how the recurrence of a point necessarily brings along local isometries near this point which may not be exendible to global isometries.

Let $v \in V$ be a recurrent point, and let $g_i \to \infty$ be local isometries such that $v_i = g_i(v) \to v$. Thus we have isometries g_i^{-1} moving v to v_i, and as v_i are close to v we may expect these isometries form part of a *connected* local isometry group G_0 acting near the point v. If the structure φ in question is a Riemannian metric, then all g_i are *uniformly bounded*, and we construct G_0 by taking convergent subsequences of g_i. This does not work for general A-structures, but a certain modification of this argument does apply (see [72] and 4.4, 4.6.A in [29]). The conclusion one obtains roughly says that the local isotropy group $\mathrm{Is}^{\mathrm{loc}}(v)$, $v \in V$, is "essentially" as "big" as the full isometry group $\mathrm{Is}(V, \varphi)$ provided V is compact or (and) admits a finite smooth G-invariant measure. An instance of such a result is Theorem 4.6.A for semisimple groups. The following is a specific example.

1.12.A. **Example.** Consider again our "very important" $\mathrm{SL}(2, \mathbf{R})/\Gamma$-example (see 0.11.A, 1.9.C). Then $\mathrm{Is}(V) = \mathrm{SL}(2, \mathbf{R})$ and also the local isotropy group $\mathrm{Is}^{\mathrm{loc}}(v) = \mathrm{SL}(2, \mathbf{R})$ acting by conjugation, for all $v \in V$.

To see another example, we state the following.

1.12.B. **Proposition** (Compare 4.4 and 4.6 in [29]). *Let (V, φ) be a compact manifold endowed with a rigid C^{an} structure φ of algebraic type (see 5.5). If $G = \mathrm{Is}(V, \varphi)$ is noncompact, then there exists a point $v \in V$ such that the local isotropy subgroup G_v is noncompact. If, moreover, there is a smooth finite G-invariant measure on V, then G_v is not compact for all points $v \in V$.*

2. Geometric structures associated with Anosov actions

2.1. We describe in this subsection certain situations where an a priori complicated dynamics may preserve a continuous (and in rare cases smooth) rigid structure.

First, we recall the definition of a *hyperbolic* action (as introduced by Anosov in 1966 [2]) of the groups $G = \mathbf{Z}$ and $G = \mathbf{R}$ on a compact manifold V. We assume the action is locally free (which is automatic for

Z) and denote by $I \subset T = T(V)$ the subbundle of the vectors tangent to the orbits. Notice that the fiber dimension of I is

$$\dim I = \dim G = \begin{cases} 0 & \text{for } G = \mathbf{Z}, \\ 1 & \text{for } G = \mathbf{R}. \end{cases}$$

We start by recalling the contracting and expanding subbundles T^+ and T^- defined as follows (compare 0.2.A):

$$T^+ = \left\{ \tau \in T(V) | \lim_{g \to +\infty} \|Dg(\tau)\| = 0 \right\},$$

$$T^- = \left\{ \tau \in T(V) | \lim_{t \to -\infty} \|Dg(\tau)\| = 0 \right\},$$

where Dg denotes the differential of the action $g: V \to V$, $g \in G$, and $\| \ \|$ refers to a fixed Riemannian metric on V.

Recall (see 0.2.A) that T^+ and T^- are not, in general, subbundles (see examples in 2.3 following the definition of Anosov action below) as the dimension of the fiber may not be constant on V. Now, we have the first Anosov axiom:

A_1. T^+ and T^- are continuous subbundles in $T(V)$, and their fiber dimensions satisfy

$$\dim T^+ + \dim T^- = \dim V - \dim G.$$

Moreover, the subbundles T^+, T^-, and I Whitney split the tangent bundle $T(V) = T^+ \oplus T^- \oplus I$. The second axiom states that the convergence to 0 in the definition of T^+ and T^- is exponential.

A_2. There exist constants $C > 0$ and $\lambda > 1$ such that

$$\|Dg(\tau)\| \leq C\lambda^{-|g|} \|\tau\|$$

for $g \geq 0$ and $\tau \in T^+$ as well as for $g \leq 0$ and $\tau \in \Gamma^-$.

2.2. **Remarks.** (a) Notice that the definition of an Anosov action does not depend on the choice of the Riemann metric since V is compact.

(b) There exist two invariant foliations S^+ and S^- on V, whose tangent bundles are T^+ and T^- respectively. These are defined by the following construction: points v_1 and v_2 in V lie in a leaf of S^+ iff $\text{dist}(g(v_1), g(v_2)) \to 0$ for $g \to +\infty$, and S^- is similarly defined for $g \to -\infty$. The leaves of S^+ are called *stable* manifolds and those of S^- *unstable*.

2.3. **Examples.** (i) Let $V = S^n$, where S^n is identified with $\mathbf{R}^n \cup \{\infty\}$ via the stereographic projection. If the action of $G = \mathbf{R}$ on S^n is given by the scaling

$$x \mapsto e^g x \quad \text{on } \mathbf{R}^n,$$

then $T^+ \subset T(S^n)$ consists of all vectors tangent to S^n outside the south pole (which corresponds to $0 \in \mathbf{R}^n$). Similarly, T^- misses the north pole.

(ii) Another example is that of the action on S^n corresponding to a parallel translation on $\mathbf{R}^n$. Here both T^+ and T^- consists of the tangent vectors outside the south pole. The proof is an easy exercise for the reader.

2.4. The simplest example of an Anosov action. Let V be the 2-dimensional torus, i.e., $V = \mathbf{R}^2/\mathbf{Z}^2$, and let f be a diffeomorphism of V whose lift to $\mathbf{R}^2$ is the linear map f_0 given by the matrix $\left(\begin{smallmatrix} 2 & 1 \\ 1 & 1 \end{smallmatrix}\right)$. Then f generates an Anosov system. In fact, the eigendirections of the corresponding automorphism of the plane $\mathbf{R}^2$ define invariant line fields on the torus: the contracting and the expanding ones. Notice that the invariant line fields are smooth in this example.

Another remark is that there exists a quadratic form on $\mathbf{R}^2$ of signature $(1, 1)$ invariant under f_0. Such a form gives rise to an f-invariant Lorentz metric on T^2, which is also invariant under the translations of T^2 (see 3.4.1(i)). We shall show later in this section (see 2.6) that the existence of such a metric is rather typical for Anosov systems. Also, notice that the action of f on T^2 is topologically transitive and even ergodic (see 0.5.B) for the Haar measure on T^2.

2.4.A. **Remark.** The Anosov property is stable under C^1-small perturbation of the action. For example, a torus diffeomorphism C^1-close to the automorphism $\left(\begin{smallmatrix} 2 & 1 \\ 1 & 1 \end{smallmatrix}\right)$ of the above example is always Anosov although its contracting and expanding direction fields may be less than C^2-smooth even in the case when the diffeomorphism is analytic (see [3]).

2.5. Example. *The stable horospherical foliation.* Let us describe the stable and unstable foliations for the geodesic flow in the unit tangent bundle $V = UT(W)$ of a complete manifold W of negative sectional curvature K.

We start with the case where W is simply connected. We take a tangent vector $\tau \in UT(W)$ at $w_0 \in W$, and let $R = \mathbf{R}_+ \subset W$ be the geodesic ray issuing from w_0 and directed by τ. The conditions $K \leq 0$ and $\pi_1(W) = 0$ imply that (see, e.g., [40])

$$(*) \qquad \operatorname{dist}_W(t_1, t_2) = t_2 - t_1$$

for all segments $[t_2 - t_1] \subset R = \mathbf{R}_+$.

Now for each $w \in R \subset W$ we take the ball $B_w(r)$ of radius $r = r(w) = \operatorname{dist}(w, w_0)$ and observe with $(*)$ that $B_w(r') \supset B_w(r)$ for $r' \geq r$. The increasing union $B_\infty = \bigcup_{w \in R} B_w(r)$ is called the *horoball* defined by τ, and can be viewed as the ball of infinite radius with the center at infinity.

The topological boundary of B_∞ is called the *horosphere* $S = S(\tau)$ normal to τ. By using $K \le 0$ it is easy to see that S is a smooth hypersurface in W passing through w_0 and normal to τ.

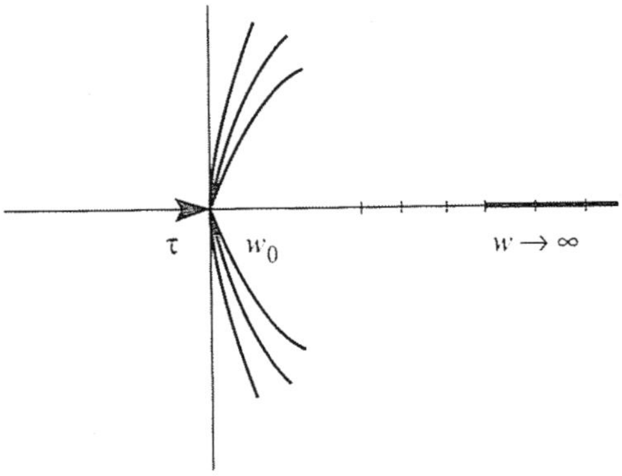

Next we consider the *normal lift* $\tilde{S}(\tau) \subset V = UT(w)$ of $S(\tau)$ where the point $\tilde{s}$ over each $s \in S(\tau)$ is represented by the unit tangent vector $\tau(s)$ normal to S and directed in the same way as τ. Since the lifted horospheres corresponding to different vectors, say $\tilde{S}(\tau_1)$ and $\tilde{S}(\tau_2)$, either coincide or, otherwise, are disjoint in V, they foliate the unit tangent bundle $V = UT(W)$. This foliation is invariant under the geodesic flow. In fact if we apply the geodesic flow g to $\tilde{S}(\tau)$, we obtain the lift of another horosphere, namely $\tilde{S}(g(\tau))$. The following diagram depicts what is happening downstairs in W.

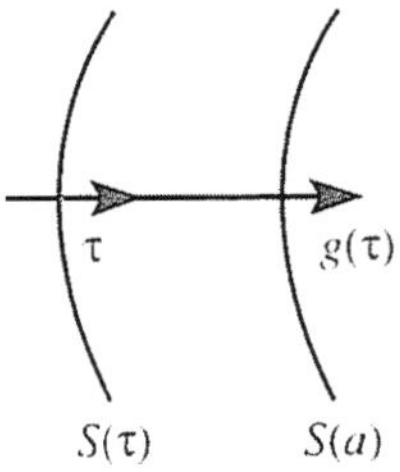

Now let W be compact and let us apply the above construction to the universal covering of W. Then we get a foliation of the universal covering which clearly is *invariant* under the deck transformation group and so defines a foliation of our compact $V = UT(W)$. It is not hard to see (for $K < 0$) that this is exactly the stable foliation for the geodesic flow. One gets the nonstable one by taking the horospheres $S(-\tau)$ for $\tau \in UT(W)$.

2.5.A. **Remark.** If W has constant negative curvature, then the horospheres in the Poincaré model are represented by the spheres tangent to the boundary S_∞ of the ball giving the Poincaré model of $W = \mathbf{H}^n$.

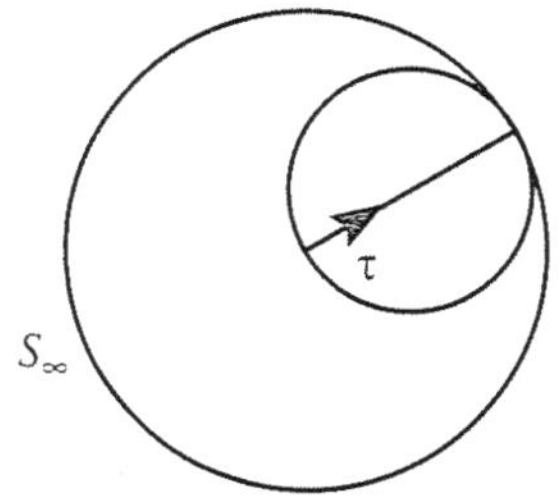

2.6. Quadratic forms on split spaces. The pair of subbundles $\pi = (T^+, T^-)$ associated to an Anosov **Z**-action (see (2.1)) is not a rigid structure as defined in 5.10 and 5.11. In fact, since T^+ and T^- are integrable, the local isometry group of π is infinite dimensional; it consists of the diffeomorphisms preserving the corresponding foliations S^+ and S^- and thus (locally) isomorphic to $\operatorname{Diff}\mathbf{R}^k \times \operatorname{Diff}\mathbf{R}^l$, where k and l are the dimensions of S^+ and S^-.

However, if an Anosov system preserves an additional structure φ, such as a symplectic (see 2.6.B below) or contact structure, then the "sum" (π, φ) may very well be rigid. In fact, by coupling π and φ we can obtain our old friend pseudo-Riemannian metric.

To do this, we start with a simple algebraic observation.

Let L be a linear space split into the direct sum $L = L' \oplus L''$, and let ω be a bilinear form defined on L. Then one can canonically construct a quadratic form, say $\overline{\omega}$, on L as follows:

$$\overline{\omega}(l, l) = \omega(l', l'') \quad \text{for all } l \in L,$$

where l' and l'' are projections of l to L' and L'', respectively.

2.6.A. **Lemma.** *Let ω be a nonsingular antisymmetric form on L, and let L', L'' be isotopic (often called Lagrangian) subspaces. Then the form ω is a nonsingular quadratic form of signature (n, n) for $2n = \dim L$, for which L' and L'' are isotropic. (We recall that a subspace in L' is called ω-isotropic if $\omega|L' \equiv 0$.)*

Proof. Take bases $\{l_i'\}$ and $\{l_j''\}$ of L' and L'' respectively in such a way that $\omega(l_i', l_j'') = \delta_{ij}$. Then, in the basis $\{l_i', l_j''\}$ of L, the form ω also has $\overline{\omega}(l_i', l_j'') = \delta_{ij}$.

2.6.B. Now, let an Anosov action of $\mathbf{Z}$ on V preserve a *symplectic* structure, that is, a closed nonsingular exterior 2-form ω. (The closeness of ω is immaterial at this point.) Then the expanding and contracting bundles T^+ and T^- are clearly ω-isotropic, and so the above lemma provides a continuous invariant pseudo-Riemannian metric φ on V, of signature (n, n) for $2n = \dim W$.

2.6.C. **Remark.** The above metric has not yet been successfully applied by anybody to study general Anosov systems. However, if one assumes that the subbundles T^+ and T^- are smooth, then the metric is also smooth, and so our theory of rigid invariant structures fully applies. For example, if one applies Theorem 0.7.D to this case, one obtains the following.

2.6.D. *Local homogeneity property.* If an Anosov $\mathbf{Z}$-action preserves a smooth symplectic structure ω on V, and the bundles T^+ and T^- are C^∞ smooth, then there is an open dense invariant subset $V_0 \subset V$ admitting a structure of locally homogeneous space, and the implied (local) Lie group G acting on V_0 preserves ω, T^+, T^-, and the metric φ.

2.6.E. **Remark.** The idea of the proof of 2.6.D can be seen in the discussion in 0.6, where we have already encountered this phenomenon in the case of a 2-dimensional manifold V and a C^2-diffeomorphism. In 0.6 we only needed C^2-smoothness, as we used the curvature of the manifold to obtain the homogeneity of all V (namely, the open dense invariant subset $V_0 \subset V$ admitting a structure of a locally homogeneous space turned out to be all of V). Thus, returning to our situation in the 2-dimensional case, we come to the following well-known fact (see [5]).

2.6.E$_1$. *If f is a C^2-smooth Anosov diffeomorphism of a smooth, compact connected and orientable surface preserving a smooth measure and having C^2-smooth stable and unstable foliations, then f is smoothly conjugate to a linear automorphism of the torus T^2.*

Recall that each automorphism of $T^2 = \mathbf{R}^2/\mathbf{Z}^2$ is given by an integral linear transformation of $\mathbf{R}^2$, that is,

$$\begin{pmatrix} x \\ y \end{pmatrix} \mapsto \begin{pmatrix} a & b \\ c & d \end{pmatrix} \begin{pmatrix} x \\ y \end{pmatrix},$$

with $a, b, c, d \in \mathbf{Z}$ and $ad - bc = \pm 1$.

2.6.E$_2$. One may hope that there is a similar classification in the higher-dimensional case (see the references at the end of 2.11).

2.7. R-actions and geodesic flows. We want to extend the previous discussion to the case of Anosov actions of the group $G = \mathbf{R}$. In fact, in this case we stand a much better chance of having an invariant rigid structure since the Anosov splitting $T(V) = I \oplus T^+ \oplus T^-$ together with a vector field X generating the one-dimensional subbundle I form, generically, a *rigid* structure (as opposed to the case $G = \mathbf{Z}$ where $\dim I = 0$). Here X denotes the field generating the action, and the notion of genericity applies to the triples (X, T^+, T^-) where the subbundles T^+ and T^- are integrable.

To see that genericity implies rigidity we consider the (one-codimensional) subbundle $T' = T^+ \oplus T^- \subset T(V)$, and let ξ be the 1-form defined by the following two conditions:

$$\ker \xi = T', \quad \xi(X) = 1.$$

We take the differential $\omega = d\xi$, and assume that the 2-form ω is nonsingular on T'. Then, as earlier, we obtain a quadratic form φ' on T'.

2.7.A. *Warning.* The notions of genericity and rigidity in the above discussion are by no means rigorous. First of all, both notions apply, strictly speaking, to smooth objects, while T^+ and T^- are in general only continuous. (One also needs C^1-smoothness to take the differential of ξ, but this is a less serious matter.)

Secondly, a 2-form on T' can be nonsingular only if $\dim T'$ is even.

Granted this, the nonsingularity can be generically ensured locally, near a fixed point, but not over all of V. Finally, one cannot honestly apply the notion of genericity to T^+ and T^-, as these come as a result of a specific (infinite) construction and cannot be changed with a sufficient freedom underlying the idea of genericity.

Yet, for all these shortcomings the above discussion put forth strong evidence in favor of (generic) rigidity of (X, T^+, T^-).

2.8. Symplectic systems. Now, let us suppose that an $\mathbf{R}$-action on a smooth manifold preserves a symplectic form ω. Notice that such an action cannot be Anosov.

In fact, since ω and the generating field X are invariant, the condition $\lim_{g\to\infty} \|Dg(\tau)\| = 0$, $g = \mathbf{R}$, for some tangent vector field τ implies that $\omega(\tau, X) = 0$. It follows that the subbundles T^+ and T^- are contained in the (one-codimensional) subbundle $\operatorname{Ker} X^*_\omega$, where X^*_ω denotes the 1-form $\tau \mapsto \omega(\tau, X)$. Since ω is antisymmetric, the field X is also contained in $\operatorname{Ker} X^*_\omega$ and so T^+, T^-, and X do not span $T(X)$ as required by the Anosov condition.

Next we recall that the Lie derivative L_X acting on the exterior forms can be expressed in terms of the exterior differential d and the interior product with X, called i_X, by the following (obvious and well-known) formula

$$L_X = i_X \circ d + d \circ i_X. \tag{$*$}$$

Then, by the definition of i_X, we have $i_X \circ \omega = X^*_\omega$ and observe that $L_X\omega = 0$ as ω is invariant under the flow generated by X. Since also $d\omega = 0$, from $(*)$ it follows that

$$0 = L_X\omega = dX^*_\omega,$$

which implies that X^*_ω is closed, and so the subbundle $\operatorname{Ker} X^*_\omega$ is integrable. Since $X \subset \operatorname{Ker} X^*_\omega$, the leaves of the resulting foliation are invariant under our **R**-action and we fix our attention on one such leaf, denoted by V. Now the bundles I, T^+, and T^- restricted to V are contained in $T(V)$ and may very well provide the Anosov splitting of the **R**-action restricted to V. If indeed the action is Anosov on V, then the restriction of ω on the 1-codimensional subbundle $T^+ \oplus T^- \subset T(V)$ is nonsingular, because ω is nonsingular, and $T^+ \oplus T^-$ is transversal and ω-orthogonal to X at the same time. With a nonsingular ω on $T^+ \oplus T^-$ we obtain as earlier a nonsingular quadratic form of signature $(n-1, n-1)$ on $T^+ \oplus T^-$ for $2n - 1 = \dim V$. This form orthogonally adds up with the form on I defined by being equal to 1 on X, and thus we get a pseudo-Riemannian metric of type $(n, n-1)$ on V.

2.9. Geodesic flows. Let us recall the symplectic description of the geodesic flow on $V = UT(W)$ for a manifold W with a Riemannian metric h. First we define the *canonical symplectic form* ω on the cotangent bundle $T^*(W)$ by $\omega = d\eta$, where η is the tautological 1-form on $T^*(W)$. Namely, this η is uniquely characterized by the following identity which must hold for all smooth 1-forms α on W, which are also viewed as sections $\alpha\colon W \to T^*(W)$,

$$(+) \qquad \alpha^*(\eta) = \alpha,$$

where α^* denotes the form on W induced from η by the map $\alpha\colon W \to T^*(W)$, and α on the right-hand side of $(+)$ is thought of as a 1-form on W.

In the case $W = \mathbf{R}^n$ one can see that

$$\eta = \sum_{i=1}^{n} y_i\, dx_i,$$

where x_i are the coordinates of $\mathbf{R}^n$, and y_i are the (impulse) coordinates in the cotangent space $T_0^*(\mathbf{R}^n)$ $(= \mathbf{R}^n)$, where we use the splitting

$$T^*(\mathbf{R}^n) = \mathbf{R}^n \times T_0^*(\mathbf{R}^n).$$

The above η can be thought of as a (universal) form on $\mathbf{R}^n$ with undetermined coefficients y_i. Then every α-section is given by n functions

$$y_i = y_i(x_1, \cdots, x_n), \qquad i = 1, \cdots, n,$$

and the (induced) α-form on $\mathbf{R}^n$ becomes

$$\sum_{i=1}^{n} y_i(x_j)\,dx_i,$$

which agrees with the tautological definition of η. The advantage of the above local formula for η is the following expression for ω,

$$\omega = d\eta = \sum_{i=1}^{n} dy_i \wedge dx_i,$$

which clearly shows ω is nonsingular.

To go further we invoke the metric h (which has not been used so far) and define the following (Hamilton) function on $V' = T^*(W)$,

$$H(Q) = \|Q\|_h^2,$$

for all covectors Q in $T^*(W)$. Then we take the ω-gradient $Y = \mathrm{grad}_\omega H$, defined by the equality,

$$i_Y\omega = dH,$$

which means $\omega(Y, \tau') = dH(\tau')$ for all tangent vectors τ' on V'. Then the above $(*)$ for the Lie derivative tells us that

$$L_Y\omega = ddH + i_Y\, d\omega = 0,$$

which makes ω invariant under the flow generated by Y.

Next, the (already used) relation

$$Y_\omega^* \overset{\text{def}}{=\!=} i_Y\omega = dH$$

shows that the levels of the function H are invariant under this flow. In particular, the flow preserves the unit cotangent bundle $UT^*W = \{H = 1\} \subset T^*(W)$.

Finally, we use the metric h to identify vectors with covectors. Thus we obtain a diffeomorphism between $T(W)$ and $T^*(W)$, which brings ω and the flow from $T^*(W)$ to $T(W)$, while the level $\{H = 1\} \subset T^*(W)$ goes to the unit tangent bundle $V = UT(W) \subset T(W)$. This is invariant under the flow as well as the level $\{H = 1\}$ in $T^*(W)$.

Now, comes the punchline: *The flow in* $V = UT(W)$ *which we have just constructed, is identical with the geodesic flow.*

To prove this one should write down the defining vector field X for the geodesic flow and identify it with Y transported to $T(W)$. This is quite easy (see, e.g., [40]). Yet, if one wants to avoid explicit formulas for the geodesic flow, one may proceed in a less formal way which is conveniently divided into three steps.

Step 1. Show that the two flows are identical in the (simplest) case o.f.: $W = \mathbf{R}$ with the standard metric. This is immediate.

Step 2. Do the same for $W = \mathbf{R}^n$ by the splitting $\mathbf{R}^n = \mathbf{R} \times \mathbf{R}^{n-1}$, where the pertinent $\mathbf{R}$ is the line in $\mathbf{R}^n$ defined by the vector $\tau \in T(\mathbf{R}^n)$ at which we want to check the equality of the two fields.

Step 3. Recall that the vector fields on $T(W)$ defining the geodesic flow is algebraically expressed in terms of the first jet of h. In fact, one uses the Γ^h_{ij}-coefficients, but we do not care about the specific shape of the formula. Since every Riemannian metric h on W is Euclidean up to the second order at every point $w \in W$, the conclusion of Step 2 extends to all (W, h).

This argument still does not quite reveal the geometric reason for the existence of an invariant symplectic (or any other) structure for the geodesic flow. Here is an alternative geometric description of this structure, which makes the invariance clear.

First we recall that every normally oriented hypersurface $S \subset W$ lifts to $UT(W)$ by the unit normal field to S defined by the coorientation (compare 2.5). The lifted hypersurfaces give us a distinguished class of $(n-1)$-dimensional submanifolds in $UT(W)$ called *base* submanifolds. Distinguishing such a class is, in fact, equivalent to giving a geometric structure to $UT(W)$. In the present case one can define this structure by taking all base submanifolds in $V = UT(W)$ passing through a given point $v \in V$ and taking the linear subspace $K_v \subset T_v(W)$ generated by all these submanifolds. This K_v happens to be of codimension one for all $v \in V$ and the resulting codimension-one subbundle $K \subset T(V)$ is called the *canonical contact structure* on $UT(W)$. In fact, one can easily see that K is identical with $T^+ \oplus T^-$ for the geodesic flow in the negative curvature case.

Now we apply the geodesic flow g_t to some base manifold $\tilde{S}$ which is the lift of a smooth hypersurface $S \subset W$. Then, at least for small t, $g_t(\tilde{S})$ equals the lift $\tilde{S}_t$ for the *equidistant* hypersurface $S_t \subset W$ having

$$\mathrm{dist}(s, S) = t$$

for all $s \in S_t$ (compare 2.5). Thus the structure defined by the base manifolds (whatever its name) is invariant under the geodesic flow.

We invite the reader to follow this discussion to the end and to relate the above geometric picture with the symplectic formalism. We also suggest one look at what happens to the round spheres $S^{n-1} \subset \mathbf{R}^n$ lifted to

$UT(\mathbf{R}^n)$ when the equidistant hypersurface reduces to a single point in the center of S^{n-1}.

2.10. Hyperbolic geodesic flows. Once we obtain an invariant symplectic structure ω for the geodesic flow (recall that this ω comes from the canonical form on $T^*(W)$) we can construct an invariant pseudo-Riemannian metric on $V = UT(W)$ in the Anosov case.

2.10.A. **Theorem** (Cartan-Kanai). *Let W be a compact manifold of negative curvature $K < 0$. Then the geodesic flow on $V = UT(W)$ preserves a continuous pseudo-Riemannian metric φ on V.*

Furthermore, if the horospherical foliations in $UT(W)$ are C^k-smooth, then φ is also C^k.

2.10.B. **Remarks.** (i) In general, the horospherical foliations are only continuous (in fact they are Hölder continuous), and so is φ.

(ii) One knows (see [34] that for $\frac{1}{4}$-pinched manifolds, $-1 < K < -\frac{1}{4}$, the foliations are C^1-smooth, and so is φ.

(iii) If we assume the foliation C^∞, then we can use the rigidity theory to obtain, in particular, an open dense locally homogeneous subset in $UT(W)$ as in the case of **Z**-actions. In fact, one expects W is a locally symmetric manifold in this case.

(iv) Many properties of geodesic flows for $K < 0$ can be seen by looking at the *ideal boundary* $\partial_\infty = \partial_\infty \tilde{W}$ of the universal covering $\tilde{W}$ of W (see [55], [15], [6]). This ∂_∞ is a topological space homeomorphic to S^{m-1}, $m = \dim W$, and the deck transformation group $\Gamma = \pi_1(W)$ acts on ∂_∞ by homeomorphisms.

The smoothness of the horospherical foliations implies the existence of a Γ-*invariant smooth structure* (i.e., the structure of a smooth manifold) on ∂_∞, and one wants to know how this structure on ∂_∞ influences the (dynamics of the) action of Γ on ∂_∞. More precisely, one has the following.

2.10.C. **Questions.** When does ∂_∞ admit a Γ-invariant C^α-structure? Here α may be a real number $\alpha = k + \varepsilon$, $0 \leq \varepsilon \leq 1$, where ε refer to the Hölder continuity of the k th derivatives. For example, does the existence of a Γ-invariant C^α-structure for $\alpha \geq 2$ imply that W is homotopy equivalent to a locally symmetric manifold? Another question (motivated by the Mostow rigidity theorem) concerns the uniqueness of Γ-invariant structures. A natural approach to these problems would be to pass from a smooth structure to a Γ-invariant rigid A-structure. For example, one may seek a (generalized) conformal structure on S_∞ and (or) a symplectic structure on $S_\infty \times S_\infty$.

2.10.D. **Remark.** If one goes back from a structure on ∂_∞ to $V = UT(W)$, one does not obtain a structure invariant under the geodesic flow but rather some *transversally invariant* structure for the 1-dimensional foliation into the orbits of this flow.

2.11. A very brief overview of some results and problems in Anosov systems. The most important and quite amazing property of Anosov systems is their *topological stability*. That is, if A' is a small C^1-perturbation of our Anosov action A on V, then there exists a homeomorphism (which is not, in general, C^1-smooth) of V close to the identity which sends each orbit of A' to an orbit of A. Notice that for $G = \mathbf{Z}$ such a homeomorphism necessarily commutes with the action, but this is not so for $G = \mathbf{R}$. This result in the complete generality is due to Anosov, but in many important cases this goes back to Birkhoff and Morse. In fact, Morse [54] has essentially proven a deeper global version of the stability theorem for compact manifolds W_1 and W_2 of negative curvature: *If* $\pi_1(W_1) = \pi_1(W_2)$, *then there exists a homeomorphism* $UT(W_1) \to UT(W_2)$ *which sends each orbit of the geodesic flow of* W_1 *to that of* W_2.

A similar global result is known (see [19], [48]) for Anosov $\mathbf{Z}$-systems on *infra-nil-manifolds* V. These systems generalize linear automorphisms of tori and exhaust all *known* Anosov $\mathbf{Z}$-actions up to topological equivalence. One may think that the infra-nil-systems give a complete list of Anosov actions. Yet one cannot even prove that *no simply connected* manifold supports an Anosov action.

There are by far more known Anosov actions for $G = \mathbf{R}$ than for $\mathbf{Z}$. This is almost entirely due to the abundance of manifolds of negative curvature. But one does not know, for example, if the existence of an Anosov $\mathbf{R}$-action on V implies that the fundamental group of V has exponential growth.

Notice that the above mentioned problems have motivated the study of the relations between $\pi_1(V)$ and $\mathrm{Is}(V)$ for V with a rigid structure. Unfortunately, there is no feedback so far except for Anosov systems with *smooth* foliations S^+ and S^- which is an extremely restrictive assumption.

The second beautiful feature of Anosov actions is the density of the periodic points (i.e., the points with compact orbits) provided the recurrent points are dense. This is due to Hedlund, Hopf, and Busemann for certain classes of hyperbolic geodesic flows, and to Anosov in the general case. One does not know if the recurrency condition is always satisfied for $G = \mathbf{Z}$.

The third basic property is the *ergodicity* of the actions preserving a smooth invariant measure. This general result is due to Sinai and Anosov, while some special cases of hyperbolic geodesic flows go back to Hedlund [33] and Hopf [37]. Nowadays one knows completely the measure theoretic structure of an Anosov system A. This structure is determined by a single invariant, the *measure entropy* of A, unless $G = \mathbf{R}$ and the system is obtained as the mapping cylinder of a **Z**-action. (See [2], [3], [13] for a more complete discussion.)

Geometric structures invariant under hyperbolic (and not only hyperbolic) actions have been studied since the earlier days of classical mechanics and differential geometry. For example, the symplectic structure for the geodesic flow was revealed by Poincaré (following Lagrange, Hamilton, Liouville, etc.). The Anosov structure for manifolds of negative curvature can be traced back to Lobachevski and Hadamard, and the invariant pseudo-Riemannian metric to E. Cartan. Interest in these structures has recently been revived in an attempt to classify (in the smooth category) Anosov's systems with the smooth stable and unstable foliations (see [7], [18], [17], [39]). A typical result in this direction is the following theorem of Hurder-Katok and Ghys: *If a compact negatively curved surface W has C^2-smooth horospherical foliations, then W has constant curvature.*

For higher-dimensional manifolds W with $K(W) < 0$ the results are less complete. For example, one knows (see [16]) that *if the horospherical foliations are C^∞, then the geodesic flow is C^∞-isomorphic to that of some manifold W' with constant curvature $K' < 0$, provided that either* $\dim W$ *is odd or* $-4 < K(W) < -1$. (Of course one expects that $K(W)$ itself is constant.)

Another result in this direction concerns arbitrary (not necessarily geodesic) Anosov flows whose *Lyapunov subbundles* are C^∞-smooth. Namely every such flow on a compact manifold is *C^∞-isomorphic* to the geodesic flow on a locally symmetric manifolds of **R**-rank = 1 (see [7]). (A priori, a Lyapunov subbundle $\tau \subset T(V)$ is only measurable. Thus the theorem applies to those flows where *every* invariant measurable subbundle is smooth.) Recently, this result was extended in [7] to all contact flows with smooth stable and unstable foliations (this more general result makes the above Lyapunov bundles irrelevant). Furthermore, the C^∞-condition was reduced to some C^k, $k < \infty$, in [32].

3. Isometries of simply connected real analytic manifolds

3.1. Now we return to a smooth A-rigid (see 1.8.A and also §5) manifold (V, φ). In this section we want to justify our earlier claim that the assumption $\pi_1(V)=0$ roots out any nontrivial dynamics of $G=\mathrm{Is}(V,\varphi)$-actions on V.

Unfortunately, all results which we state in this section need the real analyticity of φ, and we do not know what happens in the C^∞-case. We start by recalling some basic properties of $\mathrm{Is}(V, g)$ for C^{an} manifolds (the reader can consult §3 in [29] for more details).

3.2. Let φ be a C^{an} smooth rigid A-structure (see 0.4, 1.8, and also §5 for the definitions) on a compact manifold V. Then we have the following Propositions 3.2.A and 3.2.B.

3.2.A. *If V is simply connected, then the following hold:*

(i) *The group $\mathrm{Is}(V, \varphi)$ has at most finitely many connected components.*

(ii) *The isotropy subgroup $\mathrm{Is}(V, \varphi, v) \subset \mathrm{Is}(V, \varphi)$ also has finitely many connected components for all $v \in V$.*

(iii) *The topology induced from V on the $\mathrm{Is}(V, \varphi)$-orbit of each point $v \in V$ equals the quotient topology on $\mathrm{Is}(V, \varphi)/\mathrm{Is}(V, \varphi, v)$. In fact the orbits are semianalytic subsets (see 3.5.1.B) in V, and there exists at least one orbit which is a closed C^{an}-submanifold in V.*

3.2.B. *Assume as earlier $\pi_1(V) = 0$ and let $\mathrm{Is}(V, \varphi)$ preserve a smooth volume element on V. Then* (i) *all orbits of $\mathrm{Is}(V, \varphi)$ are compact;* (ii) *the group $\mathrm{Is}(V, \varphi)$ contains a closed connected normal abelian subgroup $A_0 \subset \mathrm{Is}(V, \varphi)$ such that $\mathrm{Is}(V, \varphi)/A_0$ is compact.*

The complete proof of these statements (which can be found, together with an extensive discussion of related theorems, in [29, §3.5 and 3.7]) is too technical and involved to be presented in detail here. However, in order to give some indications of why 3.2.A and 3.2.B are true, we shall briefly describe in 3.5 below a proof of 3.2.A(i) and 3.2.B(ii) valid for the special case when the structure φ in quesiton is a pseudo-Riemannian metric on V.

3.3. Remark. The reader might notice that the result 3.2.B(i) ensuring compactness of $\mathrm{Is}(V, \varphi)$-orbits under certain conditions on V and φ has already been quoted twice in these notes (compare 0.12.C where this was stated for a pseudo-Riemannian metric φ, and see also 1.11.B).

3.4. Isometries of Lorentz manifolds. Assume that (V, φ) is an n-dimensional C^{an} compact and simply connected Lorentz manifold. Then we know from Theorem 0.12.A in the Introduction that the isometry group $\mathrm{Is}(V, \varphi)$ is compact.

In fact, a key ingredient of the proof of Theorem 0.12.A (see [14]) is the compactness of $\mathrm{Is}(V, \varphi)$-orbits. This follows from Proposition 3.2.B(i) applied to the pseudo-Riemannian metric (which is an A-rigid structure) where the needed invariant measure comes from the pseudo-Riemannian volume element on V (which is obviously invariant under $\mathrm{Is}(V, \varphi)$, and finite for compact V).

The following remarks apply, strictly speaking, to Theorem 0.12.A concerning pseudo-Riemannian manifolds, but may also be used in the general discussion.

3.4.1. **Remarks.** (a) In Theorem 0.12.A real analyticity is only used for the following: each point $v \in V$ admits a neighborhood $U_v \subset V$ such that for every smaller connected $U' \subset U_v$, every Killing field on U' extends to U_v (compare with 5.15). This property for Riemannian C^{an} manifolds was proven by Nomizu [58], and his argument immediately generalizes to pseudo-Riemannian manifolds. The simple connectedness in 0.12.A is needed for the extension of local Killing fields to all of V (see 1.12).

(b) It is unknown whether Theorem 0.12.A remains true for C^∞ metrics. The major difficulty in the C^∞ case is that there may exist nonextendible local Killing fields. (For more on this point the reader can see §1.7 in [29], and also 1.12 and 5.15 in these lectures.)

(c) The Lorentz $(n-1, 1)$ condition on the signature of the metric tensor φ in 0.12.A is essential. In fact, the manifold $V = S^3 \times S^3 \times S^3$ admits an analytic metric of type $(7, 2)$ whose isometry group is $T^3 \times \mathbf{R}$ (see §5 in [14] and 1.11.D in these lectures).

(d) Note that if one allows $\pi_1(V) \neq \{0\}$, one may have a noncompact isometry group starting from dimension 2.

The simplest example is as follows (compare 2.4):

(i) Take $V = T^2 = \mathbf{R}^2/\mathbf{Z}^2$ and an automorphism $A\colon T^2 \to T^2$ which lifts to the linear map $\tilde{A}\colon \mathbf{R}^2 \to \mathbf{R}^2$ which has real eigenvalues λ_1 and λ_2, $\lambda_1 \neq \lambda_2$. For example, $\tilde{A}$ is given by the matrix $\left(\begin{smallmatrix} 2 & 1 \\ 1 & 1 \end{smallmatrix}\right)$, and then $\lambda_1 = (3+\sqrt{5})/2$, $\lambda_2 = (3-\sqrt{5}/2)$.

Let x_1 and x_2 be the corresponding eigenvectors, and define a quadratic form $\tilde{g}$ on $\mathbf{R}^2$ by $\tilde{g}(\tilde{x}_1, \tilde{x}_1) = \tilde{g}(\tilde{x}_2, \tilde{x}_2) = 0$ and $\tilde{g}(\tilde{x}_1, \tilde{x}_2) = 1$. This $\tilde{g}$ is clearly $\tilde{A}$-invariant, and so induces an A-invariant metric g on T^2. It follows that the isometry group of (T^2, g) is noncompact as it contains $\mathbf{Z}$ (generated by A) as a closed subgroup, unless $A^2 = \mathrm{Id}$.

(ii) Another very important example is that of $V = \mathrm{SL}(2, \mathbf{R})/\Gamma$ discussed in 0.11.A (see also 1.9.C).

Finally, observe that in example (i) above the fundamental group $\pi_1(T^2/\mathbf{Z}^2)$ is $\mathbf{Z}^2$, which is not too much. But in the case of

$V = \mathrm{SL}(2, \mathbf{R})/\Gamma$ the fundamental group is at least as large as Γ. One may wonder if the size of $\pi_1(V)$ influences that of the isometry group $\mathrm{Is}(V)$; some answers in this direction are presented in [29].

3.5. Outline of the proof of 3.2.A(i), (ii) and of 3.2.B for the case of a pseudo-Riemannian structure. We describe here a somewhat simplified version of the argument in [29] for the case when (V, φ) is a pseudo-Riemannian manifold of type (p, q), $p + q = \dim V$. In the original proof (see §§3.5 and 3.7 in [29]), 3.2.A and 3.2.B are derived from an appropriate form of the Frobenius theorem combined with a result of Rosenlicht (see 6.4.A and see also 5.14) on the orbit structure of algebraic actions. Here, since we work in the C^{an} category, we do not make explicit use of the Frobenius theorem.

3.5.1. Let (V, φ) be an n-dimensional connected pseudo-Riemannian manifold. Denote by $E \to V \times V$ the fibration where the fiber E_{v_1, v_2} equals the set of linear isometries between the tangent spaces $T_{v_1} \to T_{v_2}$. By using exponential coordinates at v_1 and v_2 we assign to each $e: T_{v_1} \to T_{v_2}$ a germ of a diffeomorphism of V sending $v_1 \to v_2$ and call it $\tilde{e}$. Then we define $E^r \subset E$ as the subset consisting of maps $e: T_{v_1} \to T_{v_2}$ for all v_1, v_2 such that the metric φ and the induced metric $\varphi' = \tilde{e}^*(\varphi)$ have equal Taylor coefficients of order $\leq r$ at v_1.

Next we compactify E and E^r as follows. Define a pseudo-Riemannian metric $\overline{\varphi}$ on $V \times V$ by $\overline{\varphi} = \varphi \oplus -\varphi$, and let $\overline{E} \to V \times V$ be the space of n-dimensional $\overline{\varphi}$-isotropic subspaces in $T(V \times V)$. The space E naturally embeds into $\overline{E}$ as an open dense subset. That is, each linear isometry $e: T_{v_1} \to T_{v_2}$ goes to its graph $\Gamma_e \subset \Gamma_{v_1} \times \Gamma_{v_2} = T_{(v_1, v_2)}$.

Then, we use the exponential map in the product manifold $V \times V$, and denote by $\hat{e}$ the germ at the origin of the exponential image of $\overline{e} \in \overline{E}$. This is a local n-dimensional submanifold in V. If $\overline{e} \in E \subset \overline{E}$, then the corresponding $\hat{e}$ is the graph of the germ $\tilde{e}$. Let $\overline{E}^r \subset \overline{E}$ be the set of those $\overline{e}$ where the metric $\overline{\varphi}|\hat{e}$ vanishes with order r at the origin. Clearly, $\overline{E}^r \supset E^r \subset E \subset \overline{E}$. Notice that if V is compact, then $\overline{E}$ and $\overline{E}^r$ are compact for all r.

We are now in a position to prove the following statement:

3.5.1.A. *If V is C^{an}, then $\overline{E}^r \subset E$ is a compact analytic subset.*

Proof. For every $\overline{e}$ at a point $w \in W \times W$ we denote by $J^r_{\overline{e}}$ the space of r-jets of quadratic differential forms at $w \in \hat{e} \subset V \times V$ on the submanifold $\hat{e} \subset V$. Clearly, this is a finite-dimensional vector space, and the union $J^r = \bigcup_{\overline{e}} J^r_{\overline{e}}$ has a natural structure of a real analytic vector bundle $J^r \to \overline{E}$.

This bundle J^r comes along with a section $j: \overline{E} \to J^r$, where to each $\overline{e}$ we assign the jet of the form $\varphi \oplus -\varphi$ on $V \times V$ restricted to $\hat{e}$. Clearly, j is real analytic, and the above condition "equal Taylor coefficients of order r" is equivalent to the vanishing of j. Thus $\overline{E}^r$ is represented as the zero set of an analytic section. q.e.d.

Now, note that the complement $\overline{\Sigma}^r = \overline{E}^r \backslash E^r$ consists of those $\overline{e} \subset T(V \times V)$ whose projection to $T(V)$ is *not* injective. It follows that $\overline{\Sigma}^r$ is a compact analytic subset in E^r. Since $\overline{E}^{r+1} \subset E^r$, we conclude, in the compact analytic case (see 3.5.1.B) that, $\overline{E}^{r+1} = \overline{E}^r$ for large r and similarly $\overline{\Sigma}^{r+1} = \overline{E}^r$. Hence, $E^{r+1} = E^r$ for large r (see 3.5.1.B$_2$).

Let $E^\infty = E^r$ for large r. Then using analyticity of φ one can see that $e \in E^\infty$ if and only if the germ $\tilde{e}$ is a local isometry. Next, fix a point $v_0 \in V$, and let $E(V_0) \subset E$ consist of linear isometries $T_{v_0} \to T_v$ for all $v \in V$. Notice that $E(v_0)$ is a real analytic set.

Let $E^\infty(v_0) = E^\infty \cap E(v_0)$, and observe that the isometry group $\mathrm{Is}(V, \varphi)$ embeds into $E^\infty(v_0)$ by $f \to D_{v_0} f$, where $D_{v_0} f$ stands for the differential of the isometry f at the point v_0. If V is a simply connected C^{an}-manifold, then it is not difficult to show (see [29]) that the image of $\mathrm{Is}(V, \varphi)$ in $E^\infty(v_0)$ is the union of some connected components of $E^\infty(v_0)$. In fact, this is equivalent to the following property of Killing fields proven by Nomizu in [58]: If V is a simply connected real analytic manifold, then every germ of a Killing field at v_0 extends to a Killing field on all of V. Nomizu's original result refers to the case of a C^{an} Riemannnian manifold, but this argument immediately extends to pseudo-Riemannian manifolds (compare 3.4.1(a)).

Since $E^\infty(v_0) = \overline{E}^\infty \cap \overline{E}(v_0) \backslash \overline{\Sigma}^\infty$ is a difference of two compact analytic sets, it has at most finitely many connected components (see 3.5.1.B$_3$).

Similarly, one can see that the isotropy subgroup $\mathrm{Is}_v(V, \varphi)$ has finitely many connected components. In fact, Is_v equals the intersection $E_{v,v} \cap E^\infty$ and hence is an analytic (even algebraic) subset in $E_{v,v} = \mathrm{Is}(T_v(V)) \equiv \mathrm{O}(p, q))$. (Recall that here φ is a pseudo-Riemannian metric of type (p, q).)

3.5.1.B. *Semianalytic sets.* Let us state here the properties of semianalytic sets used in the above argument. We start with

3.5.1.B$_1$. **Definition.** A subset A in a compact real analytic manifold V is called *analytic* if A equals the zero set of a system of real analytic functions on V.

It is clear that the pullback of an analytic set under an analytic map is analytic, and that the zero set of an analytic section of an analytic vector bundle is real analytic.

It is equally clear that the finite union and the finite intersection of real analytic sets are real analytic. Less obvious is the following classical result.

3.5.1.B$_2$. **Nöther property.** *Every decreasing sequence of real analytic sets*

$$A_0 \supset A_1 \supset A_2 \cdots$$

stabilizes. That is, there exists an r, such that

$$A_\infty \stackrel{\text{def}}{=} \bigcap_i A_i = A_r.$$

Next, we call a subset $A \subset V$ An-*constructive* if it is the difference of two analytic sets, i.e., if $A = A_1 - A_2$. For us an important property of An-constructive sets reads:

3.5.1.B$_3$. *Every* An-*constructive set has at most finitely many connected components.* In fact one knows that, more generally, every *semianalytic* subset in V has at most finitely many connected components when one defines semianalytic sets as the sets of solutions of finite systems of analytic inequalities.

For a more extensive treatment of the basic geometric facts in the theory of semianalytic sets we refer the reader to the work of Lojasiewicz (see [46], [47]).

3.5.2. **Useful remark.** Observe that, besides the group $\mathrm{Is}(V, \varphi)$ itself, one can find other subgroups, say, $A \subset \mathrm{Is}(V, \varphi)$, whose action on V enjoys properties 3.2.A(i), (ii) discussed in 3.5.1 above.

For example, if A is the centralizer of a system of connected subgroups in $\mathrm{Is}(V, \varphi)$, then A is the full isometry group of the structure, call it φ', in V obtained by augmenting φ with a system of Killing fields generating the subgroup in question (compare with 1.10). Since the metric φ is rigid, the structure φ' is also rigid (see 0.4(c)). Moreover, the previous discussion also applies to the structure φ' so that 3.2.A(i), (ii) hold for $\mathrm{Is}(V, \varphi')$ and its isotropy subgroup $\mathrm{Is}_v(V, \varphi')$ for all $v \in V$.

A useful application of this is the following 3.5.2.A which allows us to reduce a large part of the proof of Theorem 0.12.A to the case of an *Abelian* isometry group. (For more on this, see 3.5.5 at the end of this section.)

3.5.2.A. *Abelianization trick.* Let $(X_1, \cdots, X_m)$ be a *maximal* system of commuting Killing fields on V, and let $A \subset \mathrm{Is}(V, \varphi)$ be the connected Abelian subgroup generated by $(X_1, \cdots, X_m)$. Then the subgroup $\mathrm{Is}(V, \varphi') \subset \mathrm{Is}(V, \varphi)$ for $\varphi' = (\varphi, X_1, \cdots, X_m)$ equals the centralizer of A in $\mathrm{Is}(V, \varphi)$.

Since A is maximal, the connected identity component $\mathrm{Is}^0(V, \varphi) \subset \mathrm{Is}(V, \varphi)$ equals A.

3.5.3. We explain now why the orbits of $\mathrm{Is}(V, \varphi)$ are compact. We begin by considering the space $\mathbf{R}^{p,q}$ with the standard form

$$h_0 = \sum_{i=1}^{p} dx_i^2 - \sum_{j=p+1}^{n} dx_j^2,$$

where (p, q) is the type of φ. Consider pseudo-Riemannian metrics h on $\mathbf{R}^{p,q}$ such that $h - h_0$ together with its first derivatives vanishes at the origin, and let H^r be the space of the first r Taylor coefficients of all such h. Note, that H^r is a linear space of dimension

$$\frac{n(n+1)}{2}\left(1 + n + \cdots + \frac{(n+r-1)!}{(n-1)!r!}\right) - \frac{n(n+1)}{2} - \frac{n^2(n+1)}{2},$$

and that the group $\mathrm{O}(p, q)$ naturally (and linearly) acts on H^r.

Next, we take an orthonormal frame F at some point $v \in V$, and let $D^r(F, v, \varphi) \subset H^r$ denote the string of the first r Taylor coefficient of φ in the exponential coordinates in V corresponding to F. Changes of the frame F correspond to the above action of $\mathrm{O}(p, q)$ on H^r, and so we get a map of V into $H^r/\mathrm{O}(p, q)$, say $\mathscr{D}_\varphi^r : V \to H^r/\mathrm{O}(p, q)$. The quotient space $H^r/\mathrm{O}(p, q)$ is not a Hausdorff space. However, by the algebraic quotient theorem (see, for example, [71], [62], [57], and also 6.4.A, 5.14.B), there exists an $\mathrm{O}(p, q)$-invariant real algebraic stratification of H^r, say $H^r = H_0^r \cup H_1^r \cup \cdots \cup H_{i_r}^r$, such that each quotient space $H_i/\mathrm{O}(p, q)$, $0 \le i \le i_r$, is a manifold, and the quotient map $H_i \to H_i/\mathrm{O}(p, q)$ is a smooth fibration (compare with 5.14.B). Then one can easily show (see §3 in [29] for details) that for each r there exists a stratum H_i^r, for some $0 \le i \le i_r$, such that the pullback $V_i^r = (\mathscr{D}_\varphi^r)^{-1}(H_i^r/\mathrm{O}(p, q))$ is an open dense subset in V, and the map $\mathscr{D}_\varphi^r$ is continuous (in fact, real analytic, if φ is C^{an}) on V_i^r. Note that V_i^r is invariant under $\mathrm{Is}(V, \varphi)$.

If V is compact, real analytic, and r is sufficiently large, then by the previous discussion each "fiber" $(\mathscr{D}_\varphi^r)^{-1}(h)$ for $h \in H_i^r/\mathrm{O}(p, q)$ is a union of finitely many orbits of $\mathrm{Is} = \mathrm{Is}(V, \varphi)$ since the action of Is preserves the pseudo-Riemannian measure on V and this measure is finite for compact V; the orbit $\mathrm{Is}(v) \subset V$ for almost all $v \in V_i^r$ also admits a finite $\mathrm{Is}(V, \varphi)$-invariant measure by the classical measure decomposition theorem (see 1.5.D(ii)).

It follows that almost all orbits of $\mathrm{Is}(V, \varphi)$ in V_i^r are compact (see Remarks 3.5.4 below and 3.7.A in [29] for an explanation). Finally, one obtains by continuity that *all* orbits of $\mathrm{Is}(V, \varphi)$ are compact.

3.5.4. **Remark.** The above compactness of almost all orbits of $\mathrm{Is}(V, \varphi)$ in V_i^r is a consequence of a well-known result by Montgomery (see [53]). For the reader's sake we recall the statement:

3.5.4.A. *Consider a homogeneous space* $X = G/H$ *with a finite* G-*invariant measure, where* G *is a connected Lie group. If the isotropy subgroup* $G_x \subset G$, $x \in X$, *is connected, then* X *is compact, and therefore each maximal compact subgroup* $K \subset G$ *is transitive on* X.

By applying the above 3.5.4.A to our $\mathrm{Is}(V, \varphi)$-orbits we do not only obtain the compactness of orbits but also infer the following useful proposition.

3.5.4.B. **Proposition.** *Let* $K \subset \mathrm{Is}^0(V, \varphi)$ *be a maximal compact subgroup in the connected identity component* $\mathrm{Is}^0(V, \varphi) \subset \mathrm{Is}(V, \varphi)$. *Then the orbits of* $\mathrm{Is}^0(V, \varphi)$ *equal those of* K.

(See 3.7.A and 3.7.A_3 in [29] for a proof.)

Note that in the case when $\mathrm{Is}(V, \varphi)$ is connected Abelian, 3.5.4.A is immediate without connectedness of the isotropy subgroup. Also, in this case the orbits of $\mathrm{Is}(V, \varphi)$ equal those of the maximal torus $T \subset \mathrm{Is}^0(V, \varphi)$.

3.5.5. Finally, we observe that the discussion in 3.5.1 and 3.5.3 equally applies to the structure $\varphi' = (\varphi, X_1, \cdots, X_m)$ introduced in 3.5.2.A. To do this, one only needs to modify the definition of the space H^r by adding the Taylor coefficients of the fields $X_1, \cdots, X_m$. If the fields X_i constitute a *maximal* system of commuting Killing fields, then the connected identity component of $\mathrm{Is}(V, \varphi')$ is Abelian and thus (by a trivial case of 3.5.4.A) yields the compactness of orbits of $\mathrm{Is}(V, \varphi')$ as $\mathrm{Is}(V, \varphi')$ has only finitely many connected components. This is exactly what is used in [14] for the proof of Theorem 0.12.A. In fact, the compactness of the isometry group $\mathrm{Is}(V, \varphi)$ for Lorentz manifolds is achieved by showing that every maximal connected Abelian subgroup $A \subset \mathrm{Is}(V, \varphi)$ is compact. (See §7 in [14] for an elementary proof of the fact that if all maximal connected Abelian subgroups in a connected Lie group G are compact, then G is compact.)

4. Actions of semisimple Lie groups

4.1. When a semisimple Lie group G acts on a smooth manifold V, this action often looks as if it preserves some rigid geometric structure. In fact, any Lie group G acting on V preserves certain geometric structures related to the corresponding action of the Lie algebra $L = L(G)$ on V.

The action of L on V can be seen as a system of vector fields $X_1, \cdots, X_k$ on V corresponding to a basis $l_1, \cdots, l_k$ in L such that $[X_i, X_j]_V = [l_i, l_j]_L$ for all $i, j = 1, \cdots, k$. Note that the fields X_i in general are not preserved by the action of G. More precisely, if some $g \in G$ sends the point v to v', then the frame at v given by $\{X_1, \cdots, X_k\}$ is sent to the frame at v' given by $\{\mathrm{ad}_g X_1, \cdots, \mathrm{ad}_g X_k\}$ via the adjoint action of G on L. (We use the word "frame" even though the vector fields $X_1, \cdots, X_k$ may be dependent.) For example, if the group G is abelian, then the action of G preserves the frame. In general, one can take some adjoint-invariant polynomial function or tensor on the Lie algebra of G and this will give a geometric structure on V invariant under G. (More general structures are obtained by considering invariant functions on the spaces of jets of vector fields.) A similar kind of invariant structure arises when the linear maps $L \to T_v(V)$ given by $(l_1, \cdots, l_k) \to (X_1, \cdots, X_k)_v$ have ranks r independent of $v \in V$, for $0 \le r \le k = \dim L$.

Notice that the map $L \to T_v(V)$ is essentially the differential of our action at $v \in V$, and that its kernel, say $K_v \subset L$, invariantly depends on the point $v \in V$. That is, the map $V \to \mathrm{Gr}_r(L)$ defined by $v \mapsto K_v \in \mathrm{Gr}_r(L)$ is G-equivariant for the action of G on the Grassmann manifold $\mathrm{Gr}_r(L)$ induced by the adjoint action of G on L.

The map $V \to \mathrm{Gr}_r(L)$, which is a special case of the generalized Gauss map in the sense of [29], was introduced by Zimmer in [71] where he proves the following among other things:

4.1.A. **Theorem.** *If a noncompact simple Lie group G acts on a compact manifold V preserving a finite measure μ on V such that the above maps $L \to T_v(V)$, $v \in V$, have rank r independent of v, then either $r = 0$ or $r = k = \dim L$. In both cases the Gauss-Zimmer map $\alpha: V \to \mathrm{Gr}_r(L)$ is constant as the Grassmann manifold $\mathrm{Gr}_r(L)$ reduces to a single point.*

Proof. Consider the push-forward measure $\alpha_*(\mu)$ on $\mathrm{Gr}_r(L)$, and observe that $\alpha_*(\mu)$ is invariant under the ad_G-action. Since the later action is algebraic, we can use the following important lemma.

4.1.B. **Furstenberg-Tits lemma** (see §3.2. in [71] and 6.4.B$_4$ in these lectures). *If an algebraic action preserves a finite measure ν, then this action when restricted to the support of ν factors through an action of a compact group. In particular, if the algebraic group in question has no nontrivial compact factor group, then the action fixes* $\mathrm{supp}(\nu)$.

Now, return to the above measure $\alpha_*(\mu)$ on $\mathrm{Gr}_r(L)$ where L is simple, and note that the action of ad_g, $g \in G$, on $\mathrm{Gr}_r(L)$ has no fixed point for $1 \le r \le \dim L - 1$. Indeed, a fixed point would be an ad_g-invariant r-dimensional subspace in L, namely an ideal of L, and this is ruled out since we assumed L to be simple. This concludes the proof of 4.1.A.

4.2. Remark. The above Theorem 4.1.A is a modified version of the original Zimmer's result. In fact, Zimmer does not assume $r = \text{const}$ and proves the following by using the above argument.

4.3. Full Zimmer theorem (see [72], [70]). *Let G be a noncompact simple Lie group acting on (V, μ). Then for almost all (with respect to μ) points $v \in V$ the isotropy subgroup $G_v \subset G$ is either discrete or equals all of G.*

4.4. A natural question that arises in light of Theorem 4.3 is the following.

4.4.A. *Open problem.* Let a noncompact simple Lie group G smoothly and faithfully act on a connected manifold V, such that the action preserves a smooth finite measure. Is then the isotropy group G_v discrete for *all* $v \in V$?

4.4.B. **Example** (A. Connes). Let $\Gamma \subset G$ be a noncocompact lattice. Then G/Γ admits no smooth G-invariant compactification. (We invite the reader to find a proof on his own.)

4.4.C. **Remark.** If the action of G is faithful on each connected component of V, then the fixed point set $V_0 \subset V$ where $G_v = G$ is nowhere dense. In fact we have the following.

4.4.D. **Thurston's stability theorem** [65]. *If some group G of diffeomorphisms of V fixes a point $v_0 \in V$, and the action of G on $T_{v_0}(V)$ is trivial, then G admits a nontrivial homomorphism into $\mathbf{R}$.*

As simple groups admit no such homomorphisms, the action of our G is nontrivial on $T_{v_0}(V)$ for all v_0 in the fixed point set $V_0 \subset V$, and so V_0 is nowhere dense.

4.5. Remark. A nontrivial action of a simple Lie group G on $T_{v_0}(V)$ cannot fix a hyperplane in $T_{v_0}(V)$, since the group $\mathrm{LinAut}(\mathbf{R}^n, \mathbf{R}^{n-1})$ is solvable and thus contains no simple Lie group, where $\mathrm{LinAut}(\mathbf{R}^n, \mathbf{R}^{n-1})$ denotes the subgroup in $\mathrm{GL}(n, \mathbf{R})$ fixing $\mathbf{R}^{n-1} \subset \mathbf{R}^n$.

It follows that the codimension of V_0 is at least 2, for an appropriate notion of C^1-codimension. This can be seen even better in the case when the maximal compact subgroup $K \subset G$ is nontrivial. (Recall that the only simple Lie group where the maximal subgroup $K = \{\mathrm{Id}\}$ is the universal cover $\widetilde{\mathrm{SL}(2, R)}$ of $\mathrm{SL}(2, \mathbf{R})$). Here obviously, we have

$$V_0 = \mathrm{Fix}(G) \subset \mathrm{Fix}(K) \subset V,$$

where the set $\mathrm{Fix}(K)$ of fixed points of K is a smooth submanifold of codimension ≥ 2 in V.

It seems that $V_0 = \mathrm{Fix}(G)$ is a C^r-submanifold for C^r-actions and the action is linearizable near V_0. This is probably known to experts but the only published result (see [43]) known to the authors is more special.

4.6. Now, let us look at the action for our simple Lie group G near a *recurrent* point $v \in (V, \mu)$ (see 1.3.A) such that $g_i(v) \to v$ for some divergent sequence $g_i \in G$. These g_i act on the frame $\{X_1, \cdots, X_k\}$ by ad_{g_i}, where $k = \dim G$, and one can show that this action comes from some kind of infinitesimal action of ad_G at $v \in V$ (compare the recurrency discussion in 1.3). This was done by Zimmer in a general dynamical framework. Here, following [29] we state (without proof) a geometric version of Zimmer's theorem for G-actions preserving a rigid A-structure φ on V, when G is as earlier assumed to be noncompact simple and the action of G is faithful.

4.6.A. **Theorem.** *Let (V, φ) be a C^∞-smooth rigid A-manifold. If $G \subset \mathrm{Is}(V, \varphi)$, and the action of G preserves a finite smooth measure μ on V, then for almost all $v \in V$ the local isotropy group $\mathrm{Is}_v(V, \varphi)$ contains an isomorphic copy of the group $\mathrm{ad}\, G \subset \mathrm{Aut}\, L(G)$ (see 5.2 in [29]).*

The above result shows that locally the structure φ is highly symmetric, and this may be used often for a detailed study of (V, φ). In fact, such a study may be carried over in the special case of Lorentz manifolds as we shall see presently.

4.7. Consider a compact Lorentz manifold (V, φ) and assume that the isometry group $G = \mathrm{Is}(V, \varphi)$ contains $\mathrm{SL}(2, \mathbf{R})$ as a subgroup (see 0.13.A). We want to complement Theorem 0.13.A by giving a *complete* geometric description of such (G, V, φ). To do this, we begin with the following construction which specializes the one given in 6.9.

4.7.1. Let $\tilde{\varphi}_0$ be a bi-invariant metric on a Lie group G_0, $(\tilde{V}_1, \tilde{\varphi}_1)$ be a Riemannian manifold, and $G_1 = \mathrm{Is}(\tilde{V}_1, \tilde{\varphi}_1)$ be its isometry group. Set $\tilde{V} = G_0 \times \tilde{V}_1$, and let Γ be a discrete subgroup in the product group $G = G_0 \times G_1$. The left- and right-translations of G_0 on itself together with the (only one) action of G_1 on $\tilde{V}_1$ give rise to two actions of each of the groups G_0, $G = G_0 \times G$ and $\Gamma \subset G$ on $\tilde{V}$, which are called left and right action respectively.

Consider the left action of G_0 on $\tilde{V}$ and the right action of Γ on $\tilde{V}$. Since these actions commute, we obtain an action of the group G_0 on the quotient manifold $V = \tilde{V}/\Gamma$. Note that this action of G_0 preserves the metric φ on V which descends from the metric $\tilde{\varphi} = \tilde{\varphi}_0 + \tilde{\varphi}_1$ on $\tilde{V}$ and that the type of φ is $(m_+ + m, m_-)$, where (m_+, m_-) is the type of $\tilde{\varphi}_0$ and $m = \dim \tilde{V}_1$.

4.7.2. One obtains an invariant metric of more general kind by *warping* $\tilde{\varphi}$ on $\tilde{V}$ with an arbitrary strictly positive G_1-invariant function $\tilde{f}$ on $\tilde{V}_1$. Namely, one takes

$$\tilde{\varphi}_{\tilde{f}} = \tilde{f}\tilde{\varphi}_0 + \tilde{\varphi}_1$$

or, more pedantically,

$$\tilde{\varphi}_{\tilde{f}}(g_0, v_1) = \tilde{f}(v_1)\tilde{\varphi}_0(g_0) + \tilde{\varphi}_1(v_1)$$

(if one wants to keep track of the arguments $g_0 \in G$ and $v_1 \in \tilde{V}_1$).

Clearly $\tilde{\varphi}_{\tilde{f}}$ is Γ-invariant and thus defines a metric $\tilde{\varphi}_{\tilde{f}}$ on V, which is G_0-invariant as well as φ. Finally, we notice that the manifold V is compact if $\tilde{V}_1/G_1$ is compact and $\Gamma \subset G$ is cocompact in $G = G_0 \times G_1$.

4.8. **Remark.** One could consider a more general situation where $(\tilde{V}_1, \tilde{\varphi}_1)$ is pseudo-Riemannian. In this case one should insist on properness of the action of the isometry group G_1 on $\tilde{V}_1$ (which is automatic in the Riemannian case even if the isometry group is not compact) in order to obtain a Hausdorff quotient space $V = \tilde{V}/\Gamma$.

4.9. Splitting theorem (second version, compare 0.13.A). *Let* (V, φ) *be a compact Lorentz manifold such that* $\mathrm{Is}(V, \varphi) \supset G_0 = \mathrm{SL}(2, \mathbf{R})$. *Then* V *is obtained by the above construction for a form* $\tilde{\varphi}_0$ *on* G_0 *proportional to the Killing form and for some* $(\tilde{V}_1, \tilde{\varphi}_1)$, $\tilde{f}$ *on* $\tilde{V}_1$, *and* $\Gamma \subset G_0 \times \mathrm{Is}(\tilde{V}_1, \tilde{\varphi}_1)$.

(The reader can consult §5.4 in [29] for a discussion related to this theorem.)

Notice that the present version of the splitting theorem is equivalent to 0.13.A. Namely, the following properties (i)–(iii) claimed by 0.13.A are direct corollaries of 4.9.

(i) The isotropy subgroups of the action of G_0 are discrete for all $v \in V$ (compare 4.10(b) below).

(ii) The metric φ is nonsingular on the 3-dimensional G_0-orbits.

(iii) The normal subbundle to the orbits is integrable and the leaves are totally geodesic.

4.10. Conversely, Theorem 4.9 can be easily deduced from the above properties (i)–(iii).

4.10.1. **Remark.** (a) Property (i) above confirms (in a very special situation) our conjecture 4.4.A claiming local freedom of measure preserving actions of simple groups.

(b) The above Theorem 4.9 generalizes to pseudo-Riemannian manifolds of arbitrary type (p, q) if the isometry group contains a semisimple subgroup G_0 large enough with respect to $\min(p, q)$.

(c) There exist pseudo-Riemannian manifolds V with "insufficiently large" subgroups of isometries $G_0 \subset \mathrm{Is}(V)$ where $\tilde{V}$ do not split in the above sense. One can see this by looking at the action of a subgroup $G_0 \subset G$ on G/Γ where G/Γ is given a pseudo-Riemannian metric coming from a bi-invariant metric on G (see 6.7).

4.11. In order to get some additional insight into (semi)simple group actions we shall now consider some examples which are more complicated than the algebraic actions and those on G/Γ considered in §1.9.

4.11.1. *Induced actions* (see 6.9). Let G be a semisimple Lie group acting on G/Γ, where $\Gamma \subset G$ is a discrete subgroup, and consider an action of Γ on some connected manifold F. Then we take the quotient $V = (G \times F)/\Gamma$ for the diagonal Γ-action and observe that G naturally acts on this V. Furthermore if G/Γ and F are compact, then also V is compact. Similarly, if $\mathrm{Vol}(V/\Gamma) < \infty$ and the action of Γ on F preserves some measure, then the same is true for the action of G on V. Now, to make a practical use of the above construction one needs interesting examples of Γ-actions on F, which do not come from an action of $G \supset \Gamma$.

But the work by Zimmer suggests (see [73]) that for $\mathrm{rank}_{\mathbf{R}} G \geq 2$ (recall that G is assumed to be noncompact simple and that $\mathrm{rank}_{\mathbf{R}}$ stands for the rank of the symmetric space G/K, where $K \subset G$ is the maximal compact subgroup in G) an action of Γ extends to that of some Lie group G' with finitely many connected components if we assume, for example, that V is compact and (or) that the action of Γ preserves a smooth finite measure and $\mathrm{Vol}(G/\Gamma)$ is finite. (If the action of Γ on F admits no invariant Riemannian metric, then the above G' should be equal to G.)

On the other hand, one can construct different kinds of actions for many cases of $\mathrm{rank}_{\mathbf{R}} = 1$. For example if there is a nontrivial homomorphism $\Gamma \to \mathbf{Z}$, i.e., $H^1(\Gamma, \mathbf{Z}) \neq 0$, then every action of $\mathbf{Z}$, i.e., a single diffeomorphism of F, gives rise to a Γ-action on F.

4.11.2. Here is a potential generalization of the above "induced construction" in the case $G = \mathrm{PO}(n, 1) = \mathrm{IS}(\mathbf{H}^n)$, where $\mathbf{H}^n$ is the hyperbolic space. Let W be a compact manifold foliated into n-dimensional leaves, and assume that these leaves carry metrics of constant curvature -1. Then the above G naturally acts on the manifold V of orthonormal n-frames tangent to the leaves. In fact, V equals the space of those maps $f\colon \mathbf{H}^n \to W$ which locally isometrically send V onto a leaf. Then the action of G on $\mathbf{H}^n$ induces an action of G on V. To make this construction work, one needs foliations of curvature -1, which in general are hard to come by.

However, if $n = 2$, one can start with a 2-dimensional foliation $\mathcal{F}$ with an arbitrary metric and then use the Riemann mapping theorem to produce a new metric of constant curvature. (According to D. Sullivan this idea is due to Winkelnkemper.) To be precise, we assume that V is compact and that the foliation is hyperbolic, i.e., there exists no nonconstant conformal map of $\mathbf{R}^2$ into a leaf of the foliation. Then we consider the space $\mathcal{H}$ of *conformal* maps of $\mathbf{H}^2$ into the leaves of $\mathcal{F}$, and observe that $\mathrm{PSL}(2, \mathbf{R}) = \mathrm{IS}(\mathbf{H}^2)$ naturally acts on $\mathcal{H}$.

The space $\mathcal{H}$ is infinite dimensional but contains an invariant submanifold $V \subset \mathcal{H}$ consisting of the maps which are *covering maps* of $\mathbf{H}^2$ onto the leaves. The manifold V is diffeomorphic to a S^1-bundle over W, namely to the bundle of unit vectors which are tangent to the leaves.

The action of $\mathrm{PSL}(2, \mathbf{R})$ on V obtained in this way is continuous but not, in general, smooth. In fact smooth actions of this type seem quite rare especially if one wants a smooth invariant measure. (Note that one does have such a measure if the foliation admits a transversal measure.)

4.12. The above discussion raises the following question: Does there exist a compact *simply connected* manifold V which admits a faithful smooth (C^∞ or C^{an}) action of a simple Lie group preserving a smooth finite measure μ on V?

We know that the answer is "no," if the action is C^{an} and preserves a rigid structure (see [29] and §3 in these lectures).

4.13. We end this section with the following:

4.13.A. **Remark.** The above construction can be generalized by taking for $\mathcal{H}$ another space of maps $\mathbf{H}^2 \to W$ satisfying some system of partial differential equations.

For example, one can use harmonic maps into Riemannian manifolds W (preferably of negative curvature) and holomorphic maps into (Kobayashi hyperbolic) complex manifolds W. Of course $\mathcal{H}$ itself is too large for us, and we are interested in G-invariant submanifolds and (or) finite G-invariant measures in $\mathcal{H}$.

There is no systematic theory here, but at least there is one beautiful example. Namely, let W be the Riemann moduli space (or rather a finite nonsingular covering of the moduli space), and let V consist of those (extremal) holomorphic maps $\mathbf{H}^2 \to W$ whose lifts to the Teichmüller space $\tilde{W}$, which is the universal covering of W, are isometric embeddings with respect to the Teichmüller metric in $\tilde{W}$ (this metric equals the Kobayashi metric by Royden's theorem). The reader can see references [52] and [68] for the study of the $\mathbf{R}$-part of this action.

5. Infinitesimal geometric structures

5.1. Classically, a geometric structure φ on V is expressed in local coordinates by finitely many, say by s, functions, and these transform by certain rules depending on a particular *type* of φ when one changes the coordinates. In other words, φ is an $\mathbf{R}^s$-valued function on the space $\mathcal{U} = \mathcal{U}(V)$ of coordinate charts in V where a point in $\mathcal{U}$ is a pair (v, u), $v \in V$ and $u = (u_1, \cdots, u_n)$, for $n = \dim V$, is a local coordinate system around v, i.e., u is a locally diffeomorphic map of a small neighborhood of $v \in V$ into $\mathbf{R}^n$ sending $v \mapsto 0$. When we pass to another coordinate system u' around v, the value $\varphi(v, u')$ is expressed by certain algebraic formulas depending on the type of φ in terms of $\varphi(v, u)$ and the partial derivatives of certain orders $\leq r$ (also depending on the type of φ) at v of the coordinates u'_i viewed as functions of u_i. In particular, if u_i and u'_i only differ in order $> r$ (i.e., $u_i - u'_i$ vanishes at v along with all derivatives of order $\leq r$), then $\varphi(v, u) = \varphi(v, u')$. That is, $\varphi(v, u)$ depends only on the rth order jet (or rth order differential) of u at v.

Let us express this property with the following.

5.2. Jet language. By definitions the jet $J_v^r f$ of a smooth function f on V defined in a neighborhood of a point $v \in V$ is the equivalence class of f under the following relation: two smooth maps f and f' defined in a small neighborhood of $v \in V$ are called *r-equivalent* at v if the partial derivatives of these maps of orders $0, 1, \cdots, r$ are equal at v, where the partial derivatives are taken in a fixed coordinate system around v.

This notion of r-equivalence (and hence, of an r-jet) does not depend on the coordinate system. In fact, the chain rule shows that the partial derivatives of order $\leq r$ in one coordinate system can be expressed in terms of those in another system.

5.2.A. **Remarks.** (i) The definitions of the r-equivalence and jets immediately extend to maps $f \colon V \to \mathbf{R}^s$. Moreover, these notions can be applied to maps into an arbitrary smooth manifold X. This is done by embedding X into some $\mathbf{R}^N$ and then by noticing that the r-equivalence for maps $V \to X \hookrightarrow \mathbf{R}^N$ does not depend on the embedding.

(ii) According to our definition, the only meaningful expression involving jets which we can write so far is $J_v^r f = J_v^r f'$; this is just another way to say that f and f' are r-equivalent at v. But, if we fix local coordinates around v, we see more structure as a jet of any map is the same thing as the totality of partial derivatives of orders $\leq r$ evaluated at $v \in V$.

It follows that the space J_v^r of all r-jets, say for maps $f \colon (V, v) \to \mathbf{R}^m$, can be identified (the identification depends on a choice of local coordinates) with the Euclidean space $\mathbf{R}^{mN_0}$ for $N_0 = 1 + n + n(n+1)/2 + \cdots + (n+r-1)!/(n-1)!r!$. Furthermore, the space

$$J^r(V, \mathbf{R}^m) = \bigcup_{v \in V} J^r_v$$

has a natural structure of a smooth manifold fibered over V with the fibers J^r_v. This fibration is trivial over each coordinate chart $(U, u_1, \cdots, u_n)$ in V, as the coordinates $u_1, \cdots, u_n$ (obviously) define a splitting $J^r(U) = J^r_v \times U$, for each $v \in U$.

5.3. The fibration $\mathscr{D}^r(V) \to V$. If we apply the r-equivalence relation to all local coordinate systems u in $(v, u) \in \mathscr{U}$, we obtain a quotient of $\mathscr{U}$, call it $\mathscr{D}^r(V)$, whose points are pairs (v, δ), where $v \in V$, and δ is an element in the space of r-jets of locally diffeomorphic maps of V (or rather of a small neighborhood of $v \in V$) into $\mathbf{R}^n$ sending $v \mapsto 0$. We denote this space by $\mathscr{D}^r_v$, and then see that

$$\mathscr{D}^r(V) = \bigcup_{v \in V} \mathscr{D}^r_v.$$

Each *fiber* $\mathscr{D}^r_v$ of $\mathscr{D}^r_v(V)$ is naturally embedded into the jet space J^r_v of all jets of maps $V \to \mathbf{R}^n$ and the subset $\mathscr{D}^r_v$ in $J^r_v = \mathbf{R}^N$, $N = n(1 + n + n(n+1)/2 + \cdots + (n+r-1)!/(n-1)!r!)$ is distinguished by two conditions. The first amounts to vanishing of the first n coordinates, which correspond to the requirement $u(v) = 0$, and the second is given by nonvanishing of the Jacobian matrix at v, which reflects the locally diffeomorphic nature of u. To be more precise, we look at the tautological projections between the jet spaces $J^r_v \to J^{r'}_v$, for all $r' \leq r$ (to go from J^r to $J^{r'}$ we just forget derivatives above order r'), and we also note that the space of 1-jets of maps $(V, v) \to \mathbf{R}^n$ sending $v \mapsto 0$ is nothing else but the space of linear maps of the tangent space $T_v(V) \to \mathbf{R}^m$. Now, we first invoke the condition $u(v) = 0$ thus restricting to the subspace $\mathbf{R}^{N-n} \subset \mathbf{R}^N = J^r_v$ containing $\mathscr{D}^r_v$, and then use the above projection on the 1-jets, say

$$p_1 : \mathbf{R}^{N-n} \to \operatorname{Hom}(T_v(V), \mathbf{R}^n).$$

Now $\mathscr{D}^r_v \subset \mathbf{R}^{N-n}$ is given by the condition $\mathscr{D}^r_v = \{\delta \in \mathbf{R}^{N-n} | \operatorname{rank} p_1(\delta) = n\}$.

Notice that $\mathscr{D}^r_v$ is an open subset in $\mathbf{R}^{N-n}$ and thus has a natural structure of a smooth manifold. Furthermore, the space $\mathscr{D}^r(V)$ also has a natural structure of a smooth manifold such that the projection $\mathscr{D}^r(V) \to V$ becomes a smooth locally trivial fibration with fiber $\mathscr{D}^r_v(V)$. In fact, a coordinate system $u_1, \cdots, u_n$ in a neighborhood U of a point $v_0 \in V$ defines in an obvious way a splitting $\mathscr{D}^r(U) = \mathscr{D}^r_{v_0} \times U$.

5.4. Geometric structures. A geometric structure of order r can now be defined as a map φ of $\mathscr{U} = \mathscr{U}(V)$ into a smooth manifold Φ such that $\varphi(v, u)$ only depends on $J^r_v(u)$. Thus φ factors through the projection $\mathscr{U} \to \mathscr{D}^r(V)$ and so defines a map of $\mathscr{D}^r(V)$ to Φ, which we also denote by $\varphi: \mathscr{D}^r(V) \to \Phi$. In fact, we usually refer to a structure as to a map defined on $\mathscr{D}^r(V)$ rather than on $\mathscr{U}$, and we assume it is C^∞-smooth.

5.4.A. **Remark on $\Phi = \mathbf{R}^s$.** The study of general structures can be reduced to those where the target space Φ is the euclidean space $\mathbf{R}^s$ as our smooth manifold Φ can be embedded into some euclidean space.

5.5. *A*-structures. First we recall that $\mathscr{D}^r_v(V)$ is an *open* subset in $\mathbf{R}^{N-n}$, and thus we can speak of polynomials, rational and algebraic functions on $\mathscr{D}^r(V)$ defined as restrictions to $\mathscr{D}^r(V)$ of corresponding functions on $\mathbf{R}^{N-n}$.

Now, a structure $\varphi: \mathscr{D}^r(V) \to \mathbf{R}^s$ is called of *algebraic type* or an *A-structure* if the restriction of φ to each fiber $\mathscr{D}^r_v(V)$ is algebraic. Similarly, we define the notion of algebraic type for structures $\varphi: \mathscr{D}^r(V) \to \Phi$, where Φ is an arbitrary real algebraic manifold.

5.6. Our definition attaches no transformation law to a geometric structure φ. In fact, we do not need this for our applications. On the other hand, all natural structures come along with some rule of transformation, under the (jets of) coordinate changes.

Let us explain the meaning of this in our language. We start with describing the corresponding transformation group.

5.7. The group $\mathscr{D}^r$. Let $(V, v) = (\mathbf{R}^n, 0)$ and $\mathscr{D}^r = \mathscr{D}^r_0(\mathbf{R}^n)$. Since we can compose (locally diffeomorphic) maps $\mathbf{R}^n \to \mathbf{R}^n$ sending $0 \mapsto 0$, this induces a composition law in the jet space and the space $\mathscr{D}^r$ acquires a structure of a group. Recall that $\mathscr{D}^r$ is realized as an open subset in $\mathbf{R}^N$ and $N = n(n + n(n+1)/2 + \cdots + (n+r-1)!/(n-1)!r!)$ distinguished by the nondegeneracy of the matrix of the first $n \times n$ coordinates. The composition law $\mathscr{D}^r \times \mathscr{D}^r \to \mathscr{D}^r$ is given by a polynomial map on $\mathscr{D}^r \times \mathscr{D}^r \subset \mathbf{R}^N \times \mathbf{R}^N$ which corresponds to the chain rule (for higher derivatives). The inverse map $\delta \to \delta^{-1}$ is a *rational* map on $\mathbf{R}^N \supset \mathscr{D}^r$ with poles on $\mathbf{R}^N \backslash \mathscr{D}^r$.

The group $\mathscr{D}^r$ naturally acts on each $\mathscr{D}^r_v$, as the r-jet of the composed map $u' = \delta \circ u: (V, v) \xrightarrow{u} \mathbf{R}^n \xrightarrow{\delta} \mathbf{R}^n$ only depends on those of u and δ, and so we have an action of $\mathscr{D}^r$ on $\mathscr{D}^r(V)$. This action is free and so every orbit $\mathscr{D}_v^r$ is identical to $\mathscr{D}^r$. It is also clear that this action is exactly what remains of coordinate changes when we pass to r-jets.

5.8. $\mathscr{D}$-invariant structures. A structure $\varphi: \mathscr{D}^r(V) \to \Phi$ is called *$\mathscr{D}$-invariant* or a *$\mathscr{D}^r$-structure* if the manifold Φ is endowed with an action of the group $\mathscr{D}^r$, and φ is $\mathscr{D}^r$-equivariant.

We say in this case that the structure is of *type* Φ, where "the type" refers to the transformation law of φ encompassed by the action of $\mathscr{D}^r$ on Φ.

5.7.A. **Remarks.** (i) The type Φ of φ is not uniquely defined in terms of φ. In fact, we always can enlarge (and sometimes diminish) Φ by embedding it in a larger $\mathscr{D}^r$-space. However, the type is clear from the context. For example, if φ is a Riemannian metric on V (cf. 5.9.A(a)), then the type Φ is the space of all positive quadratic forms on $\mathbf{R}^n$ with the standard action of $\mathscr{D}^1 = \mathrm{GL}(n, \mathbf{R})$. But, if we do not care much about the positivity of φ, we look at the larger space of *all* quadratic forms as the pertinent type.

(ii) The $\mathscr{D}$-condition is not as restrictive as one may a priori think. For example, every structure φ of *algebraic type* defines in a canonical way another A-structure, say φ', which does have the $\mathscr{D}$-invariance property. To see the idea, let us consider a smooth map $p: T(V) \to \mathbf{R}$ whose restriction to each tangent space $T_v(V) = \mathbf{R}^n$ is a polynomial of degree $\leq d$ for a fixed d. Such polynomials on $\mathbf{R}^n$ form a linear space, say Φ of finite dimension $N = 1 + n + n(n+1)/2 + \cdots + (n+d-1)!/(n-1)!d!$, and $\mathscr{D}^1 = \mathrm{GL}(n, \mathbf{R})$ naturally acts on Φ. Now, each frame $\delta \in \mathscr{D}_v^1(V)$, $v \in V$, gives us an identification of $T_v(V)$ with $\mathbf{R}^n$, and so the map p on $T_v(V)$ gives us a vector, say $\varphi_p(\delta) \in \Phi$. The resulting map $\delta \mapsto \varphi_p(\delta)$ clearly is $\mathscr{D}^1$-equivariant as well as fiberwise polynomial and thus provides us with an A-structure on V in the narrow equivariant sense.

5.9. Structures as sections. Notice that the action of $\mathscr{D}^r$ on $\mathscr{D}^r(V)$ is free and proper with the orbits $\mathscr{D}_v^r(V)$, and so $\mathscr{D}^r(V)$ is the principal $\mathscr{D}^r$-bundle over V. The equivariant maps $\mathscr{D}^r(V) \to \Phi$ can be identified with the sections of the associated bundle, denoted by $\Phi(V) \to V$.

In fact, most geometric structures naturally appear as sections of such associated bundles. For example, *tensors* φ on V are the sections of (tensor) bundles which are associated to the principal $\mathrm{GL}(n, \mathbf{R})$-bundle $\mathscr{D}^1(V)$, that is, the frame bundle on V (see below for further examples).

5.9.A. **Examples.** (a) Riemannian metrics are geometric structures of the first order which have both properties A and $\mathscr{D}^1$. One sees this immediately by recalling that a Riemannian metric g is represented in local coordinates $u = u_k$, $k = 1, \cdots, n$, around $v \in V$ by $s = \frac{1}{2}n(n+1)$ smooth functions

$$g_{lk}(v) = g\left(\frac{\partial}{\partial u_l}, \frac{\partial}{\partial u_k}\right), \qquad 1 \le l \le k \le n,$$

i.e., by the components of the metric tensor at $v \in V$, and, if g'_{ij} represent g with respect to another coordinate system u', then

$$g'_{ij} = \sum_{k,l} \frac{\partial u_k}{\partial u'_i} \frac{\partial u_l}{\partial u'_j} g_{kl}.$$

(b) Affine connections on V are locally defined by Γ^k_{ij} coefficients which transform according to a certain rule under coordinate changes. The transformation rule involves first and second derivatives of the coordinates. Thus affine connections have second order. Note that if φ is an affine connection on V, then its type Φ (see 5.8) is the Euclidean space of Γ^k_{ij} coefficients with the action of the group $\mathscr{D}^2$ defined by the coordinate changes. Thus affine connections are $\mathscr{D}^2$-invariant. It is clear that they are of algebraic type.

(c) A field of k-planes on V is an A-structure of the first order. In fact, φ is a section $V \to \mathrm{Gr}_k(V)$, where $\mathrm{Gr}_k(V)$ is the Grassmann manifold of V, namely the set of all k-dimensional vector spaces of the vector spaces $T_v(V)$, $v \in V$. The full linear group $\mathscr{D}^1 = \mathrm{GL}(n, \mathbf{R})$, for $n = \dim V$, naturally acts on the fiber $\mathrm{Gr}_k(T_v(V)) = \mathrm{Gr}_k(\mathbf{R}^n)$ of the fibration $\mathrm{Gr}_k(V) \to V$. Thus, if φ is a field of k-planes on V, its type Φ is the Grassmannian $\mathrm{Gr}_k(\mathbf{R}^n)$ of the k-dimensional subspaces of $\mathbf{R}^n$ with the standard action of the group $\mathscr{D}^1$ on $\mathbf{R}^n$.

5.10. Isometries and the idea of rigidity. A diffeomorphism between smooth manifolds $f\colon V_1 \to V_2$ induces, by passing to jets of f, a diffeomorphism

$$\mathscr{D}^r(V_1) \to \mathscr{D}^r(V_2),$$

and so every structure $\varphi_2\colon V_2 \to \Phi$ induces a structure of the same type Φ on V_1. Now, given two structured manifolds (V_1, φ_1) and (V_2, φ_2) of the same type Φ one says that $f\colon V_1 \to V_2$ is an *isometry* if the induced structure on V_1 equals φ_1. The following definition may serve as a motivation to our (more elaborate) definition of rigidity in 5.11.

5.10.A. *Iso-rigidity.* A structure manifold (V, φ) is called *Iso*^{k}*-rigid at* $v \in V$ if for every germ at v of an isometry $f\colon V \to V$ fixing v, the equality of the jets

$$(*) \qquad J^k_f(v) = J^k_{\mathrm{Id}}(v)$$

implies that

$$(**) \qquad f = \mathrm{Id},$$

where Id denotes the identity map $V \to V$, and the equalities $(*)$ and $(**)$ (according to the "germ" terminology) are meant in a small neighborhood of $v \in V$.

Next, (V, φ) is called *Iso*^{k}*-rigid* if it is rigid at all $v \in V$.

The following simple proposition allows us to pass from local to global isometries.

5.10.B. **Proposition.** *Let (V, φ) be a connected rigid manifold and f_1 and f_2 be isometries $(V, \varphi) \to (V', \varphi')$, where (V', φ') is another manifold of the same type as (V, φ). If $J^k_{f_1}(v_0) = J^k_{f_2}(v_0)$ at some point $v_0 \in V$, then $f_1 = f_2$.*

Proof. By applying the Iso-definitions to $f_2^{-1} \circ f_1$ at the points $v \in V$, where $f_1(v) = f_2(v)$, one concludes that the set of those $v \in V$, where $J^k_{f_1}(v) = J^k_{f_2}(v)$ is *open* in V. On the other hand, this set is obviously closed in V as we (tacitly) assume f is C^r-smooth.

5.10.C. **Examples.** (i) One knows that Riemannian and pseudo-Riemannian metrics φ are Iso^1-rigid, as every isometry can be recaptured from the differential at a single point by using the exponential map. In fact these φ satisfy the stronger rigidity property defined in 5.11.A.

(ii) If (V, φ) has no local isometries at all, then it is trivially Iso-rigid though not necessarily rigid in the sense of 5.11.A.

5.11. Infinitesimal isometries and k-rigidity. We recall the natural action of $\mathscr{D}\mathrm{iff}(V)$ on $\mathscr{D}^r(V)$ which induces an action of $\mathscr{D}\mathrm{iff}(V)$ on i-jets of maps $\varphi: V \to \Phi$. In fact, an action of a diffeomorphism f on the ith jet of φ at a given point $\delta \in \mathscr{D}^r(V)$ only depends on the jet J^{r+i}_f at the point $v \in V$ under δ. Now we consider the group $\mathscr{D}^{r+i}(V)$ of $(r+i)$-jets of diffeomorphisms fixing v (this group is isomorphic to $\mathscr{D}^{r+i} = \mathscr{D}^{r+i}(\mathbf{R}^n, 0)$), and then define the *infinitesimal isotropy subgroup of the isometries* of a given structure $\varphi: \mathscr{D}^r(V) \to \Phi$ as the subgroup

$$\mathrm{Is}^{r+i}(v) = \mathrm{Is}^{r+i}(V, \varphi, v) \subset \mathscr{D}^{r+i}(v)$$

consisting of (jets of) diffeomorphisms fixing J^i_φ on the fiber $\mathscr{D}^r_v(V) \subset \mathscr{D}^r(V)$. That is,

$$(*) \qquad J^i_{\varphi \circ \tilde{f}} | \mathscr{D}^r_v = J^i_\varphi | \mathscr{D}^r_v,$$

where $\tilde{f} = J^r_f$. (If φ is a $\mathscr{D}^r$-structure represented by a section $V \to \Phi(V)$, then $(*)$ reduces to the corresponding jet equality at v.)

Let us observe natural maps

$$\mathrm{Is}(v) \to \mathrm{Is}^{\mathrm{loc}}(v) \to \cdots \to \mathrm{Is}^{r+i}(v) \xrightarrow{p_{r+i-1}} \mathrm{Is}^{r+i-1}(v) \to \cdots \to \mathrm{Is}^{r}(v),$$

where $\mathrm{Is}(v)$ denotes the isotropy subgroup of the isometry group $\mathrm{Is}(V, \varphi)$ at v, and $\mathrm{Is}^{\mathrm{loc}}(v)$ consists of the germs of the isometries fixing v. Also notice that one cannot define Is^{r-i} for $i > 0$ for the structures of order r, but we extend the notation p_k to $k = r$ for the projection

$$p_{r-1}\colon \mathrm{Is}^{r}(v) \to \mathscr{D}^{r-1}(v).$$

5.11.A. **Definition.** A structure φ is called *k-rigid at* v for $k \geq r-1$ if the map p_k is injective. The structure φ is *k-rigid* if it is rigid at all points $v \in V$.

5.11.A$_1$. **Remark on rigidity and Iso-rigidity.** Notice that Iso^k-rigidity states in this language that the map $\mathrm{Is}^{\mathrm{loc}} \to \mathrm{Is}^k$ is injective. It will become clear later on (see 5.13) that rigid manifolds are Iso-rigid but there are nonrigid Iso-rigid manifolds (see 5.11.B(v) below). The rigidity in the following examples will be clarified in §5.16.

5.11.B. **Examples.** (i) Let φ be a full frame field on V, i.e., a system of $n = \dim V$ independent vector fields. Then φ is a 0-rigid structure. (The Iso-rigidity is apparent in this case.)

(ii) Affine connections as well as pseudo-Riemannian metrics are 1-rigid. This can be seen either by looking at the 1-jets of exponential maps (compare 5.10.c(i)) or using (i) above (compare 5.16.C).

(iii) Conformal pseudo-Riemannian structures on V are 2-rigid for $\dim V \geq 3$. This is a classical result, probably due to E. Cartan (see 5.16.F(iv)).

(iv) Conformal structures on surfaces are *not* rigid. In fact these are not even Iso-rigid. The same applies to complex analytic structures in all dimensions and to symplectic (see 2.6.B) and contact structures.

(v) Take a generic Riemannian metric φ_0 on V and a C^∞-function ψ vanishing at a single point with infinite order. Then the structure $\varphi = \psi\varphi_0$ is Iso-rigid, as there exist no nontrivial isometries for this φ. But it is nonrigid as $\mathrm{Is}^r(V, \varphi, v_0) = \mathscr{D}^r_{v_0}(V)$, and so none of the projections $\mathrm{Is}^k \to \mathrm{Is}^{k-1}$ is injective.

This example also shows that a small perturbation of an Iso-rigid structure does not have to be Iso-rigid. In fact a small perturbation of the above φ may be identically zero in a neighborhood of v_0, which would make $\mathrm{Is}^{\mathrm{loc}}(v_0) = \mathscr{D}\mathrm{iff}^{\mathrm{loc}}(v_0)$. On the other hand, we shall see in 5.18 that the rigidity is stable under small perturbations of φ.

5.12. Isometries of rigid structures. It is immediate from the definition of rigidity that the action of the isometry group $\mathrm{Is} = \mathrm{Is}(V, \varphi)$ on $\mathscr{D}^s$ is free insofar as φ is k-rigid and $s \geq k$.

It follows that $\dim \mathrm{Is} \leq \dim \mathscr{D}^k$.

In fact one classically knows (since S. Lie) that Is has a natural structure of a Lie group such that the action of Is on V is smooth. Another important classical property is the following.

5.12.A. *Properness of the action of* Is *on* $\mathscr{D}^s(V)$ *The action of* Is *on* $\mathscr{D}^s(V)$ *is proper as well as free* (see [29] for the proof adapted to our language).

Let us indicate a converse to the above property 5.12.A.

5.12.B. *Let a Lie group G act on V. If the induced action on $\mathscr{D}^k(V)$ is free and proper, then there exists a G-invariant rigid structure φ on V which, moreover, is $\mathscr{D}$-invariant.*

Proof. Observe that if a Lie group G smoothly acts on a manifold V such that the corresponding action on $\mathscr{D}^k(V)$ for some $k = 1, 2, \ldots$ is free and proper, then the quotient space $\Phi_k = \mathscr{D}^k(V)/G$ is a smooth Hausdorff manifold. The action of $\mathscr{D}^k$ on $\mathscr{D}^k(V)$ induces a smooth action on Φ_k, and the structure $\varphi_k \colon V \to \Phi_k(V)$ corresponding to the quotient map $\mathscr{D}^k(V) \to \Phi_k$ is G-invariant.

This structure is not necessarily rigid. But one can pass to φ_{k+1} for $\Phi_{k+1} = \mathscr{D}^{k+1}(V)/G$, and this φ_{k+1} is rigid since the action of G on $\mathscr{D}^k(V)$ is free.

5.12.B$_1$. **Remark.** The structure which we have constructed in the proof of 5.12.B is by no means an A-structure. The following is a typical example.

5.12.C. **Example.** Let $G = \mathbf{Z}$, and consider an Anosov action of $\mathbf{Z}$ on a compact manifold V (see 2.1). Then the corresponding action of $\mathbf{Z}$ on $\mathscr{D}^1(V)$ is free and proper (the proof is not difficult), and the corresponding structure φ_1 is rigid as the action of $\mathscr{D}^1 = \mathrm{GL}(n, \mathbf{R})$ on $\mathscr{D}^1(V)/\mathbf{Z}$ is locally free.

5.12.D. **Remark.** Let us give another construction of an invariant structure by appealing to the following trivial fact.

If an action of G on W is proper, then there exists an invariant Riemannian metric g on W. (In fact, the existence of g is characteristic for proper actions; see [60].) Now, such a metric g on $W = \mathscr{D}^k(V)$ can easily be interpreted as a geometric structure φ on V. The rigidity of g implies that of φ. Notice that this structure is not $\mathscr{D}$-invariant.

5.13. Local integrability of infinitesimal isometries. Denote by $\mathrm{Dif}^{r+i}(v, v')$ the space of $(r+i)$-jets (of germs) of diffeomorphisms $V \to V$ mapping $v \mapsto v'$ and then, for a given structure φ of order r on V, consider the subset $\mathrm{Is}^{r+i}(v, v') \subset \mathrm{Dif}^{r+i}(v, v')$ of the jets which send $J^i_\varphi|\mathscr{D}^r_v(V) \to J^i_\varphi|\mathscr{D}^r_{v'}(V)$. The jets preserving J^i_φ are called *infinitesimal isometries of order* i.

Notice that $\mathrm{Is}^{r+i}(v, v)$ coincide with $\mathrm{Is}^{r+i}(v)$ defined in 5.11. Also observe that the group $\mathrm{Is}^{r+i}(v)$ naturally acts on $\mathrm{Is}^{r+i}(v, v')$, and this action is free and transitive. It follows that the k-rigidity of φ as defined in 5.11.A implies that the natural projection $\mathrm{Is}^{k+1}(v, v') \to \mathrm{Is}^k(v, v')$ is injective for all v and v' in V. This injectivity property explicitly states that rigidity amounts to *uniqueness* of an extension of the infinitesimal isometries of (V, φ) of order k to those of order $k+1$.

Now, an infinitesimal isometry $d \in \mathrm{Is}^i(v, v')$ is called *locally integrable* if there is an isometry $f: U \to V$ for some neighborhood $U \subset V$ of v, such that $d = J^{r+i}_f(v)$.

Global integrability means, by definition, the existence of a global isometry $f: V \to V$ such that $J^{r+i}_f(v) = d$.

Infinitesimal isometries in general are not locally integrable, and local integrability does not imply the global one.

5.13.A. **Example.** Consider a C^∞ metric φ on V, which is flat on a given open subset $U \subset V$ and has variable curvature outside U. Then for each $v \in \partial U$ the infinitesimal isometry groups satisfy

$$\mathrm{O}(n) = \mathrm{Is}^1(v) = \mathrm{Is}^2(v) = \cdots = \mathrm{Is}^k(v) = \cdots,$$

while $\mathrm{Is}^{\mathrm{loc}}(v)$ is generically trivial. Furthermore $\mathrm{Is}^{\mathrm{loc}}(v') = \mathrm{O}(n)$ for all $v' \in U$. Yet the full isometry group of V may be trivial.

The following proposition, which probably goes back to S. Lie, shows that local integrability fails only on some nowhere dense subset in V.

5.13.B. *If the structure φ is $(r+i)$-rigid, then there exist an integer i_0 (which depends only on $r+i$ and $\dim V$) and an open dense subset $U \subset V$ depending on φ such that every infinitesimal isometry in $\mathrm{Is}^{r+i_0}(u, v)$ is locally integrable for all $u \in U$ and $v \in V$.*

A well-known corollary for Riemannian metrics g is due to Singer [63].

5.13.C. **Corollary.** *If (V, φ) is infinitesimally homogeneous (namely, if $\mathrm{Is}^{r+j}(v, v')$ is nonempty for all $j = 1, 2, \ldots$ and all v and v' in V), then V is locally homogeneous (i.e., the pseudogroup of local isometries is transitive).*

The proof is based on the Frobenius theorem for totally integrable systems (see 5.17.B and also §1.6 in [29]).

To construct a local isometry sending v to v' we take some point u in the above U. Then Proposition 5.13.B ensures local isometries sending $v \to u$ and $u \to v'$. The composition of these is what we need.

5.13.D. **Remark.** One does not know how big $r+j$ is required to be for a given type of structure to insure the local homogeneity. But one knows that in the Riemannian case $r+j$ must be of order $n = \dim V$ (see [67]; also see [25, p. 165] for a related discussion).

5.14. Partition into Is^{r+j} and $\mathrm{Is}^{\mathrm{loc}}$-orbits. Now, let V be a manifold with a C^∞ (C^{an}) smooth (possibly nonrigid) structure φ of order r. Then, for each $j = 1, 2, \ldots$ we have the partition of V into the orbits under infinitesimal isometries of order j, called Is^{r+j}-orbits: two points v_1 and v_2 are in the same orbit if the set $\mathrm{Is}^{r+j}(v_1, v_2)$ (see 5.13) is nonempty. The quotient space for this Is^{r+j}-partition is denoted by V/Is^{r+j}. Similarly, we define $\mathrm{Is}^{\mathrm{loc}}$-orbits referring to the local isometry pseudogroup of V.

For general structures of *nonalgebraic* type the Is^{r+j}- and $\mathrm{Is}^{\mathrm{loc}}$-partitions may be arbitrarily complicated. For example, the partition into the orbits by an Anosov diffeomorphism is an Is^{r+j}-partition for the invariant structure constructed in 5.12.C.

On the other hand, if the structure is A, then the Is^{r+j}-partition is *regular* (at least on an open dense set in V) in the following sense.

5.14.A. **Definition.** A partition of V is called *regular* if it equals the partition into the level sets of a smooth submersion of V into some smooth manifold W.

Now, we can state the regularity property of the Is-partition.

5.14.B. **Regular partition theorem.** *Let φ be a C^∞-smooth A-structure of order r. Then for every $j \geq 0$ there exists an open dense subset $V_j \subset V$ such that the restriction of the Is^{r+j}-partition to V_j is regular. Furthermore if φ is rigid, there exists an open dense subset, say $V_\infty \subset V$, such that all Is^{r+j}-partitions, $j = 0, 1, \cdots$, are regular on V_∞, and $\mathrm{Is}^{\mathrm{loc}}$ is regular on V_∞ as well. In fact, the $\mathrm{Is}^{\mathrm{loc}}$-partition equals the Is^{r+j}-partition on V_∞ for all sufficiently large j. Moreover, this V_∞ is invariant under local (and hence global) isometries of V. (It is even invariant under infinitesimal isometries.)*

Idea of the proof. The Is^{r+j}-partition can be thought of as the partition into the levels of the map which assigns to each v the infinitesimal isometry class of φ at v of order $r+j$. If the structure φ is A, one can actually produce such a map (called the generalized Gauss map in [29]) which ranges in the quotient $\Phi'/\mathscr{D}^{r+j}$, where Φ' is some algebraic manifold acted upon the group $\mathscr{D}^{r+j}$. Then the regularity theorem for Is^{r+j}-partition follows from the algebraic stratification Theorem 6.4.A which ensures an open dense invariant subset $\Phi'' \subset \Phi'$ such that the partition into $\mathscr{D}^{r+j}$-orbits is regular in Φ''. The passage from the infinitesimal partition to the $\mathrm{Is}^{\mathrm{loc}}$ partition is then achieved by using the Frobenius theorem (see 5.17.B; also see §1 in [29]).

As an immediate corollary we get the following proposition for topologically transitive actions claimed in 0.7.A.

5.14.C. **Locally homogeneity theorem.** *If the isometry group* $\mathrm{Is}(V, \varphi)$ *is topologically transitive on* V *(i.e., if there exists a dense orbit), then there exists an open dense locally homogeneous subset in* V.

5.15. Killing fields and globalization. The essential results which we have stated so far (see 5.13.B and 5.14.B) concern local rather than global isometries where the step from "local" to "global" is obstructed by nonextendibility of isometries (compare Example 5.13.A). Now, we want to impose a *regularity condition* on the *Killing fields* of (V, φ) in order to remove this obstruction. Here, a tangent field X on some open subset $U' \subset V$ is called *Killing* if it integrates to isometries $X_t \colon U' \to U$ for the open subsets $U' \subset U$ which are relatively compact in U and where $t \in [0, \varepsilon]$ for some $\varepsilon = \varepsilon(U', X) > 0$. We call φ *regular* if the sheaf of Killing fields is locally constant (i.e., if the *dimension* of the space of Killing fields on a small connected subset in V is independent of this subset). If φ is rigid, this is equivalent to the following.

5.15.A. *Extension property.* Each point $v \in V$ admits a neighborhood $U \subset V$ such that every Killing field on every connected subset $U' \subset U$ extends to U.

5.15.B. An important consequence of the extension of Killing fields is a similar extension for the isometries in the identity components of the isometry pseudogroup (for the obvious topology in this pseudogroup). It is well known (see [58], [1]) that *rigid real analytic structures are regular.* But the regularity may easily fail for C^∞-structures. For example, every nonflat connected Riemannian manifold (V, g), where g is flat on some $U \subset V$, is *not* regular (see 5.13.A). (It is not hard to show that every Riemannian C^∞ metric g can be "regularized" by an arbitrarily small C^∞-perturbation g' such that $\mathrm{Is}(V, g) = \mathrm{Is}(V, g)$. But this is unknown for more general structures such as pseudo-Riemannian metrics.)

If V is regular and simply connected (it is enough to assume that $\pi_1(V)$ admits no nontrivial homomorphism into any finite group), then the sheaf of Killing fields is constant. If, moreover, V is compact without boundary, then global fields integrate to isometries of V. Furthermore, for every connected open set $U \subset V$, every isometry $U \to V$ in the connected component of the local isometry pseudogroup extends to an isometry of V. This allows one to globalize Proposition 5.13.B as follows.

Let $\mathrm{Dif}^{r+i}(V)$ denote the manifold of $(r+i)$-jets of germ of C^∞-diffeomorphisms $V \to V$, that is,

$$\mathrm{Dif}^{r+i}(V) = \bigcup_{v, v' \in V} \mathrm{Dif}^{r+i}(v, v'),$$

and let $\mathscr{I}^{r+i} \subset \mathrm{Dif}^{r+i}(V)$ be the corresponding union of $\mathrm{Is}^{r+i}(v, v')$. Let $\mathscr{I}_v^{r+i} = \bigcup_{v' \in V} \mathrm{Is}^{r+i}(v, v') \subset \mathscr{I}^{r+i}$, and let $\mathscr{I}_v^{r+i} = \mathscr{I}_V^{r+i}$ denote the connected component of the $(r+i)$-jet at v of the identity map $V \to V$.

Using 5.13.B one can easily show the following:

5.15.C. *If (V, φ) is regular (i.e., real analytic) compact simply connected, then for every $u \in U$ and all sufficiently large i_0 the jet map $J: f \mapsto J_f^{r+i_0}(u)$ establishes a homeomorphism of the connected component of the identity $\mathrm{Is}_0 \subset \mathrm{Is}(V, \varphi)$ onto $\mathscr{I}_u^{r+i_0}$.*

5.15.D. **Remarks.** (i) One can show (see 1.7 in [29]) that if (V, φ) is rigid *real analytic*, then infinitesimal isometries at every point $v \in V$ of sufficiently high order admit local extensions.

(ii) Let us indicate a globalization of the Regular Partition Theorem 5.14.B for regular (V, φ). (Do not confuse two notions of regularity!)

I. *f (V, φ) is regular simply connected, then there exists an open dense subset U invariant under the aciton of* $\mathrm{Is} = \mathrm{Is}(V, \varphi)$ *such that the partition of U into the orbits of the identity component* $\mathrm{Is}_0 \subset \mathrm{Is}$ *is regular.*

The proof easily follows from 5.14.B and 5.15.B.

5.16. Frame fields, rigidity and complete differential systems. We explain below how general rigid structures can be reduced to frame fields on $\mathscr{D}^r(V)$. This will provide a link of isometries with totally integrable systems needed for an application of the Frobenius theorem in the proof of 5.13.B and 5.14.B.

5.16.A. Let us look more closely at a *full frame field* φ on a manifold V, that is, a system of n independent vector fields on V for $n = \dim V$. It is clear that the infinitesimal isometry group $\mathrm{Is}^1(V, \varphi, v)$ is trivial for all $v \in V$, which amounts to 0-rigidity of φ.

Next, we observe that a structure φ of order r on $\mathscr{D}^l(V)$ naturally induces a structure on V, say φ^*, of order $r+l$. In fact, a local coordinate system $u_1, \cdots, u_n$ on V induces that on $\mathscr{D}^l(V)$, namely, the system corresponding to differentiation in these coordinates. To make sense of this, we recall that $\mathscr{D}^l(V)$ is an open subset in the space $J^l = J^l(V, \mathbf{R}^n)$ of l-jets of maps $f\colon V \to \mathbf{R}^n$. If we assume for example $l = 3$, then each coordinate $u = u^d_{ijk}$ in J^3, is the function on J^3 defined by the condition

$$u(J^3_f) = \frac{\partial^3 f_d}{\partial u_i \partial u_j \partial u_k},$$

where f_d denotes the d th component of f.

Now we have a natural embedding $\mathscr{D}^{r+l}(V) \to \mathscr{D}^r(\mathscr{D}^l)$ which gives the required operation $\varphi \to \varphi^*$. It is trivial that if φ is k-rigid, then φ^* is $(k+l)$-rigid.

An important example is the following.

5.16.B. *The full frame field φ on $\mathscr{D}^k(V)$ defines a k-rigid structure φ' on V.*

Our next example looks more familiar to differential geometers.

5.16.C. *Generalized connections.* A generalized connection of order r on V is a horizontal subbundle φ for the fibration $\mathscr{D}^r(V) \to V$.

That is, φ is an n-dimensional subbundle of $T(\mathscr{D}^r(V))$ transversal to the fibers. Such a connection can be viewed, by the above, as a structure of order $r+1$ on V. Next, observe that $\mathscr{D}^r(V) \to V$ is a principal fibration, and so the fiber $\mathscr{D}^r_v \subset \mathscr{D}^r(V)$ carries a full frame field (being the principal homogeneous space of the group $\mathscr{D}^r$). In fact, this field at a point $\delta \in \mathscr{D}^r_v$ corresponds to an r-jet of a local coordinate system at v. Then the first jet of this system is a frame at V, and the inverse of the differential brings this frame to φ over v. Thus φ defines in a canonical way a full frame field φ' on $\mathscr{D}^r(V)$. It is trivial to see that the rigidity of φ' implies that of φ. Thus generalized connections are r-rigid. In particular, ordinary affine connections are 1-rigid.

Now we see once again that pseudo-Riemannian metrics φ are 1-rigid, by (trivially) reducing this rigidity to that of the Levi-Civita connection φ' of φ.

5.16.D. The above implication

$$\text{rigidity of } \varphi' \Rightarrow \text{rigidity of } \varphi$$

is of quite general (and trivial) nature: suppose we have a structure φ of order r and φ' is obtained by some "canonical procedure" applied to φ. Here, "canonical procedure" means that the components of φ' are smooth functions in the partial derivatives of φ of order $\leq s$, or, more invariantly, that φ' is obtained from $\varphi\colon \mathscr{D}^r(V) \to \Phi$ by applying a differential operator of order s on $\mathscr{D}^r(V)$. Then it is obvious that the rigidity of φ' implies that of φ.

5.16.E. *Framed definition of rigidity.* A structure φ is *k-rigid* if there exists a full frame field on $\mathscr{D}^k(V)$ obtained from φ by a "canonical procedure."

5.16.F. **Remarks.** (i) By taking derivatives one sees that k-rigidity implies $(k+1)$-rigidity.

(ii) It is clear from the previous definition that framed rigidity implies rigidity.

The implication "rigidity" $\Rightarrow$ "framed rigidity" is not difficult, though the proof is somewhat boring.

(iii) For all natural examples of rigid structures one sees first the framed rigidity and then proceeds to rigidity. Thus the (boring) implication rigidity $\Rightarrow$ framed rigidity has little practical importance.

(iv) Frame-rigidity is very close to the notion of "structure of finite type by Cartan" where one requires a general connection of a special kind. The "finite type" terminology refers to the fact that the isometry group of such a structure is finite dimensional.

According to this terminology, structures of "infinite type," such as symplectic, foliated, etc., are those which have infinite dimensional isometry groups.

(v) The basic properties of the isometry group $\mathrm{Is}(V, \varphi)$ mentioned in 5.12 are immediate with the framed definition. Indeed, we have a frame φ' on $\mathscr{D}^r(V)$ such that isometries of (V, φ) induce those of $(\mathscr{D}^r(V), \varphi')$. These, in fact, are usual isometries for the Riemannian metric φ'' (associated to φ') with respect to which the frame φ' is orthonormal. Thus the properties of $\mathrm{Is}(V, \varphi)$ follow from the standard facts on isometries of Riemannian manifolds.

5.17. Isometries and partial differential equations. The isometry condition for a diffeomorphism $f: (V, \varphi_1) \to (V, \varphi_2)$ can be expressed by a system of partial differential equations of order r, where r is the order of the structure. In fact, the induced metric $f^*(\varphi_2)$ is expressed with J_f^r (see 5.2), and so the equality

$$f^*(\varphi_2) = \varphi_1$$

is an equation on J_f^r, i.e., a partial differential equation on f.

5.17.A. **Example.** Let the structures in question be full frame fields on V. Then the equation $f^*(\varphi_2) = \varphi_1$ has first order. At each point $v \in V$ it prescribes the differential of f, as this is uniquely defined by the condition that one given frame goes to another. Such systems of first order where the partial derivatives at each point are (smoothly) expressed in terms of the space coordinates and values of the unknown functions are called *complete.*

5.17.B. More generally, a system $\mathcal{S}$ of partial differential equations of a certain order $s+1$ is called *complete* if every partial derivative of order $s+1$ of the unknown map f can be "smoothly" expressed in terms of the derivative of order $\leq s$.

It follows that the derivatives of order $\leq s$ of every solution f of $\mathcal{S}$ satisfy, along every curve $C \subset V$, a certain system of ordinary differential equations of the first order. Therefore, if we prescribe the values of the derivatives of f of order $\leq s$ at a fixed point $v_0 \in V$ and then join v_0 with another point $v \in V$ by a curve $C \subset V$, then the solution of the partial differential equations uniquely determines $f(v) = f_C(v)$ for every solution f of $\mathcal{S}$.

If $f(v)$ does depend on the choice of C, then the system is unsolvable.

To insure solvability one imposes a *consistency* condition on $\mathcal{S}$ which amounts to a certain system of differential relations between the coefficients of the system $\mathcal{S}$.

These relations are equivalent to the (infinite) system of equations $\frac{d}{dt} f_{C_t}(v) = 0$ for all $v \in V$ and all one-parameter families C_t of curves in V between v_0 and v.

The solvability of complete integrable systems is the content of Frobenius' theorem.

5.17.C. Let us return to frame k-rigid systems (of order $r \leq k+1$), and observe that the situation here for $k > 0$ is similar to that for $k = 0$. Namely, the isometry equation

$$(*) \qquad f^*(\varphi_2) = \varphi_1$$

can be expressed by a complete system of order $k+1$. In fact, a little thought shows that the framed definition is equivalent to completeness of $(*)$.

Now, the proof of the local integrability theorem 5.13.B is reduced, by Frobenius' theorem, to verifying the integrability condition. This is of purely algebraic nature, and can be insured in our case by the constancy of the rank of a certain map (see [29]) related to equation $(*)$.

As every smooth map has constant rank on an open dense set, by applying Frobenius' theorem we obtain the desired local integrability of such a set $U \subset V$ (see 5.13.B).

5.18. Stability of rigidity. As we mentioned in 5.11.B(v) the rigidity (unlike Iso-rigidity) is stable under smooth perturbations of the structure.

In fact, a simple (but again quite boring) argument shows that rigidity is equivalent to the nonvanishing of some differential of φ. Namely, one gives the definition of rigidity as follows.

5.18.A. Denote by $\Delta^{k+r} \subset \mathscr{D}^{k+r} = \mathscr{D}_0^{k+r}(\mathbf{R}^n)$ the kernel of the natural homomorphism $\mathscr{D}^{k+r} \to \mathscr{D}^{k+r-1}$ and call a map $\psi: \mathscr{D}^{k+r} \to W$ (where W can be any manifold) *rigid* if for every left invariant vector field ∂ on $\mathscr{D}^{k+r}$ belonging to the Lie algebra of Δ^{k+1} the derivative is not identically zero.

Now, for every structure φ of order r (see 5.4) one can define the map $\mathscr{D}^k\varphi: \mathscr{D}^{k+r}(V) \to \mathbf{R}^{sN_0}$, $N_0 = 1 + n + n(n+1)/2 + \cdots$, by taking all partial derivatives of $\varphi(v, u)$ with respect to the coordinate system $u = (u_1, \cdots, u_n)$.

Then we have the following.

5.18.B. **Stable rigidity criterion.** *If the above map $\mathscr{D}^k\varphi$ is rigid, then φ is k-rigid.*

6. Examples of A-rigid actions

6.1. We present here basic examples of A-rigid actions. Most of them have already appeared throughout these lectures, but we have brought all of them together for the convenience of the reader.

6.2. Compact groups. The easiest actions from our point of view are those of compact Lie groups G on V. Every such action is rigid. In fact, starting from an arbitrary (noninvariant) metric φ one gets an invariant one by averaging our G, where the averaged metric is

$$\overline{\varphi} = \int_G (g\varphi)\, dg;$$

here dg is the Haar measure on G.

Notice that for a *generic* metric φ, $\mathrm{Is}(V, \overline{\varphi}) = G$ provided $\dim V > \dim G$ as a simple argument shows.

The basic topological property of compact group actions is the following.

6.2.A. **Compact stratification theorem.** *Let G be a compact Lie group acting on a manifold V. Then there exists a stratification* (*see* 1.5.A) *of V into G-invariant locally closed subsets V_i, $i = 1, \cdots, s$, such that the orbits of the action of G on V_i are mutually isomorphic* (*i.e., the isotropy subgroups G_v, $v \in V$, are mutually conjugate*) *for v running over V_i. Moreover, the quotient space V/G is a smooth manifold, and the quotient map $V \to V/G$ is a smooth fibration.*

This is well known, and the proof is not difficult (see, e.g., [11]).

6.3. Algebraic groups. After compact group actions the next remarkable class of examples is that of algebraic actions of algebraic groups on algebraic manifolds, which were introduced in 1.6.

As in 1.6, we concentrate, insofar as examples are concerned, on *algebraic* subgroups in the full matrix group $GL(n, \mathbf{R})$ acting on $P^{n-1} = P(\mathbf{R}^n)$ in the usual way.

Rephrasing what was said in 1.6, we recall that a subgroup $G \subset GL(n, \mathbf{R}) \subset \mathbf{R}^{n^2}$ is *algebraic* if the closure of G in $\mathbf{R}^{n^2}$ is the zero set of a system of polynomials on $\mathbf{R}^{n^2}$.

6.3.A. **Example.** Let G be a one-parameter group (i.e., G is isomorphic to $\mathbf{R}$ as a Lie group) of diagonal transformations. That is, the elements of G are matrices

$$\begin{pmatrix} t^{\alpha_1} & & & 0 \\ & t^{\alpha_2} & & \\ & & \ddots & \\ 0 & & & t^{\alpha_n} \end{pmatrix}, \qquad t \in \mathbf{R}^{\times}_{+},$$

where $\alpha_1, \cdots, \alpha_n$ are given real numbers. Then G is algebraic if and only if α_i/α_j is *rational* for all $i, j = 1, \cdots, n$.

The proof of this is an easy exercise.

6.4. Basic facts on algebraic actions. We collect here some standard facts on algebraic groups which are used in several places in these lectures. Some of these are restatements of what has been stated earlier. For a more complete discussion the reader can see [71], [38], [10].

6.4.A. **Algebraic stratification theorem** (see [62] and compare 5.14.B). *Let G be a real algebraic group algebraically acting on a real algebraic manifold V. Then there exists a stratification $V = \bigcup_{i=1}^{s} V_i$, where each $V_i \subset V$ is a locally closed G-invariant algebraic submanifold, such that all orbits $G(v)$, $v \in V_i$, are closed subsets in V_i of dimension n_i. Furthermore, the quotient V_i/G has a natural structure of a smooth algebraic manifold such that the projection $V_i \to V_i/G$ is an algebraic map as well as a C^{an}-fibration.*

We state without proof another standard theorem.

6.4.B. **The orbit theorem.** *Let G be a real algebraic group algebraically acting on a real algebraic manifold V. Let $\overline{G(v)}$ denote the closure of the orbit $G(v) \subset V$ and $\partial G(V) = \overline{G(v)} - G(v)$. Then $\partial G(v)$ is a semialgebraic set of dimension* $\dim \partial G(v) < \dim G(v)$. *In particular, $\partial G(v)$ is contained in an algebraic set of dimension* $< \dim G(v)$.

6.4.B$_1$. **Corollary.** *If V is compact, then for every $v \in V$, $\overline{G(v)}$ contains a compact orbit. In fact, if $d_{\min}$ is the minimum of the dimension* *of the orbits which are contained in* $\overline{G(v)}$, *then each orbit in* $\overline{G(v)}$ *whose dimension equals* $d_{\min}$ *is compact.*

6.4.B$_2$. **Example.** Let $G = \mathbf{R}^n$. This G admits no compact *algebraic* homogeneous space of positive dimension. In fact, if G/H is compact and $H \neq G$, then H has infinitely many components and hence is non-algebraic. It follows that each algebraic action of $G = \mathbf{R}^n$ on a compact algebraic manifold must have a fixed point. The same is true for the group $(\mathbf{R}_+^\times)^n$ (which is isomorphic to $\mathbf{R}^n$ as a Lie group but not as an algebraic group) and also applies to complex algebraic solvable subgroups in $\mathrm{GL}(n, \mathbf{C})$. (The latter is the famous theorem of Borel-Lie.)

Next, we recall from 5.3 the principal bundle $\mathscr{D}^r(V) \to V$ whose fiber $\mathscr{D}_v^r$ over $v \in V$ consists of the r-jets (of germs) of local coordinate systems in V around v, and state the following.

6.4.B$_3$. **Jet properness theorem.** *Let* G *be a real algebraic group algebraically acting on a real algebraic manifold* V. *If* G *acts on* V *faithfully, then there exists an* r *such that the action of* G *on* $\mathscr{D}^r(V)$ *is free and proper. Moreover,* $\mathscr{D}^r(V)/G$ *is a smooth algebraic manifold for which the projection* $\mathscr{D}^r(V) \to \mathscr{D}^r(V)/G$ *is a smooth algebraic map.*

Notice that the map $\mathscr{D}^r(V) \to \mathscr{D}^r(V)/G$ provides a G-invariant A-structure on V. It is easy to see that this structure is rigid for large r, and so the algebraic actions are included in the class of rigid A-actions (see 1.8.A and also 5.5, 5.10, 5.11).

6.4.B$_4$. **Diagonal action theorem.** *If the action of* G *on* V *is faithful, then there exists an open dense subset*

$$U \subset W = \underbrace{V \times V \times \cdots \times V}_{r}$$

which is invariant for the diagonal action of G *on the* r *th Cartesian power* W *of* V *for some (sufficiently large)* r *such that the action of* G *on* U *is free and proper.*

From this one can immediately deduce the following property which we have already seen in 4.1.B.

6.4.B$'_4$. **Corollary.** *Let* μ *be a* G*-invariant Borel probability measure on* V, *and let* $V_\mu \subset V$ *be the Zariski closure of the support of* μ. *Then the action of* G *on* V_μ *factors through a compact action. In particular, if* $V_\mu = V$ *(e.g.,* $\operatorname{supp}\mu = V$*), then* G *is compact.*

6.4.B$_5$. **Finite volume property** (Compare 1.11.C.). Let $\Gamma_g \subset V \times V$ denote the graph of the action $g: V \to V$, $g \in G$. Then if V is compact, $\mathrm{Vol}(\Gamma_g) \leq \mathrm{const}$, where Vol denotes the n-dimensional volume ($n = \dim V$) for a fixed Riemannian metric in $V \times V$.

Idea of the proof. Algebraically embed $V \times V$ into P^N (for large N), and notice that the algebraic degree of $\Gamma_g \subset V \times V$ in P^N is bounded by a constant, say d. By the definition of degree (or Bezout theorem) the number of intersection points of Γ_g in P^N with a generic $(N-n)$-dimensional subspace P^n in P^N is at most d. Then by Crofton's formula, $\mathrm{Vol}(\Gamma_g)$ equals the average $\#(P^n \cap \Gamma_g)$ over all $P^n \subset P^N$. q.e.d.

Observe that in the above finite volume property the crux of the matter is that "const" may depend on V and on a choice of the metric in $V \times V$, but not on $g \in G$.

The above 6.4.B$_5$ is closely related to another important feature of the graphs Γ_g. Namely, we have the following.

6.4.B$_6$. **Hausdorff closure property.** If some compact subset $K \subset V \times V$ lies in the closure of the set of the graphs $\{\Gamma_g\}$, $g \in G$, then $\dim K \leq n = \dim V$, where the space of the subsets in $V \times V$ is given the Hausdorff topology corresponding to the Hausdorff metric: $\mathrm{dist}(A_1, A_2)$ for two subsets A_1, A_2 defined as the minimal ε such that the ε-neighborhoods of A_1 and A_2 satisfy:

$$U_\varepsilon(A_1) \supset A_2 \quad \text{and} \quad U_\varepsilon(A_2) \supset A_1.$$

Notice that attaching these Hausdorff limits K to the set of graphs $\{\Gamma_g\}$ provides an interesting compactification of the transformation group G, where the limit set $K \subset V \times V$ can be viewed as the "graph of an ideal transformation" of V.

6.4.C. **Remarks.** In fact, 6.4.B$_6$ above shows that the topological properties of algebraic actions are somewhat similar to those of compact group actions. Namely, in the algebraic case, where the group is noncompact, it can be naturally compactified.

A similar remark also applies to the finite volume property 6.4.B$_5$. Namely, in the compact case the sup of the pointwise norms of the differential Dg of the transformations $g: V \to V$ are uniformly bounded as g ranges over G. In the algebraic case we have bounded $\mathrm{Vol}(\Gamma_g) \subset V \times V$ which can be viewed as a kind of integral norm of $\mathscr{D}g$.

6.5. Diagonal one-parameter actions. Let us look again at the diagonal group

$$\begin{pmatrix} t^{\alpha_1} & & 0 \\ & \ddots & \\ 0 & & t^{\alpha_n} \end{pmatrix}, \qquad t \in \mathbf{R}_+^\times,$$

acting on P^{n-1}.

Here we allow arbitrary reals $\alpha_1, \cdots, \alpha_n$, so the action is not necessarily algebraic, and we assume without loss of generality that

$$0 < \alpha_1 \leq \alpha_2 \leq \cdots \leq \alpha_n.$$

Now, let us restate some elementary properties of diagonal matrices in geometric terms.

(a) The fixed points of the action are $(1, 0, 0, \cdots, 0)$, $(0, 1, 0, \cdots, 0)$, $\cdots$, $(0, 0, \cdots, 0, 1)$.

(b) Among these points a special role is played by the two points $(1, 0, \cdots, 0)$ and $(0, 0, \cdots, 1)$ corresponding to $\min_i \alpha_i$ and $\max_i \alpha_i$. Namely, there exists an open dense invariant set $U \subset P^{n-1}$, such that the closure of each orbit $\overline{G(v)} \subset P^n$, $v \in U$, contains these two points. In fact one may take $U = \{x_1, \cdots, x_n\}$, $x_i \neq 0$, $i = 1, \cdots, n$.

(c) Each coordinate subspace is G-invariant (coordinate subspaces are given by the equations $x_{i_j} = 0$, for given $i_1 < i_2 < \cdots < i_j < \cdots < i_n$) and these are the only closed invariant subspaces.

6.5.A. **Diagonal action of $(\mathbf{R}_+^\times)^k$ on P^{n-1}** (see 1.6.C(i)). A diagonal action of $(\mathbf{R}_+^\times)^k$ on the projective space P^{n-1} is defined by a homomorphism

$$(\mathbf{R}_+^\times)^k \to A = \left\{ \begin{pmatrix} t_1 & & 0 \\ & \ddots & \\ 0 & & t_n \end{pmatrix} \right\},$$

where A is the subgroup of the diagonal matrices in $\mathrm{GL}(n, \mathbf{R})$. More precisely, such an action is given by a matrix

$$\begin{pmatrix} m_1 & & & 0 \\ & m_2 & & \\ & & \ddots & \\ 0 & & & m_n \end{pmatrix},$$

where each m_i, $i = 1, \cdots, n$, is a monomial:

$$m_i = t_1^{\alpha_{1,i}} t_2^{\alpha_{2,i}} \cdots t_k^{\alpha_{k,i}}, \qquad (t_1, \cdots, t_k) \in (\mathbf{R}_+^\times)^k.$$

To simplify the notation, let $k = 2$. Then our action is given by the matrices

$$\begin{pmatrix} t_1^{\alpha_1} t_2^{\beta_1} & & & 0 \\ & t_1^{\alpha_2} t_2^{\beta_2} & & \\ & & \ddots & \\ 0 & & & t_1^{\alpha_n} t_2^{\beta_n} \end{pmatrix}, \qquad (t_1, t_2) \in (\mathbf{R}_+^\times)^2.$$

6.5.A$_1$. Another example of diagonal action for $k = n-1$ is that where the action is given by

$$\begin{pmatrix} t_1 & & & 0 \\ & t_2 & & \\ & & \ddots & \\ 0 & & & t_n \end{pmatrix}, \qquad (t_1, \cdots, t_n) \in (\mathbf{R}_+^\times)^n.$$

Here $G = (\mathbf{R}_+^\times)^{n-1} = A/\Delta$, where Δ equals $\mathbf{R}_+^\times$ diagonally embedded in $A = (\mathbf{R}_+^\times)^n$, and $\Delta = \mathbf{R}_+^\times$ appears as the group of diagonal matrices

$$\begin{pmatrix} t & & & 0 \\ & t & & \\ & & \ddots & \\ 0 & & & t \end{pmatrix},$$

which trivially acts on P^{n-1}, so that one looks at the action not really of $\mathrm{GL}(n, \mathbf{R})$ but of $\mathrm{GL}(n, \mathbf{R})/\Delta$ on P^{n-1}.

Notice that this action is *universal* in the sense that every diagonal matrix is contained in the group

$$A = \left\{ \begin{pmatrix} t_1 & & 0 \\ & \ddots & \\ 0 & & t_n \end{pmatrix} \right\} \subset \mathrm{GL}(n, \mathbf{R}).$$

The geometry of the action of this $G = A/\Delta = (\mathbf{R}_+^\times)^{n-1}$ on $V = P^{n-1}$ is easy to understand. For example, the orbit $G(v)$ of a point $v = (x_1, \cdots, x_n)$ where all (homogeneous) coordinates are > 0 equals the set of *all* points with positive coordinates. This set can be identified with the open $(n-1)$-symplex. (In fact, the set $\{x_i > 0\}$ in P^{n-1} is identical with the usual symplex $\Delta^{n-1} = \{x_i > 0 \mid \sum_{i=1}^n x_i = 1\}$ in $\mathbf{R}^n$.) Then the closure $\overline{G(x)}$ of the orbit gives the closed symplex, and $\partial G(v) = \overline{G(v)} - G(v)$ equals the boundary of the symplex. Furthermore, the k-dimensional orbits correspond to the k-faces of our symplex in an obvious way.

Now, turn to the general diagonal action of $G = (\mathbf{R}_+^\times)^k$ on P^{n-1} and state the following classical theorem.

6.5.B. **Convexity theorem.** *The closure of each orbit $G(v) \subset P^n$ is a finite union of orbits. Moreover, there exists a convex polyhedron $H \subset \mathbf{R}^n$ and a homeomorphism $\overline{G} \to H$ such that each orbit $G(v)$ goes onto the interior of some face of H.*

Idea of proof (see [30] for details). Start with the following map $M: P^{n-1} \to \mathbf{R}^n$, $M: (x_1, \cdots, x_n) \to (x_1^2, x_2^2, \cdots, x_n^2)$ where the homogeneous coordinates are normalized by $\sum_{i=1}^n x_i^2 = 1$. Then P^{n-1} goes to the standard symplex $\Delta^{n-1} \subset \mathbf{R}^n$. If $G = (\mathbf{R}_+^\times)^{n-1} = A/\Delta$, then it is obvious that each orbit goes to some face of Δ^{n-1}.

6.5.C. Now, let us turn to the general case, but to keep the notation simple, let $k = 2$ and

$$G = \begin{pmatrix} t_1^{\alpha_1} t_2^{\beta_1} & & \\ & \ddots & \\ & & t_1^{\alpha_n} t_2^{\beta_n} \end{pmatrix}, \qquad (t_1, t_2) \in (\mathbf{R}_+^\times)^2.$$

Denote by $S \subset \mathbf{R}^2$ the set of pairs $(\alpha_1, \beta_1), \cdots, (\alpha_n, \beta_n)$, and let $C = \operatorname{Conv}(S) \subset \mathbf{R}^2$ be the convex hull of S. This C is a finite polygon with at most n vertices. We may assume, by permuting the coordinates if necessary, that the first k points $(\alpha_1, \beta_1), \cdots, (\alpha_k, \beta_k)$ are the vertices while the remaining ones lie inside (or on the open edges of) C.

Notice that these k points play a special role (similar to $\max \alpha_i$ and $\min \alpha_i$ in 6.5(b)). Namely, each of the n points $(1, 0, \cdots, 0), (0, 1, \cdots, 0, 0), \cdots, (0, 0, \cdots, 1)$ in P^{n-1} is fixed under the diagonal action. But the first k, $(1, 0, \cdots, 0), (0, 1, \cdots, 0), \cdots, (\underbrace{0, \cdots, 1}_{k}, \cdots, 0)$, which correspond to the vertices of C with our notation, are *attractive* in some neighborhood in P^{n-1}. This means there exists an open dense $U \subset P^{n-1}$, such that the closure of each orbit $\overline{G(v)}$, $v \in U$, contains these k points. On the other hand, there exists no such *open* U for the remaining fixed points.

Now, let L denote the (unique) affine map $\Delta^{n-1} \to \mathbf{R}^2$ which sends the k th vertex of Δ^{n-1} to $(\alpha_k, \beta_k) \in \mathbf{R}^2$, and consider the composed map $L \circ M \colon P^{n-1} \to \mathbf{R}^2$ for the above $M \colon P^n \to \Delta^{n-1}$. One can show (see [30]) that each G-orbit in P^{n-1} homeomorphically goes under $L \circ M$ onto an open face of the convex polygon C, and thus we obtain the claimed correspondence between orbits and convex sets (see [30] for details).

6.5.D. We conclude the discussion of algebraic groups by explaining the conformal action of $\mathrm{O}(n+1, 1)$ on S^n (see 0.10.A). We start with the standard embeddings

$$S^n \subset \mathbf{R}^{n+1} \subset P^{n+1},$$

and let $G \subset \mathrm{SL}(n+2, \mathbf{R})$ be the group of projective transformations which map S^n into itself.

The action of G on S^n preserves round subspheres as these are the intersections of S^n with linear subspaces. Thus this action is conformal.

Next, we represent $S^n \subset P^{n+1}$ by a cone in $\mathbf{R}^{n+2}$. Namely, we take the cone defined by the equation

$$x_1^2 + x_2^2 + \cdots + x_{n+1}^2 = x_{n+2}^2,$$

and observe that the linear transformations of $\mathbf{R}^{n+1}$ preserving this cone form, up to the scalars, the orthogonal group $O(n+1, 1)$ for the quadratic form

$$x_1^2 + x_2^2 + \cdots + x_{n+1}^2 - x_0^2.$$

Thus we get the required conformal action of $O(n+1, 1)$ on S^n.

Let us change the coordinates in order to have the form

$$x_1^2 + x_2^2 + \cdots + x_n^2 + yz.$$

Now, we have an interesting diagonal 1-parameter group, $\mathbf{R}_+^\times$, namely

$$G = \begin{pmatrix} 1 & & & & 0 \\ & 1 & & & \\ & & \ddots & & \\ & & & t & \\ 0 & & & & t^{-1} \end{pmatrix}$$

acting conformally on S^n.

This group acts by north pole-south pole transformations (see 2.3) which can be seen geometrically if one writes $S^n = \mathbf{R}^n \cup \{\infty\}$, where the point $0 \in \mathbf{R}^n \subset S^n \subset P^{n+1}$ corresponds to the eigenspace spanned by $(0, 0, \cdots, 1)$, and ∞ corresponds to $(0, 0, \cdots, -1)$. Then G acts by scaling on $\mathbf{R}^n$, $x \mapsto t \cdot x$. This action has two fixed points on S^n. The south pole corresponds to $0 \in \mathbf{R}^n$, and the north pole to ∞. If $s \in S^n \setminus \{\text{south pole}\}$, then $ts \to$ north pole for $t \to +\infty$ and if $s \in S^n \setminus \{\text{north pole}\}$, then $ts \to$ south pole for $t \to -\infty$.

We suggest that the reader write in matrices the transformation corresponding to the parallel translation of $\mathbf{R}^n \subset S^n = \mathbf{R}^n \cup \{\infty\}$.

6.6. Twisted torus action (Compare 1.11.D). Now we want to describe a simple class of actions which look very much like algebraic actions though they are not algebraic.

Let T^n freely act on V, and let $\mathcal{G} \subset \mathrm{Diff}(V)$ be the corresponding gauge group: each $g \in \mathcal{G}$ is a fiber preserving diffeomorphism $V \hookleftarrow$, where each fiber $T_v = T^n$ goes into itself by a translation in T^n. Thus g is determined by a function, say $\varphi\colon V \to T^n$, such that $g_\varphi(v) = \varphi(v) \cdot v$ for the action $t \cdot v$ of T^n of V. Note that $\mathcal{G} \supset T^n$, and we are interested in connected finite-dimensional subgroups $G \subset \mathcal{G}$ which contain the torus

$T^n \subset \mathcal{G}$. Such a subgroup G can be given by k vector fields on V, for $k = \dim G - n$ tangent to the T^n-orbits. These fields on each orbit should correspond to some field coming from the Lie algebra $\mathbf{R}^n = L(T^n)$, and our fields are, in fact, determined by maps $f_j \colon W \to \mathbf{R}^n$, $j = 1, \cdots, k$, for $W = V/T^n$.

The following proposition shows that the action of G is rigid in most cases. For example, it is rigid if the maps f_j are real analytic.

6.6.A. **Proposition.** *The following conditions are equivalent*:

(1) *The action of G is A-rigid.*

(2) *The action is rigid.*

(3) *The induced action of G on $\mathcal{D}^r(V)$ is free for some r.*

(4) *There exists an r, such that: if a linear combination $f = \sum_{j=1}^{k} \lambda_j f_j$ has $\partial^I f(v_0) = 0$ for $|I| = 1, \cdots, r$, where v_0 is some point in V (and the partial derivatives are taken in some local coordinates around v_0), then f is constant on the connected component of v_0 in V.*

Sketch of the proof. Obviously, $(1) \Rightarrow (2)$ while $(2) \Rightarrow (3)$ by 5.12. Let us show $(1) \Leftrightarrow (4)$. First, observe that condition (4) is equivalent to the action of G to be locally free on $\mathcal{D}^r(V)$, and hence $(3) \Rightarrow (4)$ and $(1) \Rightarrow (4)$.

Next, to see that $(4) \Rightarrow (1)$, notice that under condition (4), the action of the isotropy subgroup G_v on $\mathcal{D}_v^r(V)$ is locally free for all $v \in V$, and that this implies that the action is in fact free. Then one can see that the embedding $G_v \subset \mathcal{D}_v^r$ is algebraic, i.e., it is injective and the image is an algebraic subgroup in $\mathcal{D}_v^r$. That is, we have a fibration $\overline{\mathcal{D}}^r \to V$ with algebraic fibers $\mathcal{D}_v^r/G_v$, and T^n naturally acts on $\overline{\mathcal{D}}^r$.

Now, observe that G-invariant functions on $\mathcal{D}^r(V)$ can be identified with T^n-invariant ones on $\overline{\mathcal{D}}^r$. These are easy to construct (for example by averaging over T^n) as T^n is compact. In fact, there exists a system of such functions algebraic on each fiber $\mathcal{D}_v^r/G_v$, $v \in V$, such that the isometry group of the corresponding structure $\mathcal{D}^r(V) \to \mathbf{R}^s$ (see 5.4) equals G. The details are left to the reader.

6.6.B. **Examples.** (i) (Compare 1.11.D.) Take $V = S^3$ and $G = S^1 \times \mathbf{R}$ acting on S^3 by twisted rotations as described in 1.1.1.D.

(ii) Here our manifold is $V = \mathbf{R}^{n-m} \times T^m$ equipped with the standard (product) affine structure, and our group G is the group of *affine* transformations of V mapping each torus $x \times T^m$ into itself by some rotation of T^m. Clearly, G is isomorphic to $T^m \times \mathbf{R}^{m(n-m)}$.

6.6.C. **Remark.** The actions described in 6.6 are *not* algebraic unless $G = T^m$.

In fact, they violate the diagonal action property (see 6.4.B_4) and the finite volume property (see 6.4.B_5) of algebraic actions as explained in 1.11.D and 1.11.E.

6.7. Homogeneous actions. Most dynamically significant rigid actions appear as we have already mentioned in 0.5.A in the homogeneous framework. Namely, V is the homogeneous space for G, and the acting group may be G itself or a subgroup H of G. If the action of G on V is faithful, then it is easy to see that the corresponding action on $\mathscr{D}^r(V)$ is free and proper for large r. Hence the action is rigid. In fact, it is not hard to show that the action is A-rigid.

These homogeneous and subhomogeneous (i.e., of subgroups $H \subset G$) actions on V include the algebraic actions as P^n is an $\mathrm{SL}(n+1, \mathbf{R})$-homogeneous space.

A twisted torus action also appears in the homogeneous framework according to 6.6.B(ii). Notice that in the algebraic and the twisted torus actions the isotropy group $G_v \subset G$ is connected and has at most finitely many components. But one obtains by far more interesting examples if the isotropy group G_v has infinitely many components. The key case here is where the isotropy group is a discrete subgroup, say $\Gamma \subset G$. Let us indicate a particularly interesting invariant structure φ on $V = G/\Gamma$ coming from a bi-invariant structure $\tilde{\varphi}$ on G as it descends to G/Γ.

6.7.A. **Example.** Let G be semisimple, and $\tilde{\varphi}$ correspond to the Killing form on G. Then φ (like $\tilde{\varphi}$) is a pseudo-Riemannian G-invariant metric on $V = G/\Gamma$.

Notice that the isotropy group G_v is conjugate to Γ for all $v \in V$ while the local isotropy group $\mathrm{Is}^{\mathrm{loc}}(V, \varphi, v)$ contains extra transformations coming from $G \times G$ acting on G by left- and right-translations. It follows, in the semisimple case, that $\mathrm{Is}^{\mathrm{loc}}$ is locally isomorphic to the group G itself. More generally, one can easily show that if G has no compact factor group, then for every G-invariant A-structure φ on V one has the Lie algebra $L(G)$ inside $L(\mathrm{Is}^{\mathrm{loc}}(v), \varphi)$. Notice that the above discussion is justified by the existence of many interesting discrete subgroups Γ in G (see 1.9.D). The first example is our famous $G = \mathrm{SL}(2, \mathbf{R})$ with the lattice $\Gamma = \mathrm{SL}(2, \mathbf{Z}) \subset G$ (see 1.9.D(ii)).

6.8. Besides pseudo-Riemannian metrics one may have other interesting invariant structures. For example, if $G = \mathrm{GL}(m, \mathbf{R})$, then the flat affine structure on G induced from $\mathbf{R}^{m^2} \subset \mathrm{GL}(m, \mathbf{R})$ is bi-invariant. Thus we get compact *affine flat manifolds* of the form $\mathrm{GL}(m, \mathbf{R})/\Gamma$ acted upon by $\mathrm{GL}(m, \mathbf{R})$ preserving the affine structure. Then passing to

PSL(m, **R**) one obtains similar examples with *flat projective structures*. Notice that this structure for $m \geq 3$ does not come from the Killing form (an exercise to the reader).

6.9. One can elaborate the previous examples by "twisting" them with proper actions. Namely, let another group, say G, freely and properly act on V_1, and let $\tilde{\varphi}_1$ be an invariant Riemannian metric (see 5.12.D). Then take a discrete subgroup $\Gamma \subset G \times G_1$ and observe that G acts on $(G \times V_1)/\Gamma$ preserving the (local) product structure $\varphi \times \varphi_1$ for any bi-invariant structure $\tilde{\varphi}$ on G (compare 4.7.1).

6.10. Locally homogeneous actions. Next, after homogeneous spaces come *locally homogeneous* ones. Standard examples of such manifolds are affine flat, conformally flat, projectively flat manifolds.

The isometry group $\mathrm{Is}(V)$ of a locally homogeneous space V can be arbitrarily complicated if no compactness (or finiteness of volume) assumption is imposed on V.

For example, for any countable group Γ one can take a domain $U \subset \mathbf{R}^4$ with $\pi_1(U) = \Gamma$, and then Γ isometrically acts on the universal covering V of U.

A more convincing example is as follows. Start with some Riemannian metric g_0 on a surface V, such that $\mathrm{Is}(V, g_0) = \Gamma$, and then conformally change g_0 to get a complete metric g on V with *constant* negative curvature. Thus $\mathrm{Is}(V, g) \supset \Gamma$ and it is easy to make $\mathrm{Is}(V, g) = \Gamma$ as well.

If V is compact, and (or) the action of $\mathrm{Is}(V)$ preserves a smooth volume element, then the action of the identity component $\mathrm{Is}_0(V) \subset \mathrm{Is}(V)$ on V can be understood almost as well as in the homogeneous case. But the action of discontinuous (e.g., discrete) groups on V appears more difficult to comprehend. The problem arises from the action of $\mathrm{Is}(V)$ on the fundamental group of V which makes it impossible to lift the action to the universal covering of V. On the other hand, all *known* examples of interesting discrete actions on locally homogeneous spaces are rather simple and essentially of arithmetic origin, and these can be essentially reduced to the standard action of $\mathrm{SL}(n, \mathbf{Z})$ on the torus $\mathbf{R}^n/\mathbf{Z}^n$. For example the action on a nilmanifold coming from automorphisms of the corresponding (nilpotent) Lie group is in this category.

6.11. We conclude with a particularly nice example which has homogeneous origin but is not (even locally) homogeneous.

6.11.A. **Example.** Let V be the unit tangent bundle of a complete (e.g., compact) locally symmetric space X. Consider $G = \mathbf{R}$ acting by the geodesic flow. If $\operatorname{rank} X = 1$ (e.g., X has constant sectional curvature), then V admits a natural locally homogeneous structure compatible with the flow. But for $\operatorname{rank} X \geq 2$ the manifold V is not (at least in a natural way) locally homogeneous, but is partitioned into locally homogeneous "fibers" of generic codimension $k = \operatorname{rank} X$. In fact, the quotient space of this partition can easily be identified with S^{k-1}/W, where S^{k-1} is the unit tangent sphere to some flat subspace in X, and W is the Weyl group.

6.11.A$_1$. **Remark.** The pleasant feature of this example is the existence of an invariant rigid A-structure, which can be built using the stable and the unstable foliations (as in 2.2(b)) together with the natural action of $\mathbf{R}^k$ commuting with our $G = \mathbf{R}$. It would be interesting to classify (locally homogeneous) Riemannian manifolds where the geodesic flow admits a rigid A-structure. Now, after the rows of all these rigid beauties we bring forth

6.12. An example of a nonrigid action. Take a small ball $B^n \subset V$, $n = \dim V$, and let f_1 and f_2 be generic C^∞-diffeomorphisms of V into itself (here V is any smooth manifold) sending B^n into itself and fixing the complement $V \backslash B^n$. The group generated by f_1 and f_2 obviously is free. Next take a diffeomorphism $f: V \to V$ such that the images $f^j(B^n)$ are disjoint for the iterates f^j of f. It is clear that the group Γ generated by f, f_1, and f_2 contains an infinite product of free groups as a subgroup. It follows (see 0.4.A and §4 in [29]) that Γ admits no rigid actions at all on any compact manifold or on a noncompact manifold with an invariant probability measure.

What we do not know, however, is the answer to the following question.

Question. Does Γ admit a faithful *real analytic* action on some compact manifold?

7. Some open problems on rigid structures and transformation groups

7.1. Rigidity problem for a geometric structure induced by a nondiffeomorphic map. Let V and W be n-dimensional manifolds with geometric structures of the same type, called φ on V and ψ on W, and let $f: V \to W$ be a smooth map which is an immersion (i.e., locally diffeomorphic) on an open dense subset $U \subset V$. Can the structures ϕ and ψ be A-rigid in the case where f is *not* an immersion on all of V?

7.1.A. Let, for example, ψ be a Riemannian metric on W, and φ be a quadratic form on V induced by f. Does the rigidity of φ imply that f is locally diffeomorphic?

7.1.B. Let (W, ψ) be a C^∞-smooth almost complex manifold and (V, φ) be obtained by blowing up a point $w_0 \in W$. Notice that this blown-up manifold admits a natural *continuous* almost complex structure which is our φ. Now, let φ be C^∞-smooth for some C^∞-structure on V. Does it follow that the structure φ is infinitesimally integrable (and hence nonrigid) at w_0?

7.1.B$'$. Let (W, ψ) be as above, and (V, φ) be a ramified cover of W, where the ramification locus is a smooth codimension 2 submanifold $W_0 \subset W$ whose tangent subbundle $T(W_0) \subset T(W)$ is invariant under the almost complex structure ψ (that is, an automorphism of $T(W)$ with square $-\mathrm{Id}$). Here again we ask if smoothness of (V, φ) implies infinitesimal integrability of ψ at W_0.

7.2. Extension problem for local C^{an}-isometries. Let (V, φ) and (W, ψ) be connected A-rigid n-dimensional C^{an}-manifolds of the same type, where we additionally assume V is simply connected and W is compact without boundary. Let $U \subset V$ be an open connected subset, and $f_0: (U, \varphi) \to (W, \psi)$ be a locally isometric C^{an}-immersion. Under which conditions does f_0 extend to a locally isometric immersion $f: (V, \varphi) \to (W, \psi)$? Some condition is indeed needed here as the following example shows.

7.2.A. **Example.** Let φ and ψ be flat affine connections, such that (V, φ) equals the affine space $\mathbf{R}^n$, while (W, ψ) is a compact *noncomplete* manifold. Then there is no global isometry $(V, \varphi) \to (W, \psi)$, while the local isometries are abundant.

Keeping in mind this example one may suggest that the answer to the extension question is "yes" in the case where W is *simply connected.* What is, probably, more relevant is the global extendibility of local Killing fields on (W, ψ) as is explained below.

7.2.B. Suppose φ and ψ are pseudo-Riemannian metrics of the same type, and let $I \subset \mathrm{Gr}_n(V \times W)$ denote the set of the n-planes which are tangent to the graphs of the local isometries $(U, \varphi) \to (W, \psi)$ for all $U \subset V$ (compare 3.5). If some local isometry $f_0: U \to W$ is not extendible to a boundary point $v_0 \in \partial U$, then the differential of f_0 blows up as $u \to v_0$. It follows that I is *not* a closed subset in Gr_n. In other words some sequence of graphs of local isometries converges to a totally geodesic isotropic n-dimensional submanifold $M \subset (V \times W, \varphi \oplus -\psi)$ which is vertical in the sense that its projection to V has everywhere rank $\leq n-1$.

Now we recall that the space of germs of n-dimensional totally geodesic submanifolds in $V \times W$ naturally *embeds* into the Grassmann bundle

$\mathrm{Gr}_n(V \times W)$, and the image is a *real analytic* subset in $\mathrm{Gr}_n(V \times W)$ (see 3.5.1). It follows that $M = M_0$ is included in a C^{an}-family of such submanifolds M_t for $t \in [0, 1]$, where each M_t for $t > 0$ is the graph of a local isometry $V \to W$. Therefore, W (as well as V) has a nonzero Killing field. Moreover, the local isometry pseudogroup of W is noncompact.

If $\pi_1(W) = 0$, then the local isometries extend to global ones on W. With these one probably can show that local isometries of V to W can be globally extended as required.

A similar argument may work for affine connections but, for example, the case of conformal pseudo-Riemannian structures looks more difficult.

Notice that the extension problem makes sense in the C^∞-case if we put it as follows.

7.2.C. Given an isometry $f_0 \colon (U, \varphi) \twoheadrightarrow (W, \psi)$ for an open subset $U \subset V$. Does it smoothly extend to the closure of U in V? This question can be viewed as a generalization of Problem 7.1.

As we have seen above the solution of the extension problem strongly depends on the structure of the closure of the set of graphs of (local) isometries $V \to W$. In the pseudo-Riemannian case every subvariety $M \subset V \times W$ in this closure (here "closure" refers to the Hausdorff topology in the space of subsets in $V \times W$) is smooth but, in general, such M may have singularities. For example, the closure of the graphs of the conformal transformations of the standard sphere S^n contains the unions $(s_1 \times S^n) \cup (S^n \times s_2)$ for all pairs of points s_1 and s_2 in S^n. This leads to the question on the possible structure of the singularities of M for general A-rigid structures. Here it may be useful to look not only at the graphs of the isometries themselves but also at the graphs (and their Hausdorff limits) of higher order jets of isometries. (Notice that all these questions apply to limits of graphs of solutions of quite general partial differential equations.)

7.2.D. **Extension to the boundary.** The seemingly easiest case of the extension problem is that where V and W are compact manifolds with boundaries and f_0 is an isometry of the *interior* U of V to that of W. The question is whether f_0 extends to the boundary ∂V. The answer is unknown even if the structures in question are pseudo-Riemannian metrics and the boundaries of V and W are totally geodesic. (If the boundary of V contains no geodesic segment, then the extension is possible by an easy limit argument.)

7.2.E. The answer to the following question needs better understanding of limits of graphs of isometries.

Let $M \subset V \times W$ be an irreducible (semi)analytic subset such that at a generic point $z \in M$ (where M is nonsingular) M is the graph of a local isometry (for given A rigid C^{an}-structures on V and W). The question is whether M is everywhere nonsingular and equals the graph of a local isometry of V to W at *every* point in M.

The earlier discussion provides the positive answer for pseudo-Riemannian structures and for the Riemannian conformal structure, but for more general A-rigid structures the answer is unknown.

7.3. Problems concerning different notions of (weak) local homogeneity of a geometric structure. We start with a list of different notions of local homogeneity of a structured manifold (V, φ) given in the order of the decreasing strength.

(Ho_1)(V, φ) admits a locally transitive action of a local Lie group.

(Ho_2) For every two points v_1 and v_2 in V there exists a local isometry of some neighborhood U_1 of v_1, say $f\colon U_1 \to V$, such that $f(v_1) = v_2$.

(Ho_3) For every two points v_1 and v_2 there exist arbitrarily small isometric neighborhoods U_1 of v_1 and U_2 of v_2, where the implied isometry $f\colon U_1 \to U_2$ does not have to send $v_1 \mapsto v_2$.

(Ho_4) There exists an open dense subset $U \subset V$ which satisfies one of the above (Ho_1), (Ho_2), or (Ho_3).

(Ho_5) There exists a dense subset $X \subset V$ such that the above (Ho_2) or (Ho_3) is satisfied for all pairs of points v_1 and v_2 in X.

(Ho_6) There exists a dense subset in $V \times V$, such that either (Ho_2) or (Ho_3) is satisfied for the pairs (v_1, v_2) in this subset.

The general problems here are as follows.

7.3.A. For a rigid (or A-rigid) structure φ of a given type and smoothness class C^k decide which of the implications $\mathrm{Ho}_{i+1} \Rightarrow \mathrm{Ho}_i$ hold true.

7.3.B. Let φ be a rigid C^k-structure of a given type. Which of the Ho_i-conditions imply that φ is real analytic?

These problems are already interesting for Riemannian and pseudo-Riemannian structures (compare 5.13.D and p. 165 in [25]). Also notice that the major difficulty appears when the smoothness class C^k is not very high and the infinitesimal techniques do not apply. In fact the symmetries of nonsmooth structures may be quite pathological, as in the case of the geodesic flows preserving continuous (and sometimes C^1-smooth) pseudo-Riemannian metrics (see 2.10). On the other hand, a slight strengthening of the smoothness condition probably makes a locally homogeneous structure (satisfying some Ho_i) real analytic. In fact, this may be already true on the infinitesimal level: if a C^2-rigid structure φ is C^k-infinitesimally homogeneous for $k \gg 2$, then one expects φ to be C^{an}. For example, if a (pseudo-)Riemannian metric φ has the same curvature tensor at all points of V, then φ should be C^{an} since the "constant curvature tensor" equation seems transversally elliptic.

7.3.C. *The* Is-*partition problem.* One may try a nonhomogeneous generalization by asking what is the possible topological structure of the partition of V into the local isometry orbits of (V, φ). As it stands, this question appears too general but one may hope for a meaningful answer in some particular cases.

7.4. Isometrices of Lorentz manifolds. Let (V, φ) be a compact simply connected Lorentz C^k-manifold. Is then the isometry group $\text{Is}(V, \varphi)$ compact? By D'Ambra's theorem this is true for C^{an}-manifolds, but one expects this already for C^0.

If $\text{Is}(V, \varphi)$ is noncompact, then V admits a totally geodesic foliation $\mathcal{S}$ of codimension one whose leaves are projection from $V \times V$ of totally geodesic submanifolds $M \subset V \times V$ which are limits of graphs of isometries g_i of (V, φ) for some divergent sequences $g_i \in \text{Is}(V, \varphi)$.

The existence and the structure of $\mathcal{S}$ can be seen with the following definitions.

7.4.A. **Definitions.** An n-dimensional subspace $T \subset T(V \times V)$ for $n = \dim V$ is called *asymptotic* (or $\{g_i\}$-asymptotic to be precise) if it is the limit of the tangent spaces T_i to the graphs $\Gamma_i = \Gamma_{G_i} \subset V \times V$ at some points $x_i \in \Gamma_i$. That is $T_i = T_{x_i}(\Gamma_i)$ and $T = \lim_{i \to \infty} T_i$ in the Grassmann space $\text{Gr}_n(V \times V)$. (Notice that this definition makes sense for arbitrary C^1-maps g_i between two manifolds.)

If the maps g_i preserve an affine connection in V (as is the case for Lorentzian isometries), then the asymptotic space $T = T_x$, $x = (v, v')$, exponentiates to a complete totally geodesic submanifold Γ in $V \times V$ which is, by definition, the limit of the graphs Γ_i with the reference points $x_i = (v_i, v_i')$ for $v_i' = g_i(v_i)$.

If g_i preserve a pseudo-Riemannian metric φ in V, then the intersection of Γ with the (vertical) fibers $v \times V$ are φ-isotropic since the graphs Γ_i are $(\varphi \oplus -\varphi)$-isotropic. In particular these intersections in the Lorentz case are at most 1-dimensional. In fact, if the sequence g_i diverges (i.e., it is nonprecompact in $\text{Is}(V, \varphi)$) and V is connected, then for each $v \in V$ the intersection $\Gamma \cap (v \times V)$ is either empty or 1-dimensional. Then the projection of Γ to the first component V is a totally geodesic hypersurface denoted $\Lambda \subset V$, which is called an *asymptotic leaf* for $\{g_i\}$. If two asymptotic leaves meet, the intersection cannot be transversal as at the transversal intersection point $v \in V$ the norm of the difficulties Dg_i would be (by an easy argument) uniformly bounded for $i \to \infty$, which is impossible under our assumption $g_i \to \infty$. It follows in the Lorentz case (where the leaves have codimension one) that different leaves do not meet at all.

Now, we observe that for an arbitrary countable subset N in V there exists a subsequence g' of g_i, such that for each $v \in N$ there is a $\{g'_j\}$-asymptotic leaf Λ containing v. Then the closures of these (codimension 1) leaves form our foliation $\mathscr{S}$ in the case N is chosen *dense* in V.

7.4.B. **Remarks.** (i) The above *asymptotic foliation* $\mathscr{S}$ is by no means unique. Yet the space of these $\mathscr{S}$ form a *compact* space naturally acted upon by $\mathrm{Is}(V, \varphi)$. The compactness follows from an (obvious) uniform Lipschitz bound for the tangent subbundle $S = T(\mathscr{S}) \subset T(V)$ of an arbitrary totally geodesic foliation $\mathscr{S}$ of codimension one in V.

(ii) The above *asymptotic subbundle* S is somewhat similar to the stable subbundle T^+ discussed in 0.2.B. In fact there is a more general notion of a subbundle with a "restricted growth" which embraces S as well as T^+.

(iii) Suppose V is noncompact but the isometries g in question keep invariant some compact subset $V_0 \subset V$. Then one uses the action of (the differentials of) g on the bundle $T(V)$ restricted to V_0, and one may speak of asymptotic subbundles $S \subset T(V)|V_0$. The corresponding totally geodesic submanifolds Λ obtained by exponentiating (the fibers of) S may go outside V_0 but yet they are mutually disjoint in V (because $\mathrm{codim} = 1$).

Let us apply all this to a subgroup $G \subset \mathrm{O}(n, 1)$ acting on $V = \mathbf{R}^{n,1}$ and fixing the origin $V_0 = \{0\}$. Then we recognize the space of asymptotic subbundles as the limit set of G acting (by conformal transformations) on the sphere S^{n-1} which can be described in two equivalent ways.

(a) *Usual description.* S^{n-1} is the set of isotropic rays in $\mathbf{R}^{n,1}$.

(b) *Our description.* S^{n-1} consists of the hyperplanes $H \subset \mathbf{R}^{n,1}$ on which the form $\sum_{i=1}^{n} x_i^2 - x_{n+1}^2$ is singular.

7.4.C. Now we see that the space A of asymptotic foliations (or subbundles) on a (compact) Lorentz manifold generalizes the classical limit set in the theory of Kleinian groups. This leads us to the following.

Questions. How far does the Kleinian group theory extend to the action of $G = \mathrm{Is}(V, \varphi)$ (and more generally of $G = \mathrm{Is}(V, V_0, \varphi)$) on A? When is A a (smooth) manifold? When is the action minimal? Is every $g \in G$ elliptic, parabolic, or hyperbolic? etc. The answer to these questions should eventually lead to a classification of the compact Lorentzian manifolds V with noncompact isometry groups and also of noncompact manifolds admitting compact invariant subsets and/or finite invariant measures.

Another group of questions concerns a generalization of the above construction of $\mathcal{S}$ and A to (more) general manifolds V with affine connections. Here one can distinguish the case of *flat* manifolds V where asymptotic leaves tend to be parallel in V. It is also useful to consider (arbitrarily fine) partitions of V into convex (or nearly convex) subsets, and then apply $g \in \mathrm{Is}(V)$ to these partitions (and to small convex subsets in V in general) for $g \to \infty$. Thus one may obtain in the limit certain foliations which are kind of duals of $\mathcal{S}$.

Finally one may try to extend the discussion to nonaffine (e.g., conformal) structures by looking at the limits of the graphs $\Gamma_i = \Gamma_{g_i} \subset V \times V$ and their lifts to the iterated Grassmann manifolds $G' = \mathrm{Gr}_n(V \times V)$, $G'' = \mathrm{Gr}_n(G')$, etc. This evokes the compactification problem for the space of solution of totally integrable (and more general elliptic) systems of partial differential equations.

7.5. Isometric actions of $T^k \times \mathbf{R}^l$. We have seen in 3.2.B that noncompact groups acting on compact real analytic pseudo-Riemannian manifolds (V, φ) essentially reduce to the actions of $G = T^k \times \mathbf{R}^k$, such that the G-orbits equal the T^k-orbits. This leads to the following.

Questions. Let numbers p, q, k, and l is given. Does there exist a pseudo-Riemannian manifold V of type (p, q) whose isometry group G equals $T^k \times \mathbf{R}^l$, such that the orbits of G equal those of the maximal torus $T^k \subset G$? When can one find such a V with prescribed topological properties, such as $\pi_1(V) = 0$ or V being homeomorphic to S^n for $n = p + q$? (See [14] for partial results.)

An analogous but simpler question is where we are already *given* an action of the torus T^k on V and seek an invariant pseudo-Riemannian metric φ whose isometry group contains a closed subgroup isomorphic to $\mathbf{R}^l$ whose action preserves the T^k-orbits (and act on every such orbit by some "rotation" of T^k; compare the twisted torus action in 6.6).

A useful invariant of the above actions of $T^k \times \mathbf{R}^l$ is the map which assigns to each point $v \in T$ the isotropy subgroup of this point in $T^k \times \mathbf{R}^l$. The level sets of such a map are totally geodesic in V and one is led to the problem of (local and global) description of totally geodesic partitions (and of their singularities) arising from such maps and more general maps with similar singularities. (The singularities appear in the above picture if the action of T^k is *not* free.) In fact, partitions into (complete) totally geodesic submanifolds of variable dimension are quite interesting for standard manifolds such as the spheres and projective spaces.

7.6. Pseudo-Riemannian Lichnerowicz conjecture. Let (V, φ) be a compact pseudo-Riemannian manifold such that the action of the full group of conformal transformations of (V, φ) admits no smooth invariant measure. Is then (V, φ) conformally flat?

The standard conformally flat manifold is the isotropic cone V in $\mathbf{R}^{p+1, q+1}$ with the natural (p, q)-conformal structure, acted upon by $O(p+1, q+1)$ (with no invariant measure). If $\min(p, q) = 1$, then there are finite coverings of this V with similar properties, and one believes these are the only examples.

7.7. Nonstable manifolds and geometric invariants. Let the diffeomorphism group $\mathscr{D}$ of a (compact) manifold V act on the space Φ of geometric structures φ on V of a given type. Then the isotropy subgroup $\mathscr{D}_\varphi \subset \mathscr{D}$ is the same thing as the isometry group $\mathrm{Is}(V, \varphi)$, and noncompactness of $\mathrm{Is}(V, \varphi)$ can be thought of as the first manifestation of *nonproperness* of the action of $\mathscr{D}$ on Φ at $\phi \in \Phi$. A phenomenon similar to nonproperness is *nonstability* (according to the terminology of algebro-geometric invariant theory) which means that the quotient space $\Phi/\mathscr{D}$ is non-Hausdorff at the point $\overline{\varphi} \in \Phi/\mathscr{D}$ under φ.

To see how badly non-Hausdorff $\Phi/\mathscr{D}$ may be one should look at the space Φ_0 of flat pseudo-Riemannian tori $\mathbf{R}^{p,q}/L$ for the lattices $L \subset \mathbf{R}^{p,q}$. Two generic (irrational) tori (T^n, φ_1) and (T^n, φ_2) are not isometric (i.e., lie on different orbits of $\mathscr{D} = \mathrm{Diff}\, T^n$) but they admit arbitrarily small (rational) perturbations φ_1' and φ_2' which are isometric (by a linear map $T^n \to T^n$ sending φ_1' to φ_2').

The properties of *nonstable* rigid manifolds (V, φ) are similar to those having $\mathrm{Is}(V, \varphi)$ noncompact. The problem is to prove specific theorems in this direction. In particular, one expects a "classification" of "sufficiently nonstable" A-rigid manifolds (V, φ).

The nonstability phenomenon makes it hard to produce geometric invariants of structured manifolds (V, φ). This is well seen if we compare Riemannian and pseudo-Riemannian manifolds. A compact Riemannian manifold (V, φ) has plenty of geometric invariants, such as $\mathrm{Vol}(V, \varphi)$, $\mathrm{Diam}(V, \varphi)$, $\mathrm{Inj\,Rad}(V, \varphi)$, spectral invariants of the Laplace operator on functions and forms, etc. Now, if φ is an indefinite quadratic form, then the only obvious invariant is $\mathrm{Vol}(V, \varphi)$. Of course, one can produce some invariants using the curvature but these are *not continuous* in the C^0-topology on the space of metrics. Probably, the nonstability makes impossible any meaningful *general* theory of rigid non-Riemannian structures. (A good example is the virtual nonexistence of pseudo-Riemannian geometry in spite of the joint efforts of mathematicians and physicists.) However one may expect such a theory for certain "stable classes" of structures.

Example. Call a pseudo-Riemannian structure φ of type (p, q) *tame* if there exist closed forms ω_+ of degree p and ω_- of degree q such that ω_+ does not vanish on the p-dimensional subspaces $T \subset T(V)$ on which φ is positive, and ω_- does not vanish where φ is negative.

A special case of a tame manifold (V, φ) is that where there exist transversal foliations S_+ and S_- on V of dimensions (p, q) admitting smooth transversal measures, such that φ is positive on the leaves of S_+ and negative on those of S_-.

Notice that tame manifolds may be unstable; yet, one expects a rich geometry for such manifolds and/or similar (?) classes of other rigid manifolds.

Remarks. (a) The above notion of tameness is borrowed from (almost) complex geometry.

(b) Stability, it seems, may have other sources than tameness. For example, the C^k-smooth conformal Riemannian structures on a compact manifold V are stable for every $k = 0, 1, \dots$. Namely, if we take the C^k-space Φ_k of conformal C^k-structures on V, then the quotient $\Phi_k/\mathscr{D}$ is Hausdorff for every $k = 0, 1, \cdots$, as a simple argument shows. (Notice that the isotropy group $\mathscr{D}_\varphi$ may be noncompact if $V \approx S^n$, but the Hausdorff property is valid just the same.)

7.8. Problems concerning local Lie group actions and Lie algebra actions. Consider a Lie algebra L and a faithful C^∞-action of L in a (small) neighborhood of a point $v \in V$. Equivalently, one may speak of the local Lie group G corresponding to L acting on V near v. We want to have a "practical" criterion for the existence of an invariant A-rigid structure (of a certain type) defined in some neighborhood of V. A perfect answer would be in terms of the infinitesimal data at v.

7.8.A. An extreme case of nonrigidity is that where the action fixes an open subset U whose closure contains v. In this case the differential of the action is trivial at v, and by Thurston's stability theorem [65] the Lie algebra L is necessarily solvable. On the other hand every solvable algebra L seems to admit such an action which may be constructed with a smooth deformation of a trivial action to a faithful one. (For the purpose of constructing an example it is sufficient to "degenerate" a faithful *linear* action.)

7.8.B. On the other end of the spectrum one finds the action where the isotropy subalgebra $L_v \subset L$ is semisimple. Then, at least in the C^{an}-case (and probably in the C^∞-case as well) the action of L_v is linearizable by the Kushnirenko theorem (see [43]). It immediately follows that there exists an L_v-invariant submanifold transversal to the L-orbit $L(v) \subset V$ and meeting this orbit exactly at v (where the L-orbit means the orbit of the corresponding group action). Therefore, there exists an L-invariant projection of a small neighborhood u of $L(v)$ to $L(v)$. Furthermore, one can easily show that the fibration $U \to L(v)$ admits an L-invariant connection which can then be turned into an A-rigid structure on U.

7.8.C. The above discussion also shows that the action in 7.8.B can be *globalized* in the following sense. There exist a manifold V' and an action of a Lie group G with the Lie algebra L, such that the action of L near some point $v' \in V'$ is C^∞-conjugate to the original action of L in V near v. It would be interesting to give more general criteria for the existence of a globalization, where L_v is not necessarily semisimple. For example, let L be compact semisimple, and let L_v integrate to a *closed* subgroup in the simply connected Lie group corresponding to L. Is then the action globalizable? In particular, is every local action of su(2) globalizible?

7.8.D. The globalization problem becomes significantly more difficult if one requires additional properties of V' and the action of G on V', such as

(i) compactness of V' and/or topological transitivity of the action;

(ii) the existence of a smooth invariant measure μ on V' such that $\mu(V') < \infty$ and/or ergodicity of the action.

Notice that the globalization problem in cases (i) and (ii) is already interesting for *linear* actions of L on $V = \mathbf{R}^n$.

A necessary condition for (ii) for A-rigid action is given in [29]. A refinement of that to a necessary and sufficient condition faces a difficulty similar to the following more classical problem. Let $V' = G/H$ be a homogeneous space where G and H are connected Lie groups. Does there exist, for given G and H, a discrete subgroup $\Gamma \subset G$ which acts properly on V' such that V'/Γ is compact? (Instead of the compactness one may require $\mathrm{Vol}(V'/\Gamma) < \infty$ for some G-invariant measure on V'.) One can expect a comprehensive answer in the case of a nilpotent group G. On the other hand, if G is semisimple and H is noncompact, one expects the negative answer apart from a few special examples (compare [7]).

7.8.E. We conclude the discussion of the local actions of L with the following remark of a general kind. The actions of some groups (such as $\mathbf{R}$) are quite "soft," and a meaningful study usually requires some genericity condition. Other groups (e.g., compact) act in a rigid fashion, and one studies individual actions rather than representatives of generic classes. The question is how one can combine the two approaches for the study of local actions of general Lie groups and algebras.

7.9. Problems concerning the topology of G-manifolds. Here, a G-manifold means a connected manifold V with a smooth (or real analytic) faithful action of a connected Lie group G. The general question concerns the relations between the structure of the action (and G itself) and the topology of V. This has been extensively studied for compact Lie groups but almost nothing is known for noncompact G. Let us indicate specific questions.

7.9.A. Given G, what is the minimal dimension of a *compact* G-manifold V? One also asks this question in the presence of a smooth G-invariant measure on V.

7.9.B. Let $H \subset G$ be the maximal compact subgroup, and V be an H-manifold of sufficiently large dimension $n = n(G)$. Does the action of H on V extend to that of G? (To avoid local trouble one may assume H acts freely on V.)

7.9.C. Which (compact) manifolds V admit stratified (rigid) actions of G?

A great number of examples come from algebraic geometry. Namely if $G \subset \mathrm{GL}(n+1)$ is an algebraic group, one takes the Zariski closure V_0 of some G-orbit of the action on $\mathbf{R}P^n$ and then takes some equivariant nonsingular resolution of V_0 for V. Thus one may produce many stratified actions of $\mathbf{R}^2$ on surfaces.

7.9.C1. Does every compact n-dimensional manifold V admit a faithful stratified C^{an}-action of $\mathbf{R}^n$ or, more generally, of a given simply connected solvable group G? When, moreover, may one have such an action with *finitely many* orbits and/or with *only one* n-dimensional orbit?

7.9.C2. Let $V_0 = G/H$ be a homogeneous space. Does there exist a compact G-manifold V containing V_0 as an open (and dense) subset? Can one take V compact with boundary such that $\operatorname{Int} V = V_0$? (Of course, the answer may be "yes" or "no" depending on G/H.)

7.9.D. It seems that G-spaces are easier to construct for groups G of subexponential growth (e.g., for nilpotent groups G) than for the exponential growth. The first "difficult" group is the (solvable) group G of the affine automorphisms of the line. The Lie algebra of this group G is 2-dimensional with X and Y satisfying $[X, Y] = 2X$. If X and Y are vector fields on V satisfying this relation, then the diffeomorphisms generated by Y, say $f_\lambda: V \to V$, send X to the multiple λX for all $\lambda > 0$. This makes the subgroup $H = \mathbf{R}$ generated by X very special (parabolic). For example, the action of H has "polynomial growth," that is, the norm of the differential of h satisfies the bound $\|\mathscr{D}_h\| \leq h^d$ for some $d > 0$ and all sufficiently large $h \in H = \mathbf{R}$.

7.9.E. If G is a *semisimple* group, the list of known compact G-spaces is rather limited, but yet there are no known topological restrictions on V (apart from those which come from the maximal compact subgroup). Compact G-manifolds look especially rigid if there is a smooth G-invariant measure on V. In this case one expects (on the basis of the known examples) that the fundamental group $\pi_1(V)$ is roughly as large as G. Moreover, one might think that the lifted action of G to the universal covering of V should be proper.

7.9.E1. If the $\mathbf{R}$-rank of G is ≥ 2, and G is simple, then one may even hope that every compact G-manifold with a smooth measure and ergodic action is isomorphic to a standard homogeneous example (see §6). On the other hand for $\mathbf{R}$-rank $= 1$ (and especially for $G = \mathrm{SL}_2\mathbf{R}$) one expects a variety of nonhomogeneous examples.

The notion of a G-manifold with semisimple G of $\mathbf{R}$-rank $= 1$ is similar to that of a foliation with a metric of negative curvature along the leaves. Here G-invariant measures should be replaced by *transversal measures*, and in the presence of these the underlying manifold must be (as one naively believes) rather special. For example, the fundamental group is expected to have exponential growth.

7.10. Problems concerning the structure of Diff. Let V be a compact smooth manifold, and let $\mathrm{Diff}^r(V)$ denote the group of C^r-diffeomorphisms of V for $r = 1, 2, \cdots, \infty$, an and also $r = \mathrm{alg}$, which means the group of real algebraic maps for a fixed (local) real algebraic structure on V. Notice that $\mathrm{Diff}^{\mathrm{alg}}$ is *not* an algebraic group. In fact $\mathrm{Diff}^{\mathrm{alg}}$ is infinite dimensional for all V of positive dimension as the algebraic degree of $f \in \mathrm{Diff}^{\mathrm{alg}}$ may be arbitrarily large.

7.10.A. Now we want to look at Diff as if it were a linear group (i.e., a subgroup in some GL_N). For example we want to define the notion of *(quasi)-unipotency* of an $f \in \mathrm{Diff}$. Here are some possibilities.

(U_1) The spectrum of the differential D of f acting on the linear space of vector fields on V (of given regularity class) consists of the single point $\{1\}$.

(QU_1) The spectrum of the above D lies on the unit circle.

(U_2) The *adjoint orbit* of f, that is, $\{\varphi f \varphi^{-1}\} \subset \text{Diff}$ for all $\varphi \in \text{Diff}$, contains the identity in its closure.

(QU_2) For every $\varepsilon > 0$ there exists a Riemannian metric g on V, such that

$$-\varepsilon g \leq g - f^*(g) \leq \varepsilon g.$$

Here it is useful to think of the space G of the Riemannian metrics on V as an infinite-dimensional manifold of *nonpositive curvature*, where f acts as an isometry. Then the above expresses the following property of the *displacement function* $\delta(g) = \text{dist}(g, fg)$,

$$\inf_{g \in G} \delta(g) = 0.$$

(Of course, the specific content of this discussion depends on the choice of a particular distance function in G. Geometrically the most attractive is the one corresponding to the natural Riemannian metric on G.)

The elementary geometry of nonpositive curvature tells us that the above condition $\inf \delta(g) = 0$ is equivalent to

$$i^{-1}\text{dist}(g, f^i g) \to 0 \quad \text{for every } g \in G \text{ and } i \to \infty,$$

where f^i is the ith iterate of f. Now, we observe that $\text{dist}(g, f^i g)$ is bounded (for most natural definitions of dist) by const $\log\|Df^i\|$, and so the above (dist $\to$ 0)-property is insured if $\|Df^i\|$ has subexponenital growth. This bound on growth is strengthened by our next definition.

(QU_3) The norm of the differential of the ith iterate of f, that is $\|Df^i\|$, grows at most polynomially in i (i.e., $\leq \text{const}^i$) for $i \to \infty$. The same growth condition can be applied to the higher-order differentials of f^i, and for $f \in \text{Diff}^{\text{alg}}$ one may require the polynomial growth of the algebraic degree $\deg f^i$.

Notice that the specific norm $\|Df^i\|$ depends on the choice of some Riemannian metric on V, but the *equivalence class* of the sequence $\|Df^i\|$ is independent of this choice. (Here and below two positive functions (or sequences) $a(i)$ and $b(i)$ for some variable $i \in I$ are called *equivalent* if the ratio $a(i)/b(i)$ is contained in an interval $0 < \varepsilon \leq a(i)/b(i) \leq \delta < \infty$ for all $i \in I$.) Similarly, the algebraic degree is defined up to such equivalence.

What one wants to know is the relation between these definitions and their immediate variations. Also one asks for simple criteria for (quasi-)unipotency.

7.10.B. **Example.** Suppose there exists $\varphi \in \mathrm{Diff}$, such that $\varphi f \varphi^{-1} = f^2$. Then clearly f satisfies (QU_3). On the other hand, if $f \neq \mathrm{Id}$, then φ itself, probably, does not satisfy (QU_3). (This is clear if f is contained in a one-parameter group generated by a field X on V, such that φ transforms X to $2X$. In this case the spectrum of df contains 2, which is incompatible with (QU_3).)

7.10.C. Property (QU_3) leads to the following.

Growth problem. What is the possible equivalence class of the sequence $\|Df^i\|$? For example, if $\|Df^i\|$ grows very slowly (e.g., like log log log i), does it then follow that $\|Df^i\|$ is, in fact, bounded for $i \to \infty$? (A. Katok pointed out to us that the constructions in [$2\frac{1}{2}$] provide diffeomorphisms f with a slow growth of $\|Df^i\|$. Yet such diffeomorphisms must be quite rare.)

This question may be easier for the differentials $D_r f^i$, $r \geq 2$, and for alg $\deg f^i$. Another simplified question is whether a slow growth allows a small *periodic* perturbation f' of f. For example, let f be generated by a nonzero vector field X on V. Does the slow growth imply V admits a nontrivial circle action? In particular, does the simplicial volume of V vanish? (Compare [22].) In fact, the following strong recurrency property (which goes along with the slow growth condition) may suffice: The one-parameter subgroup generated by X is *not* closed in Diff.

7.10.D. Notice that the (strong) recurrency pattern of f^i can be measured by $\mathrm{dist}_r(f^i, \mathrm{Id})$ for the (naturally) defined C^r-distance on Diff_r. Here again one may ask what are possible equivalence classes of the sequences $\mathrm{dist}_{r_{f^i}}(f^i, \mathrm{Id})$. Of course, most diffeomorphisms f generate *discrete* (cyclic) subgroups in Diff, and so the above dist_r never approaches zero. However, if we take several diffeomorphisms $f_1, \cdots, f_k$ of V, then (the equivalence class of) the induced distance function on the free group on k-generators, $F_k \to \mathrm{Diff}$, becomes quite interesting as it approaches zero.

7.10.E. Take a (Riemannian) metric g on V, act on it by f^i, and let

$$g_j = \sum_{i=0}^{j} f^i g.$$

(Sometimes one takes $\sup_i$ rather than $\sum_i$, but the asymptotic properties of $\sup_i$ and $\sum_i$ are essentially the same.) Then we pick up some geometric invariant of the metric space (V, g_i) and obtain, by sending $i \to 0$, an invariant of f.

Examples. (a) Let N_ε^j denote the minimal number of ε-balls needed to cover (V, g_j). Then one defines the *topological entropy*

$$\operatorname{ent}_\varepsilon f = \limsup_{j\to\infty} j^{-1} \log N_\varepsilon^j ,$$

and

$$\operatorname{ent} f = \lim_{\varepsilon\to 0} \operatorname{ent}_\varepsilon f.$$

(b) The volume and the diameter of (V, g_j), say Vol_j and Diam_j, lead to the invariants closely related to the entropy, that is,

$$\limsup_{j\to\infty} j^{-1} \log \mathrm{Vol}_j ,$$

and

$$\limsup_{j\to\infty} j^{-1} \log \mathrm{Diam}_j$$

(c) There is an enumerable amount of various geometric invariants besides the volume and diameter to which the above applies. Here are some which look especially attractive.

The *k-dimensional width* of (V, g_j) for $k = 0, 1, \cdots, n = \dim V$ (see [23], [28]).

The *spherical radius* of (V, g_j), that is, the minimal radius of S^n which receives a distance decreasing map $(V, g_j) \to S^n$ of nonzero degree.

The *spectral invariants* of (V, g_j), such as the number N_λ^j of eigenvalues in the interval $[-\lambda, \lambda]$ of some operator $\mathscr{D}$ associated to g_j. For example, $\mathscr{D}$ may be the Laplace operator on functions, the operator $d+\delta$ on forms or the Dirac operator. Then the *spectral entropy* is

$$\lim_{\lambda\to\infty} \limsup_{j\to\infty} j^{-1} \log N_\lambda^j .$$

An important feature of the spectral invariants is their essential *linearity*. They can be expressed solely in terms of the linear operators $\mathscr{D}$ and f^* (induced by f) on an appropriate Hilbert space $\mathscr{H}$.

Given this variety of invariants of f one wants to know the basic relations between them. Some relations follow from pure geometry where the special origin of g_j plays no role. For example, the spherical radius gives a lower bound for N_λ for $\mathscr{D} = d + \delta$ and for the Dirac operator but not for the Laplace operator (see [24]). Furthermore, the work by Yomdin on the Shub entropy conjecture (see [26] and references therein) suggests further relations which do take into account the nature of g_j (which reflects, for example, the degree of regularity of f).

7.10.F. Let us consider the Cartesian product $W = V \times \mathbf{Z}_+$ with the *maximal* metric g_+, such that

(i) for every $v \in V$ the embedding

$$\mathbf{Z}_+ = (v \times \mathbf{Z}_+) \to V \times \mathbf{Z}_+$$

is distance nonincreasing;

(ii) for every $j \in \mathbf{Z}_+$ the embedding $(V, g_i) \to (V \times j) \subset V \times \mathbf{Z}_+$ is also distance nonincreasing.

Then the geometric invariants of $W = (V \times \mathbf{Z}_+, g_+)$ give us invariants of f. This is especially interesting for one-parameter group actions (rather than $\mathbf{Z}$ generated by f), as a continuous time reparametrization of an action does not change the *quasi-isometry* class of g_+. Then the quasi-isometry invariants of g_+ give us invariants of the (one-dimensional) orbit foliation. An example of such an invariant is the L_p-*cohomology* group $L_pH^k(W, g_+)$, where $W = V \times \mathbf{R}_+$ in the one-parameter group case. Here the L_p-cohomology is defined with ε-coverings where ε eventually goes to zero.

Example. If we apply the above to the geodesic flow of a compact manifold X of negative curvature, then the space W is quasi-isometric to the universal covering $\tilde{X}$ of X. In this case the L_p-cohomology can be equally defined with differential forms on $\tilde{X}$, and some computations of these can be found in [61]. Yet, even here our knowledge of L_pH^k is far from complete.

Remark. Forgetting the parametrization and/or passing from a group action to that of the (local) pseudogroup formed by the return maps in a neighborhood of a given point corresponds to the passage from the metrical to the *conformal* geometry. In the case of a space Y of negative curvature (as the above $\tilde{X}$) this conformal geometry lives at the ideal boundary (or the sphere at infinity) of Y.

7.10.G. The above constructions provide us with certain *invariants* of Diff that are functions Diff $\to \mathbf{R}$ which are invariant under conjugation. However, these functions usually are not continuous (though some of them are semicontinuous, as, for example, the topological entropy by the Yomdin theorem). In fact, it is hard to produce continuous invariants of Diff as the space Diff/conjugations is highly non-Hausdorff. Yet this space is Hausdorff at some points (e.g., at those corresponding to structurally stable diffeomorphisms), and one asks when two given diffeomorphisms f_1 and f_2 can be distinguished by a continuous invariant Diff $\to \mathbf{R}$. But even if f_1 and f_2 are "continuously indistinguishable," i.e., if there are arbitrarily small perturbations f_1' and f_2' which are mutually conjugate, may one still study f_1 and f_2 by means of "slightly discontinuous" invariants if the "amount" of the conjugate pairs (f_1', f_2') close to (f_1, f_2) is "not too great"?

7.10.H. *The symplectic case.* Let $V = (V, \omega)$ be a symplectic manifold with $H^1(V; \mathbf{R}) = 0$. Then every symplectic vector field ∂ on V is given by a (Hamilton) function $H: V \to \mathbf{R}$ which is unique up to an additive constant. This H provides nontrivial invariants of ∂. The simplest is the oscillation of H,

$$\operatorname{osc} H = \sup H - \inf H.$$

Further invariants are those of the push-forward measure $H^*(\mu)$ on the real line, where μ is the measure on V associated with the top exterior power of the (symplectic) form ω. Also there are more sophisticated symplectic (*width* or capacity) invariants of the subsets $H^{-1}[a, b] \subset V$ for $[a, b] \in \mathbf{R}$.

For example if $\partial \neq 0$, then also $\operatorname{osc} H \neq 0$ and so the one-parameter subgroup generated by ∂ cannot be unipotent in the sense of (U_2), in the case where V is compact. In particular, such a V admits no symplectic action of any noncompact semisimple Lie group G and also of the (solvable) group of the affine automorphisms of $\mathbf{R}$.

Hofer (see [36]) recently found a remarkable extension of the above oscillation from the Lie algebra to the group Symp Diff. His work shows, in particular, that in many cases the connected identity component of Symp Diff(V, ω) contains no (U_2)-element besides Id.

7.11. Problem concerning finitely generated subgroups in Diff. We want to find specific algebraic properties of a finitely generated group Γ which admits a faithful (or at least nontrivial) C^r-action on a compact manifold V, for a given $r = 1, 2, \cdots, \infty$, an, alg. One may ask the same question with an additional restriction on the action such as the following.

(i) $\dim V \leq k$.

(ii) The action is *rigid*.

(iii) $\Gamma \subset \mathrm{Diff}^r$ is a *discrete* subgroup.

(iv) The actions of each $\gamma \in \Gamma$ is (quasi-)unipotent in the sense of some (QU)-property discussed in 7.10.

(iv)$'$ The group $\Gamma \subset$ Diff is (quasi-)unipotent as a whole in an appropriate sense. For example, one may assume that for each $\varepsilon > 0$ there exists a Riemannian metric g on V which is moved at most by ε under the action of the generators of Γ. Alternatively one may impose a bound on the size of $\|D\gamma\|$ in terms of the word length of $\gamma \in \Gamma$.

7.11.A. The properties of $\Gamma \subset \mathrm{Diff}$, one may expect by the analogy with linear groups, are as follows.

(a) residual finiteness of Γ or at least the existence of a single subgroup Γ' of finite index > 1;

(b) the solvability of the word problem;

(c) nongenericity of Γ.

Let us explain what genericity means. Take some generators $\gamma_1, \cdots, \gamma_p$, and let $r_1, r_2, \cdots, r_q, \ldots$ be an infinite sequence of relations where r_q is a "randomly chosen" word of sufficiently large length l_q. If l_q is $\geq \mathrm{const}.\log q$, then the group $\Gamma = \{\gamma_1, \cdots, \gamma_p | r_1, r_2, \ldots\}$ is known to be quite large (e.g., having exponential growth, nonamenable, etc., see [27]). On the other hand, Γ has no nontrivial linear representation. Indeed, representations $\Gamma \to \mathrm{GL}_N$ given by solutions of the algebraic(!) equations imposed by $r_1 = 1, r_2 = 1, \ldots$ on matrices $M_i = 1$, $i = 1, \cdots, p$ in GL_N, and an easy argument shows that, generically, there is only one solution $M_i = 1$, $i = 1, \cdots, p$. If we fix N, then we only need finitely many relations. In fact one may think that generic groups $\Gamma = \{\gamma_1, \cdots, \gamma_p | r_1, \cdots, r_q\}$, where $p \geq 2$ and q is much greater than p, admit no monomorphisms into $\mathrm{GL}_\infty = \bigcup_N \mathrm{GL}_N$.

7.11.B. Let us see what the chances are of having (a), (b), and (c) for $\Gamma \subset \mathrm{Diff}$. The residual finiteness property seems hard to get even for A-rigid actions.

Property (b) needs an algorithm to decide when a "given" diffeomorphism equals the identity.

Property (c), that is, nongenericity of Γ, seems the most realistic of the three. In fact, it appears quite easy for $\mathrm{Diff}^{\mathrm{alg}}$ which is, after all, a countable union of algebraic varieties. On the other hand one can probably show without much work that the p-tuples of C^r-diffeomorphisms $f_1, \cdots, f_p$, satisfying sufficiently many generic relations $r_i = \mathrm{Id}$ (thought of as difference equations), form a "rather rare" subset in the space $\underbrace{\mathrm{Diff}^r \times \cdots \times \mathrm{Diff}^r}_{p}$, provided r is sufficiently large. (Many generic groups Γ embed into $\mathrm{Diff}^0(S^4)$ and some may even embed into $\mathrm{Diff}^1(S^4)$. In fact these groups often appear as fundamental groups of singular 2-dimensional spaces of negative curvature which can be "thickened" to 5-dimensional manifolds with 4-dimensional ideal boundaries $\approx S^4$. Thus one gets $\Gamma \subset \mathrm{Diff}^0 S^4$. Furthermore, whenever one can insure the 1/4-pinching of the curvature, one obtains $\Gamma \subset \mathrm{Diff}^1 S^4$.) What looks difficult, however, is to show that this rare subset reduces to a single point, namely, $\{f_i = \mathrm{Id}$, $i = 1, \cdots, p\}$.

7.11.C. One can generalize the above discussion by replacing the free group on p generators by an arbitrary group Γ_0, and then consider factor groups Γ of Γ_0 obtained by adding relations $r_1 = 1, r_2 = 1, \cdots$, for some $r_1, r_2, \ldots$ in Γ_0. If Γ_0 is *hyperbolic*, then, generically, Γ also is hyperbolic and roughly of the same size as Γ_0 (see [27]). But if Γ_0 is nonfree, Γ is even less likely to lie in Diff. For example, if Γ_0 is a lattice in $\mathrm{Sp}(n, 1)$, then even the presence of a single relation may suffice. Moreover, one may expect that every noninjective homomorphism $\Gamma_0 \to$ Diff has finite image. (For homomorphisms $\Gamma_0 \to \mathrm{GL}_N$ this is true by Corlette's theorem [12].)

7.11.D. Many examples of groups with the nonsolvable word problem are obtained with the use of amalgamated products and the HNN-construction. It seems that these constructions often lead to groups Γ nonrepresentable in Diff, even if the word problem is solvable.

Example. Take Γ generated by $\gamma_1, \gamma_2, \gamma_3$ with two relations $\gamma_2\gamma_1\gamma_2^{-1} = \gamma_1^2$ and $\gamma_3\gamma_2\gamma_3^{-1} = \gamma_2^2$. The first relation suggests that γ_2 is hyperbolic, while the second implies that γ_2 is parabolic (compare 7.10.B) which makes an embedding $\Gamma \subset$ Diff very unlikely. (By the discussion in 7.10.B this is the case if γ_1 is contained in a one-parameter subgroup which is conjugated by γ_2 into itself.)

7.11.E. Let us indicate some properties of Diff which may be used to restrict $\Gamma \subset$ Diff. First of all, the algebraic topology provides a bound on finite subgroups in Diff. For example the group S_∞ of permutations of the integers with compact support embeds in no Diff. On the other hand S_∞ embeds into some *finitely generated* (even finitely presented) group Γ which then also does not embed into Diff. (This class of examples was pointed out to me by Michel Herman many years ago.)

7.11.E1. An important property of linear groups G (and Lie groups in general) is the following commutation inequality for the distance $\|g\| = \mathrm{dist}(g, \mathrm{Id})$,

$$(*) \qquad \|[g_1, g_2]\| \le \mathrm{const}\|g_1\| \cdot \|g_2\|,$$

which holds true for all g_1 and g_2 in G. This inequality implies that successive commutators of small (i.e., close to Id) elements in G converge to zero, and so *every discrete subgroup* $\Gamma \subset G$ *generated by small elements is nilpotent.* (This is due to Zassenhaus.)

If we try to generalize $(*)$ to the C^r-distance in Diff^r, we encounter the loss of one derivative,

$$(+) \qquad \|[g_1, g_2]\|_{r-1} \le \mathrm{const}\|g_1\|_r \cdot \|g_2\|_r,$$

which makes the above conclusion unlikely for Diff^∞. Yet some (solvable, if not nilpotent) version of that may remain valid for $\mathrm{Diff}^{\mathrm{an}}$ where the loss of derivatives is compensated by the quadratic shape of the right-hand side of $(+)$. (Notice that the sequence of functions $f_{i+1}(t) = (df_i/dt)^2$ converges to zero if it starts with a "sufficiently small" analytic function f_0.) In particular, this may be useful in the study of C^{an}-deformations of the trivial representation $\Gamma \to \mathrm{Diff}$.

7.11.E2. **Remark.** If the trivial representation of Γ to Diff admits an arbitrarily small C^1-deformation to a faithful (i.e., injective) representation $\Gamma \to \mathrm{Diff}^1$, then by Thurston's stability theorem every finitely generated subgroup $\Gamma' \subset \Gamma$ admits a nontrivial homomorphism into $\mathbf{R}$.

7.11.E2. If Γ admits a *rigid* action on V, then Γ can be *uniformly metrically embedded* (*placed* in the sense of [29]) into some Lie group. This gives a nontrivial restriction on (the "size" of) the groups Γ in $\mathrm{Diff}^{\mathrm{rigid}}$. Thus one asks if there is a similar restriction for the existence of a discrete (rather than rigid) representation $\Gamma \hookrightarrow \mathrm{Diff}$. If so, one could insure nondiscreteness of certain groups $\Gamma \subset \mathrm{Diff}$, and then try to apply the consideration of 7.11.E1 to the elements $\gamma \in \Gamma$ close to $\mathrm{Id} \in \mathrm{Diff}$.

References

[1] A. M. Amores, *Vector fields of a finite type G-structure*, J. Differential Geometry **14** (1979) 1–6.

[2] V. D. Anosov, *Geodesic flows on closed Riemannian manifolds with negative curvature*, Proc. Inst. Steklov **90** (1967), 1–235.

[2 $\frac{1}{2}$] D. V. Anosov & A. B. Katok, *New examples in smooth ergodic theory. Ergodic diffeomorphisms*, Trans. Moscow Math. Soc. **23** (1970), 1–35.

[3] V. I. Arnold & A. Avez, *Problèmes ergodiques de la mécanique classique*, Gauthier-Villars, Paris, 1966.

[4] L. Auslander, L. Green & F. Hahn, *Flows on homogeneous spaces*, Annals of Math. Studies, No. 53, Princeton University Press, Princeton, NJ, 1963.

[5] A. Avez, *Anosov diffeomorphisms*, Proc. Internat. Sympos. Topological Dynamics (C. Auslander and W. H. Gottschalk, eds.) Benjamin, New York, 1968.

[6] W. Ballmann, M. Gromov & V. Schroeder, *Manifolds of nonpositive curvature*, Progress in Math., Vol. 61, Birkhäuser, Basel, 1985.

[7] Y. Benoist, P. Foulon & F. Labourie, *Flots d'Anovos à distributions de Liaponov différentiables*, Preprint, École Polytechnique, Palaiseau.

[8] G. D. Birkhoff, *Dynamical systems*, New York, 1927.

[9] A. Borel, *Density properties for certain subgroups of semisimple Lie groups without compact factors*, Ann. of Math. **72** (1960), 179–188.

[10] ——, *Linear algebraic groups*, Benjamin, New York, 1969.

[11] G. E. Bredon, *Introduction to compact transformation groups*, Academic Press, New York, 1972.

[12] K. Corlette, *Archimedian superrigidity and hyperbolic geometry*, University of Chicago, Preprint, March, 1990.

[13] I. P. Cornfeld, S. V. Fomin & Ya. G. Sinai, *Ergodic theory*, Springer, Berlin, 1980.

[14] G. D'Ambra, *Isometry groups of Lorentz manifolds*, Invent. Math. **92** (1988) 555–565.

[15] P. Eberlein & B. O'Neill, *Visibility manifolds*, Pacific J. Math. **46** (1973) 45–110.

[16] F. Ferer, *Geodesic flows on manifolds of negative curvature with smooth horospherical foliations*, Preprint, Math. Sci. Res. Inst.

[17] R. Feres & A. Katok, *Invariant tensor fields of dynamical systems with pinched Lyapunov exponents and rigidity of geodesic flows*, Ergodic Theory and Dynamical Systems **9** (1989) 427–433.

[18] P. Foulon & F. Labourie, *Flots D'Anosov à distributions de Liapunov différentiables*, C. R. Acad. Sci. Paris **309** (1989) 255–260.

[19] J. Franks, *Anosov diffeomorphisms on tori*, Trans. Amer. Math. Soc. **145** (1969) 117–124.

[20] S. Franks & R. Williams, *Anomalous Anosov flows in global theory of dynamical systems*, Lecture Notes in Math., Vol. 819, Springer, Berlin, 1980, 158–174.

[21] H. Ghys, *Flots d'Anosov dont les feuilletages sont différentiables*, Ann. Sci. École Norm. Sup. **20** (1987) 251–270.

[22] M. Gromov, *Volume and bounded cohomology*, Inst. Hautes Études Sci. Publ. Math. **56** (1983) 213–307.

[23] ——, *Filling Riemannian manifolds*, J. Differential Geometry **18** (1983), 1–147.

[24] ——, *Large Riemannian manifolds*, Lecture Notes in Math., Vol. 120, Springer, Berlin, 1986, 108–122.

[25] ——, *Partial differential equations*, Springer, Berlin, 1986.

[26] ——, *Entropy, homology and semialgebraic geometry (after Y. Yomdin)*, Sem. Bourbaki, Juin 1986, Asterisque 145-146k, Soc. Math. France, 1987, 225–241.

[27] ——, *Hyperbolic groups*, Essays in Group Theory (S. M. Gersten, ed.), Math. Sci. Res. Inst. 75–265. Publ. 8, Springer, Berlin, 1987.

[28] ——, *Width and related invariants of Riemannian manifolds*, On the Geometry of Differentiable Manifolds (Rome 1985), Asterisque 163–164, Soc. Math. France, 1988, 93–109.

[29] ——, *Rigid transformation groups*, Géométrie Différentielle (Bernardétrie and Choquet-Bruhat, eds.), Travaux encours, Hermann, Paris, **33** (1988) 65–141.

[30] ——, *Convex sets and Kähler manifolds*, Proc. Differential Geometry and Topology, Villa Gualino (To) (F. Tricerri, ed.), World Scientific, Singapore, to appear.

[31] J. Hadamard, *Sur l'itération et les solutions asymptotiques des équations différentielles*, Bull. Soc. Math. France **29** (1901) 224–228.

[32] B. Hasselblatt, *A bootstrap for regularity of the Anosov splitting*, Preprint.

[33] G. A. Hedlund, *The dynamics of geodesic flows*, Bull. Amer. Math. Soc. **45** (1939) 241–246.

[34] M. Hirsch & C. Pugh, *Smoothness of horocycle foliations*, J. Differential Geometry **10** (1975) 225–238.

[35] G. Hochschild, *The structure of Lie groups*, Holden Day, San Francisco.

[36] H. Hofer, *On the topological properties of symplectic maps*, Preprint.

[37] E. Hopf, *Statistik der geodätischen Linien in Mannigfaltigkeiten negativer Krümmung*, Ber. Verh. Sächs. Akad Wiss., Leipzig **91** (1939), 261–340.

[38] J. E. Humphreys, *Linear algebraic groups*, Springer, New York, 1975.

[39] M. Kanai, *Geodesic flows of negatively curved manifolds with smooth stable and unstable foliations*, Ergodic Theory and Dynamical Systems **8** (1988) 215–241.

[40] W. Klingenberg, *Riemannian geometry*, Studies in Math., Vol. 1, Walter de Gruyter, Berlin, 1982.

[41] S. Kobayashi, *Transformation groups in differential geometry*, Ergebnisse Math. Grenzgeb., Vol. 70, Springer, Berlin, 1972.

[42] J. L. Koszul, *Lectures on groups of transformations*, Lectures on Math. & Physics, Vol. 20, Tata Inst. Bombay, 1965.

[43] A. G. Kushnirenko, *Linear equivalent action of a semisimple Lie group in the neighborhood of a stationary point*, Functional Anal. Appl. **1** (1967) 89–90.

[44] J. Lelong Ferrand, *Transformations conformes et quasi-conformes des variétés riemanniennes compactes*, Acad. Roy. Belg. Cl. Sci. Mem. Coll. in-8°, 2e Ser., (1971) 1–44.
[45] A. Lichnerowicz, *Geometrie des groupes de transformation*, Dunod, Paris, 1958.
[46] S. Lojasiewicz, *Sur les ensembles semialgébriques*, Sympos. Math. 3, Ist. Naz. Alta Matem, **233** (1970).
[47] ——, *Ensembles semi-analytiques*, Preprint.
[48] B. B. Mandelbrot, *The fractal geometry of nature*, Freeman, New York, 1982.
[49] R. Mañe, *Ergodic theory and differentiable dynamics*, Springer, Berlin, 1987.
[50] A. Manning, *There are no new Anosov diffeomorphisms on tori*, Amer. J. Math. **96** (1974) 422–429.
[51] G. A. Margulis, *Formes quadratiques indéfinies et flots unipotents sur les espaces homogènes*, C. R. Acad. Sci. Paris Série I Math. **304** (1987) 249–253.
[52] H. Masur, *Interval exchange transformations and measured foliations*, Ann. of Math. **115** (1982) 169–200.
[53] D. Montgomery, *Simply connected homogeneous spaces*, Proc. Amer. Math. Soc. **1** (1950) 457–469.
[54] M. Morse, *A one to one representation of geodesics on a surface of negative curvature*, Amer. J. Math. **43** (1921) 33–51.
[55] G. D. Mostow, *Strong rigidity of locally symmetric spaces*, Annals of Math. Studies, no. 78, Princeton University Press, Princeton, NJ, 1973.
[56] ——, *Discrete subgroups of Lie groups*, Advances in Math. **15** (1975) 112–123.
[57] D. Mumford, *Hilbert's* 14*th problem, the finite generation of subrings such as rings of invariants*, Proc. Sympos. Pure Math., Vol. 28, Amer. Math. Soc., 1976, 431–443.
[58] K. Nomizu, *On local and global existence of killing vector fields*, Ann. of Math. **72** (1970) 105–112.
[59] M. Obata, *Conformal transformation of Riemannian manifolds*, J. Differential Geometry **4** (1970) 311–333.
[60] R. S. Palais & C. Terng, *Critical point theory and submanifold geometry*, Lecture Notes in Math., Vol. 1353, Springer, Berlin, 1989.
[61] P. Pansu, *Cohomologie L^p des variétés à courbure négative, cas du degré un*, Rend. Sem. Mat. di Torino, to appear.
[61 $\frac{1}{2}$] M. Ratner, *On Raghunathan's measure conjecture*, to appear.
[62] M. Rosenlicht, *A remark on quotient spaces*, An. Acad. Brasil Ciênc. **35** (1963) 483–489.
[63] I. M. Singer, *Infinitesimally homogeneous spaces*, Comm. Pure Appl. Math. **13** (1960) 685–690.
[64] D. P. Sullivan, *A counterexample to the periodic orbit conjecture*, Inst. Hautes Études Sci. Publ. Math. No. 46 (1976) 5–14.
[65] W. Thurston, *A generalization of the Reeb stability theorem*, Topology **3** (1974) 214–231.
[66] T. tom Dieck, *Transformation groups*, Studies in Math., Vol. 8, Walter de Gruyter, Berlin, 1987.
[67] F. Tricerri & L. Vanhecke, *Curvature homogeneous Riemannian manifolds*, Ann. Sci. École Norm. Sup. **22** (1989), to appear.
[68] W. A. Veech, *The Teichmüller geodesic flow*, Ann. of Math. **124** (1986) 441–530.
[69] R. J. Zimmer, *Kazdan groups acting on compact manifolds*, Invent. Math. **75** (1984), 425–436.
[70] ——, *Semisimple automorphism groups of G-structures*, J. Differential Geometry **19** (1984) 117–123.
[71] ——, *Ergodic theory and semisimple groups*, Birkhäuser, Boston, 1984.
[72] ——, *On the automorphism group of a compact Lorentz manifold and other geometric manifolds*, Invent. Math. **83** (1986) 411–426.

[73] ——, *Action of semisimple groups and discrete groups*, Proc. Internat. Congr. Math., 1986.
[74] ——, *Automorphism groups and fundamental groups of geometric manifolds*, Preprint, 1990.
[75] ——, *Groups generating transversals to semisimple Lie group actions*, Preprint, 1990.
[76] ——, *Superridgidity, Ratner's theorem, and fundamental groups*, Preprint, 1990.

INSTITIT DES HAUTES ETUDES SCIENTIFIQUES, FRANCE
UNIVERSITA DI CAGLIARI, ITALY

SURVEYS IN DIFFERENTIAL GEOMETRY
1 (1991) 113-199

FLIPS, FLOPS, MINIMAL MODELS, ETC.

JÁNOS KOLLÁR

One of the most interesting and profound developments of algebraic geometry in the past ten years is the "minimal model program," also called "Mori's program." The aim of the program, originating in [78], can be summarized as follows:

Let X be a smooth projective algebraic variety. One would like to construct another algebraic variety X' such that:

(i) X' is obtained from X by a series of simple "surgery type" operations, and

(ii) the global structure of X' is simple.

The usefulness of the program depends on how well we understand the "surgery type" operations and how simple the structure of X' is.

In dimension one the program does not exist; every smooth compact curve is as simple as possible. In dimension two one recovers the construction of minimal models of smooth surfaces, which has already been done by the Italian geometers around the turn of the century.

For a long time it was believed that a similar program is impossible in higher dimensions. The main reason behind this belief was that X' cannot be chosen to be smooth. Only beginning with the works of Reid [96] and Mori [78] did it become clear that by allowing certain singularities the local structure becomes only a little more complicated while the global structure becomes much simpler.

After this conceptual obstacle was removed, the hardest part of the program turned out to be to show the existence of certain special "surgery type" operations. This was finally completed in dimension three by Mori [82] and is still unknown in higher dimensions.

There have been several survey articles recently about the program. [54] is aimed at a very general readership. [120] assumes some familiarity with algebraic geometry while [49] is aimed at those who wish to become experts. The booklet [13] grew out of a seminar aimed at advanced graduate

Received April 13, 1990. Partial financial support was provided by the National Science Foundation under grant numbers DMS-8707320 and DMS-8946082 and by an A. P. Sloan Research Fellowship.

students; Chapters 8–13 present the simplest known proof of (1.4.2–7). [57] is intended as an introduction to [82].

In this article I try to be self-contained without reproducing earlier surveys. §1 discusses the structure of Mori's program mostly without proofs. The most difficult "surgery type" operations are discussed in §2. These are called flip and flop. After a general discussion of the codimension two surgery problem in §2.1 a detailed description of flops is presented in §2.2. The description is very explicit, and this is useful for applications. §2.3 contains some examples and results concerning flips. See [57] for a more detailed introduction.

The applications of the program are presented in §3 together with several open problems and conjectures.

The remaining sections are more independent of the main body of the program and of each other. Each one presents old and new results, conjectures, and speculations centered around a question which originated in the minimal model program.

By [78] every step in the construction of X' is related to some simple geometric configuration inside X. However, it is not clear when a similar-looking configuration corresponds to a step of the program. This leads to some very interesting questions and examples which are discussed in §4.

It turns out that flips and flops also play a crucial role in understanding proper but nonprojective varieties. This approach leads to simplifications of several results and to numerous new problems. These can be found in §5.

§6 is devoted to conjectures about deformations of rational surface singularities that grew out of studying minimal models of threefolds. These conjectures are quite interesting themselves, and a conceptual proof of them may lead to a better understanding of flips.

The groundfield is the field $\mathbb{C}$ of complex numbers, unless the contrary is stated at the beginning of a section.

I would like to thank H. Clemens, A. Corti, J.-M. Hwang, T. Luo, K. Matsuki, Y. Miyaoka, S. Mori, M. Reid, N. Shepherd-Barron, H. Sterk, J. Stevens, and D. Toledo for very useful corrections and comments.

0. Notation and terminology

1. A line bundle will always be locally free in the Zariski topology. This is important for spaces like $\mathbb{C}^2 - 0$ since $\mathrm{Pic}_{\mathrm{Zariski}}(\mathbb{C}^2 - 0) = 0$ but $\mathrm{Pic}_{\mathrm{Euclidean}}(\mathbb{C}^2 - 0)$ is infinite dimensional.

2. A line bundle L on an n-dimensional proper irreducible variety or complex space X is called *big* if $h^0(X, L^{\otimes m}) > \text{const} \cdot m^n$ for $m \gg 0$.

3. A line bundle L on a scheme X is called *nef* if for every compact curve $C \subset X$ we have $\deg(L|C) \geq 0$.

4. A normal variety X has $\mathbb{Q}$*-factorial* singularities if for every codimension-one subvariety $V \subset X$ there is an integer $m > 0$ such that mV is locally definable by one equation. The main consequence of this is that every codimension-one subvariety will have a cohomology class in $H^2(X, \mathbb{Q})$.

5. *Terminal* and *canonical* singularities are defined in (1.3.2). *Du Val* singularities are defined in (1.3.3.2).

6. *Minimal models* and *relative Fano models* are defined in (1.4.8).

7. *Curve neighborhoods* are defined in (2.1.1) and the *opposite* in (2.1.5).

8. *Flops* are defined in (2.2.1) and *flips* in (2.3.1).

9. If $I \subset \mathcal{O}_X$ is an ideal sheaf, then $B_I X$ denotes the blow-up of I. If $Z \subset X$ is a closed subvariety, then $B_Z X$ denotes the blow-up of the ideal sheaf of Z.

10. The word *morphism* will be reserved for maps that are everywhere defined. In diagrams they will be denoted by a solid arrow: $\rightarrow$. Maps that need not be everywhere defined will be indicated by a broken arrow: $\dashrightarrow$.

11. Let X be a smooth proper variety or complex manifold. Let K_X denote the canonical line bundle. For every $m > 0$ the global sections of $K_X^{\otimes m}$ define a meromorphic map

$$X \dashrightarrow I_m(X) \subset \mathbb{P}(H^0(X, K_X^{\otimes m})).$$

If m is sufficiently large and divisible, then $\dim I_m(X)$ is independent of m. This number is called the *Kodaira dimension* of X and is denoted by $\kappa(X)$. We set $\kappa(X) = -\infty$ if $I_m(X) = \emptyset$ for every $m > 0$.

12. If $(x_1, \cdots, x_n)$ is a coordinate system on $\mathbb{C}^n$, then the symbol $\mathbb{C}^n/\mathbb{Z}_m(a_1, \cdots, a_n)$ denotes the quotient of $\mathbb{C}^n$ by the group action $(x_1, \cdots, x_n) \mapsto (\zeta^{a_1}x_1, \cdots, \zeta^{a_n}x_n)$, where ζ is a primitive mth root of unity.

1. Construction of minimal models

1.1. Introductory remarks. Let C be a smooth proper algebraic curve over $\mathbb{C}$ (equivalently, a compact Riemann surface). C can be endowed with a metric of constant curvature, and one has the following classification according to the sign of the curvature ($\mathbb{H}$ denotes the upper half plane):

curvature	structure
positive	$\mathbb{CP}^1$
zero	$\mathbb{C}/\mathbb{Z}^2$
negative	$\mathbb{H}/\pi_1(C)$

In higher complex dimensions there are several possible curvatures to consider. The holomorphic bisectional curvature is a very strong invariant, and there are only few varieties where it can be semidefinite.

1.1.1. Theorem [77], [107]. *Let X be a compact complex Kähler manifold. Assume that T_X admits an Hermitian metric whose holomorphic bisectional curvature is everywhere positive. Then $X \cong \mathbb{P}^n$.*

Much less is known about manifolds that admit an Hermitian metric whose holomorphic bisectional curvature is everywhere negative. It was conjectured in [122, Problem 35] that their universal cover is always Stein. The following result indicates that this may not be true:

1.1.2. Theorem [68], [76]. *There are simply connected algebraic surfaces X such that Ω^1_X is ample.*

For such a surface X some symmetric power of T_X admits an Hermitian metric whose holomorphic bisectional curvature is everywhere negative. However it is not clear whether T_X itself admits such a metric.

One can take the trace of holomorphic bisectional curvature to get the Ricci curvature of the tangent bundle. I prefer to think of it as the curvature of the determinant of the tangent bundle. These two approaches are not completely equivalent since a metric on $\det T_X$ may not always be liftable to a metric on T_X.

In algebraic geometry it is customary to consider the canonical bundle:

$$K_X \stackrel{\text{def}}{=} (\det T_X)^*$$

The sheaf of local sections of the canonical bundle is called the dualizing sheaf and is denoted by ω_X. I will try to be systematic, and say canonical bundle when I mean a line bundle and dualizing sheaf when I mean a sheaf.

Dualizing changes the sign of the curvature, creating the possibility of confusion.

Even in dimension two it is not true that every variety has a semidefinite canonical bundle, but the exceptions are easy to enumerate:

1.1.3. Theorem. *Let X be a smooth proper algebraic surface. If K_X does not admit a metric h whose curvature $\Theta = \overline{\partial}\partial \log h$ is semipositive, then there is a morphism $f: X \to Y$ which is one of the following types:*

(1.1.3.1) *Y is a smooth surface, and X is obtained from Y by blowing up a point.*

(1.1.3.2) *Y is a smooth curve, and X is a $\mathbb{P}^1$ bundle over Y.*

(1.1.3.3) *Y is a point and $X \cong \mathbb{P}^2$.*

In all cases there is an embedded copy of $\mathbb{P}^1 \cong C \subset X$ such that $(\sqrt{-1}/2\pi)\int_C \Theta < 0$.

These cases are very different in nature. (1.1.3.2) and (1.1.3.3) are very precise global structural statements. One can hardly wish for more. (1.1.3.1) merely identifies (and removes) a small part of X and gives no global information. The corresponding morphism $f: X \to Y$ is called the contraction of a (-1)-curve. This is the simplest example of an algebraic surgery operation.

The contraction (1.1.3.1) introduces a new surface Y which is simpler than X since $b_2(Y) = b_2(X) - 1$. We can apply (1.1.3) to Y and continue if possible. This gives the following:

1.1.4. Theorem. *Let X be a smooth proper algebraic surface. Then there is a sequence of contractions $X \to X_1 \to \cdots \to X_n = X'$ such that X' satisfies exactly one of the following conditions:*

(1.1.4.1) *$K_{X'}$ admits a metric whose curvature is semipositive.*

(1.1.4.2) *X' is a $\mathbb{P}^1$-bundle over a curve C.*

(1.1.4.3) *$X' \cong \mathbb{P}^2$.*

Based on this presentation and using good hindsight we can formulate the aim of the minimal model program:

1.1.5. Hope. Let X be a smooth, projective algebraic variety. Then there are certain "elementary surgery operations" such that repeated application of them produces a variety X' and

either: $K_{X'}$ admits a metric whose curvature is semipositive,

or: there is a structure theorem for X'.

This has been achieved only in dimension three, and there are many surprises along the way.

1.2. Extremal rays on smooth varieties. Curvature assumptions are very difficult to handle in algebraic geometry. The following observation leads to a slightly different notion, which is easier to deal with.

Let L be a line bundle on a complex manifold M with metric h and curvature Θ, and let $C \subset M$ be any proper curve. Then

$$c_1(L) \cap C = \frac{\sqrt{-1}}{2\pi} \int_C \Theta.$$

We will denote this number by $C \cdot L$. In particular, if Θ is semipositive, then $C \cdot L \geq 0$ for every C.

1.2.1. Definition. A line bundle L on a variety X is called nef if $C \cdot L \geq 0$ for every compact curve $C \subset X$. (This replaces the earlier confusing terminology "numerically effective".)

It is conjectured that for the canonical line bundle being nef is equivalent to admitting a metric with semipositive curvature. In general however these two notions are slightly different.

The intersection product $C \cdot L$ depends only on the homology class of C but not on C itself. With this in mind we introduce:

1.2.2. Definition.

(1.2.2.1) Let X be a smooth projective variety over $\mathbb{C}$. The cone of curves of X—denoted by $NE(X)$—is the convex cone generated by the homology classes of effective curves in $H_2(X, \mathbb{R})$, where N stands for "Numerical equivalence", and E for "Effective".

The closed cone of curves or the Kleiman-Mori cone of X—denoted by $\overline{NE}(X)$—is the closure of $NE(X)$ in $H_2(X, \mathbb{R})$.

The definitions of course make sense for any complex manifold X.

(1.2.2.2) If X is a singular variety, then instead of $H_2(X, \mathbb{R})$ one can use

$$N_1(X) \stackrel{\text{def}}{=} \frac{\{1 - \text{cycles with real coefficients}\}}{\{\text{numerical equivalence}\}} \cong (\operatorname{Pic} X \otimes \mathbb{R})^*.$$

For smooth varieties (or for varieties with rational singularities) over $\mathbb{C}$ there is a natural linear embedding $N_1(X) \hookrightarrow H_2(X, \mathbb{R})$, and under this identification the corresponding cones are the same.

(1.2.2.3) If L is a line bundle or a Cartier divisor, then taking cap product determines a linear map

$$L^{\cap}: N_1(X) \to \mathbb{R}.$$

Reformulating (1.2.1) we obtain that L is nef iff $L^{\cap}|NE(X)$ is semipositive.

1.2.3. Definition.

(1.2.3.1) Let $V \subset \mathbb{R}^m$ be a convex cone. A subcone $W \subset V$ is called extremal if

$$u, v \in V, u + v \in W \Rightarrow u, v \in W.$$

Informally: W is a face of V.

(1.2.3.2) A one-dimensional subcone is called a ray. A ray is of the form $\mathbb{R}^+ v$ for some $v \in V$.

(1.2.3.3) Let X be a smooth proper algebraic variety. A ray $\mathbb{R}^+[Z] \subset \overline{NE}(X)$ is called a K_X-negative extremal ray if it is extremal and $K_X^{\cap}: \mathbb{R}^+[Z] \to \mathbb{R}$ is strictly negative on $\mathbb{R}^+[Z] - \{0\}$. If no confusion is likely, we will use "extremal ray" to mean a K_X-negative extremal ray.

The first step toward (1.1.5) is to understand the part of $\overline{NE}(X)$ where K_X^n is negative. It turns out to have a relatively simple structure:

1.2.4. Theorem [78, 1.4]. *Let X be a smooth projective variety (any dimension) over an algebraically closed field. Then the extremal rays of the closed cone of curves $\overline{NE}(X)$ are discrete in the open halfspace $\{z \in N_1(X) | z \cdot K_X < 0\}$. If $R \subset \overline{NE}(X)$ is an extremal ray, then there is a rational curve $C \subset X$ such that $[C] \in R$.*

Once an extremal ray is identified as the source of the trouble, one would like to use it to construct a map as in (1.1.3). In dimension three a complete description is known:

1.2.5. Theorem [78, 3.1–5]. *Let X be a smooth projective threefold over $\mathbb{C}$ (or an algebraically closed field of characteristic zero). Let R be an extremal ray of the closed cone of curves. Then the following holds:*

(1.2.5.1) *There is a normal projective variety Y and a surjective morphism $f: X \to Y$ such that an irreducible curve $C \subset X$ is mapped to a point by f iff $[C] \in R$; one can always assume that $f_*\mathcal{O}_X = \mathcal{O}_Y$, and then Y and f are unique up to isomorphism.*

The following is a list of all the possibilities for f and Y.

(1.2.5.2) *First case: f is birational.*

Let $E \subset X$ be the exceptional set of f. One has the following possibilities for E, Y, and f:

(1.2.5.2.1) *E is a smooth minimal ruled surface with typical fiber C, and $C \cdot E = -1$; Y is smooth, and f is the inverse of the blowing up of a smooth curve in Y.*

(1.2.5.2.2) *$E \cong \mathbb{P}^2$, and its normal bundle is $\mathcal{O}(-1)$. Y is smooth, and f is the inverse of the blowing up of a point in Y.*

In the remaining cases Y has exactly one singular point P, and f is the inverse of the blowing up of P in Y. Let $\widehat{\mathcal{O}}_{P,Y}$ be the completion of the local ring of $P \in Y$. Then the following hold:

(1.2.5.2.3) *$E \cong \mathbb{P}^2$, and its normal bundle is $\mathcal{O}(-2)$. $\widehat{\mathcal{O}}_{P,Y} \cong \mathbb{C}[[x, y, z]]^{\mathbb{Z}_2}$, where $\mathbb{Z}^2$ denotes the ring of invariants under the group action $(x, y, z) \mapsto (-x, -y, -z)$.*

(1.2.5.2.4) *$E \cong Q$, where Q is a quadric cone in $\mathbb{P}^3$, and its normal bundle is $\mathcal{O}_{\mathbb{P}^3}(-1)|Q$. $\widehat{\mathcal{O}}_{P,Y} \cong \mathbb{C}[[x, y, z, t]]/(xy - z^2 - t^3)$.*

(1.2.5.2.5) *$E \cong Q$, where Q is a smooth quadric surface in $\mathbb{P}^3$, the two families of lines on Q are numerically equivalent in X, and its normal bundle in $\mathcal{O}_{\mathbb{P}^3}(-1)|Q$. $\widehat{\mathcal{O}}_{P,Y} \cong \mathbb{C}[[x, y, z, t]]/(xy - zt)$.*

(1.2.5.3) *Second case: f is not birational.*

Then we have one of the following cases:

(1.2.5.3.1) $\dim Y = 2$; Y *is smooth and* f *is a flat conic bundle (i.e., every fiber is isomorphic to a conic in* $\mathbb{P}^2$*).*

(1.2.5.3.2) $\dim Y = 1$; Y *is a smooth curve, and every fiber of* f *is irreducible. The generic fiber is a smooth surface* F *such that* $-K_F$ *is ample.*

(1.2.5.3.3) $\dim Y = 0$ *and* X *is a Fano variety (i.e.,* $-K_X$ *is ample) and* $b_2(X) = 1$.

1.2.6. Comments.

(1.2.6.1) The second case should be considered a fairly complete structural description. Conic bundles are well understood. If $\dim Y = 1$, then there is a complete list of the possible fibers. Fano varieties have been completely classified by Fano-Iskovskikh [36], [37].

(1.2.6.2) The cases listed under (1.2.5.2) should constitute the desired "elementary surgery operations." The first two cases are as expected, but the last three are unexpected and create a serious problem since Y is singular. Thus, we cannot continue as in (1.1.4).

The realization that singularities must appear in three-dimensional minimal models was apparently made by Ueno [116, Chapter 16]. At that time this was interpreted as a sign that there are no minimal models in dimension three. The crucial conceptual step of allowing singularities was taken by Reid [96] and Mori [78]. Choosing the right class of singularities is a technical but very important part of the program.

1.3. Terminal and canonical singularities.

1.3.1. Guiding principles. We want to investigate varieties X for which K_X is not nef. In order to do this, K_X should exist and being nef should make sense.

The usual definition of K_X works over the smooth locus of X. If X is normal (a harmless assumption), then $\operatorname{codim}(\operatorname{Sing} X) \geq 2$, hence K_X has a well-defined homology class in $H_{2\dim X-2}(X, \mathbb{Z})$. However, because of the singularities there is no product between $H_{2\dim X-2}$ and H_2. Thus, the symbol $C \cdot K_X$ makes no sense in general.

If $K_{X-\operatorname{Sing} X}$ extends to a line bundle over X, then its first Chern class is in $H^2(X, \mathbb{Z})$ and we can take the cap product with $[C] \in H_2(X, \mathbb{Z})$. For the singularity given in (1.2.5.2.3), this condition is not satisfied because of the group action. However, $K_{X-\operatorname{Sing} X}^{\otimes 2}$ will extend to a line bundle over X. Thus, we can still define a first Chern class $c_1(K_X) \in H^2(X, \mathbb{Q})$, and this is also satisfactory.

For smooth varieties the plurigenera

$$P_m(X) = \dim \Gamma(X, K_X^{\otimes m}) \qquad (m \geq 0) \tag{1.3.1.1}$$

are birational invariants, and they are the most important discrete birational invariants that we know. The birational invariance is implied by the following more local result:

Let $f\colon X \to Y$ be a proper birational morphism between smooth varieties, and let $E_i \subset X$ be the f-exceptional divisors. Then

$$K_X = f^*K_Y \otimes \mathscr{O}_X\left(\sum a_i E_i\right), \qquad a_i > 0 \text{ for every } i. \tag{1.3.1.2}$$

This can be reformulated as

$$f_*(\omega_X^{\otimes m}) = \omega_Y^{\otimes m} \quad \text{for every } m > 0. \tag{1.3.1.3}$$

Observe that $a_i \geq 0$ would be sufficient to conclude the birational invariance of plurigenera.

1.3.2. Definition. An algebraic variety X is said to have canonical (resp. terminal) singularities if the following three conditions are satisfied:

(1.3.2.1) X is normal.

(1.3.2.2) $K_{X-\mathrm{Sing}\,X}^{\otimes m}$ extends to a line bundle over X for some $m > 0$. This unique extension will be denoted by $K_X^{[m]}$. The smallest such m is called the index of K_X.

(1.3.2.3) Let $f\colon X' \to X$ be a resolution of singularities, and let $E_i \subset X$ be the f-exceptional divisors. Let

$$K_{X'}^{\otimes m} = f^*(K_X^{[m]}) \otimes \mathscr{O}_{X'}\left(\sum a_i E_i\right) \qquad (m = \mathrm{index}(K_X)).$$

Then

$$X' \text{ has canonical singularities} \Leftrightarrow a_i \geq 0 \quad \text{for every } i,$$

$$X' \text{ has terminal singularities} \Leftrightarrow a_i > 0 \quad \text{for every } i.$$

(1.3.2.4) For arbitrary i one can define

$$\omega_X^{[i]} \stackrel{\text{def}}{=} \text{double dual of } \omega_X^{\otimes i}.$$

This is a torsion free sheaf, locally free iff $\mathrm{index}\,X | i$. If X has canonical singularities, and $f\colon X' \to X$ is a resolution of singularities, then

$$f_*(\omega_{X'}^{\otimes i}) = \omega_X^{[i]} \quad \text{for } i \geq 0.$$

(1.3.2.5) Another useful consequence of the definition is the following. Let $U \subset X$ be any open set (in the Euclidean topology). Let

$$f(dz_1 \wedge \cdots \wedge dz_n)^{\otimes m} \in \Gamma(U - \mathrm{Sing}\,U,\, K_{U-\mathrm{Sing}\,U}^{\otimes m}).$$

Then for any compact $K \subset U$

$$\int_K |f|^{2/m}\, dz_1 \wedge \cdots \wedge dz_n \wedge d\bar{z}_1 \wedge \cdots \wedge d\bar{z}_n < \infty.$$

In dimension two it is easy to get a complete list by looking at the minimal resolution:

1.3.3. Proposition.

(1.3.3.1) *A two-dimensional terminal singularity is smooth.*

(1.3.3.2) *Two-dimensional canonical singularities are exactly the DuVal singularities (also called rational double points), and are given by equations:*

$$(1.3.3.3)\qquad \begin{aligned} &A_n\colon xy + z^{n+1} = 0;\\ &D_n\colon x^2 + y^2 z + z^{n-1} = 0;\\ &E_6\colon x^2 + y^3 + z^4 = 0;\\ &E_7\colon x^2 + y^3 + yz^3 = 0;\\ &E_8\colon x^2 + y^3 + z^5 = 0. \end{aligned}$$

In dimension three there is a complete list of terminal singularities. See [100] for a very nice survey.

1.3.4. Theorem [98], [17] , [85], [79], [61], [109].

(1.3.4.1) *Three-dimensional terminal singularities are isolated.*

(1.3.4.2) *A three-dimensional hypersurface singularity is terminal iff it is isolated and can be given by an equation*

$$g(x, y, z) + th(x, y, z, t) = 0,$$

where g *is one of the equations from* (1.3.3.3).

(1.3.4.3) *Every other three-dimensional terminal singularity is the quotient of a hypersurface terminal singularity (called the index-one cover) by a cyclic group. The typical case is*

$$(xy + f(z^n, t) = 0) \subset \mathbb{C}^4/\mathbb{Z}_n(1, -1, a, 0), \quad \textit{where } (a, n) = 1.$$

The exceptional cases can be written as

$$(x^2 + f(y, z, t) = 0) \subset \mathbb{C}^4/\mathbb{Z}_n(a, b, c, d) \quad \textit{for some } n \leq 4.$$

There is a complete list of the possibilities.

(1.3.4.4) *Every terminal singularity can be deformed into a collection of terminal cyclic quotient singularities* $\mathbb{C}^3/\mathbb{Z}_n(1, -1, a)$, *where* $(a, n) = 1$.

1.4. Extremal rays on singular varieties. The original deformation theoretic arguments of Mori [77], [78] do not seem to work for singular varieties. Substantially new ideas are required to extend the results. The new approach relies very heavily on vanishing theorems, and therefore it works in characteristic zero only. On the other hand, applied even in the smooth case, it gives results not accessible by the previous method; namely, it proves that extremal rays can always be contracted in any dimension. Since the proofs are reviewed in [13, Lectures 8–13], we restrict ourselves to stating the theorems and to some comments.

1.4.1. Definition.

(1.4.1.1) Let $x \in \mathbb{R}$. We define $\ulcorner x \urcorner$ to be the smallest integer $\geq x$.

(1.4.1.2) Let X be an algebraic space. Let $D = \sum a_i D_i$ be a formal sum of distinct irreducible divisors on X with rational coefficients. Let

$$\ulcorner D \urcorner = \ulcorner \sum a_i D_i \urcorner \overset{\text{def}}{=} \sum \ulcorner a_i \urcorner D_i .$$

(1.4.1.3) The fractional part of D is the collection of those D_i such that a_i is not an integer.

1.4.2. Vanishing Theorem [42], [117]. *Let X be a smooth, proper algebraic space. Let $D = \sum a_i D_i$ be a nef and big $\mathbb{Q}$-divisor. Assume that the fractional part of D has only normal crossing singularities. Then*

$$H^i(X, K_X \otimes \mathcal{O}_X(\ulcorner D \urcorner)) = 0 \quad \textit{for } i > 0 .$$

Comments. If the a_i are integers, and D is ample, then this is the Kodaira vanishing theorem. Thus one can express the result as follows: if a divisor is close to being ample, then Kodaira vanishing still holds.

The above technical formulation seems artificial, but divisors of the form $\ulcorner D \urcorner$ appear very frequently. We will see an example in (5.3.8).

1.4.3. Nonvanishing Theorem [105]. *Let X be a nonsingular projective variety. Let D be a nef Cartier divisor and let G be a $\mathbb{Q}$-divisor such that $\ulcorner G \urcorner$ is effective. Suppose that $aD + G - K_X$ is ample for some $a > 0$, and that the fractional part of G has only simple normal crossings. Then, for all $m \gg 0$,*

$$H^0(X, mD + \ulcorner G \urcorner) \neq 0 .$$

Comments. The divisor G is here for the purpose of certain applications. The proof does not simplify if we assume that $G = 0$. By (1.4.2) the higher cohomologies vanish, thus

$$H^0(X, mD + \ulcorner G \urcorner) = \chi(X, mD + \ulcorner G \urcorner) .$$

Therefore, the claim is that a certain expression involving Chern classes is not zero. In dimension three one can understand the precise form of this expression and prove the result in several cases. In higher dimensions however this approach seems to fail.

1.4.4. Basepoint-Free Theorem [6], [44], [45], [97]. *Let X be a proper algebraic space with only canonical singularities. Let D be a nef Cartier divisor such that $aD - K_X$ is nef and big for some $a > 0$. Then $|mD|$ has no basepoints for all $m \gg 0$.*

Comments. In the literature this result is stated for projective varieties only. The proof, however, analyzes only various proper modifications, and these can always be chosen to be projective.

1.4.5. Rationality Theorem [45], [53]. *Let X be a projective variety with only canonical singularities. Let H be an ample Cartier divisor, and let*

$$r = \max\{t \in \mathbb{R} \mid H + tK_X \text{ is nef}\}.$$

Assume that K_X is not nef (i.e., $r < \infty$). Then r is a rational number of the form u/v where $0 < v \leq (\operatorname{index} X)(\dim X + 1)$.

1.4.6. Cone Theorem [45], [97]. *Let X be a projective variety with only canonical singularities. Then the extremal rays of the closed cone of curves $\overline{NE}(X)$ are discrete in the open halfspace $\{z \in N_1(X) \mid z \cdot K_X < 0\}$.*

1.4.7. Contraction Theorem [44], [97]. *Let X be a projective variety with only $\mathbb{Q}$-factorial terminal (resp. canonical) singularities. Then the following hold:*

(1.4.7.1) *For every extremal ray $R \subset \overline{NE}(X)$ there is a contraction map $f: X \to Y$ such that an irreducible curve $C \subset X$ is mapped to a point by f iff $[C] \in R$. One can always assume that $f_*(\mathscr{O}_X) = \mathscr{O}_Y$, and then f and Y are unique.*

(1.4.7.2) *We have the following possibilities for f and Y:*

(1.4.7.2.1) *f is birational, and the exceptional set is an irreducible divisor. Then Y again has $\mathbb{Q}$-factorial terminal (resp. canonical) singularities. Such a contraction is called divisorial.*

(1.4.7.2.2) *f is birational, and the exceptional set has codimension at least two in X. In this case $K_{Y-\operatorname{Sing} Y}^{\otimes m}$ never extends to a line bundle over Y for $m > 0$. Such an f is called a small extremal contraction.*

(1.4.7.2.3) $\dim Y < \dim Y$. *The general fiber F has negative canonical class. Such a contraction is called a Fano contraction.*

Proof. We will show that in case (1.4.7.2.2) $K_{Y-\operatorname{Sing} Y}^{\otimes m}$ never extends to a line bundle over Y. Assume the contrary. Then $K_X^{[m]}$ and $f^* K_Y^{[m]}$ are two line bundles on X, and are isomorphic outside the exceptional set. Since the exceptional set has codimension at least two, these line bundles are isomorphic. On the other hand, if $[C] \in R$, then,

$$\deg(K_X^{[m]} | C) < 0 = \deg(f^* K_Y^{[m]} | C).$$

This is a contradiction.

Comments. (1.4.7.2.3) should be considered a structure theorem. It describes X in terms of the lower-dimensional varieties F and Y. Of course it may happen that $\dim Y = 0$ which means that K_X is ample. Even in dimension three these singular Fano varieties are not fully understood.

(1.4.7.2.1) is very satisfactory. The new variety Y has the same properties as X, but the rank of H^2 is one less. We can continue the above process with Y as in (1.1.4). So far there is no complete description of divisorial contractions in dimension three, but it seems to be attainable with a finite amount of work.

(1.4.7.2.2) is the bad news, since Y does not have canonical singularities. In this case something new must be done. This new operation is called flip. Currently its existence is known in dimension three only . We will discuss it in detail in the next section. For now assume that we can do something, and let us formulate the main theorem.

1.4.8. Definition. Let Z be a projective variety with $\mathbb{Q}$-factorial terminal singularities.

(1.4.8.1) Z is called a minimal model if K_Z is nef.

(1.4.8.2) Z is called a relative Fano model if there is an extremal ray R such that the corresponding contraction $f: Z \to Y$ maps onto a lower-dimensional variety. Thus if F is a general fiber, then $-K_F$ is ample.

1.4.9. Minimal Model Theorem for Threefolds [82]. *Let X be a smooth projective three-dimensional algebraic variety over $\mathbb{C}$. Then a succession of divisorial contractions and flips transforms X into a projective variety X' which has the following properties*:

(1.4.9.1) *X' and X are birationally equivalent.*

(1.4.9.2) *X' has only $\mathbb{Q}$-factorial terminal singularities.*

(1.4.9.3) *Either X' is a minimal model or X' is a relative Fano model.*

This X' is not unique, but only one of the alternatives in (1.4.9.3) *can occur.*

Proof. Starting with a smooth threefold X we define inductively a series of threefolds as follows. Let $X_0 = X$. If X_i is already defined, we consider K_{X_i}. If K_{X_i} is nef, then let $X' = X_i$. If K_{X_i} is not nef, then we contract an extremal ray. If we obtain a Fano contraction, then again we set $X' = X_i$. If the contraction $f_i: X_i \to Y_i$ is divisorial, then we set $X_{i+1} = Y_i$. If the contraction is small, then we set $X_{i+1} =$ the flip of f_i. All that remains is to prove that the process will terminate.

A divisorial contraction decreases $\dim H^2$ by one, so we can have only finitely many of these. A flip leaves $\dim H^2$ unchanged. Shokurov [105] proves that a flip "improves" the singularities, and this easily implies that any sequence of flips is finite.

2. Flip and Flop

Studying threefolds one frequently encounters the situation where the "bad set" is a curve and one wants to change the threefold in codimension two only. There are at least three such examples:

1. In (1.4.7.2.2) we saw that in trying to construct the minimal model of a threefold one may encounter a contraction $f: X \to Y$ which contracts only finitely many curves $C_i \subset X$. These curves have negative intersection with K_X, and therefore one would like to get rid of them. Blowing up introduces a whole family of curves that have negative intersection with the canonical class. Thus we need to search for some other operation that changes X in codimension two only. This operation will be the "flip."

2. A threefold X may have several different minimal models X_i. It turns out that the induced bimeromorphic map $X_i \dashrightarrow X_j$ is an isomorphism in codimension one. Thus the difference between X_i and X_j is in finitely many curves only. One would like to understand such maps by factoring them into a sequence of "elementary" maps. "Elementary" may mean for instance that only one irreducible curve is changed. These are the so-called "flops."

3. A threefold X may have a divisor D which is ample except on finitely many curves. One can get rid of these curves by blowing up. However, the pullback of D will not be ample, only the proper transform. Thus, for instance, $h^0(mD)$ will change in this process. One could try to make D ample by changing X in codimension two only. This will not affect $h^0(mD)$. This situation arises very naturally for nonprojective threefolds and will be discussed in detail in §5.

A large part of the difficulty in three-dimensional geometry comes from these codimension-two surgery problems. It is not too hard to find examples that show that even under quite reasonable conditions codimension-two modifications do not exist. There are some theorems that assert the existence of codimension-two modifications under very strict restrictions. These theorems are very hard, and have important consequences in three-dimensional geometry.

2.1. Curve surgery on threefolds. The aim of this subsection is to discuss general facts about algebraic curve surgery on threefolds. The two most important special cases, flips and flops, will be discussed in detail in subsequent subsections.

2.1.1. Definition.

(2.1.1.1) A three-dimensional curve neighborhood is a pair $C \subset X$, where C is a proper connected curve, and X is the germ of a normal threefold along C. We will frequently think of X as an analytic representative of the germ. For some purposes, especially if one wants to consider positive characteristic as well, one can think of X as a formal scheme along C. This would require changing some definitions.

(2.1.1.2) A three-dimensional curve neighborhood $C \subset X$ is called contractible if there is a morphism $f: (C \subset X) \to (P \in Y)$ satisfying the following properties:

(i) Y is the germ of a normal singularity around the point P;
(ii) $f(C) = P$;
(iii) $f: X - C \to Y - P$ is an isomorphism.

f and Y are uniquely determined by $C \subset X$. f is called the contraction morphism of $C \subset X$.

(2.1.1.3) $C \subset X$ is called a three-dimensional irreducible curve neighborhood if C is irreducible.

(2.1.1.4) Two three-dimensional curve neighborhoods $C_i \subset X_i$ are called bimeromorphic if there is an isomorphism $(X_1 - C_1) \cong (X_2 - C_2)$. If both neighborhoods are contractible to $P_i \in Y_i$, then $(Y_1 - P_1) \cong (Y_2 - P_2)$, thus in fact $Y_1 \cong Y_2$. Therefore, two contractible curve neighborhoods $f_i: X_i \to Y_i$ are bimeromorphic iff $Y_1 \cong Y_2$.

2.1.2. Proposition. *Let Z be a normal threefold, and let $C \subset Z$ be a connected proper curve, $C \subset X$ be the germ of Z along C, and $C' \subset X'$ be a three-dimensional curve neighborhood bimeromorphic to $C \subset X$. One can patch $Z - C$ and X' along $X - C \cong X' - C'$ to get a new threefold $C' \subset Z'$. Then*

(2.1.2.1) *Z' is proper iff Z is, and*

(2.1.2.2) *the composite map $\phi: Z \dashrightarrow Z - C \cong Z' - C' \hookrightarrow Z'$ is bimeromorphic.*

Proof. ϕ is meromorphic in codimension one, and is therefore meromorphic by the Levi extension theorem. The properness of Z' is clear.

2.1.3. Definition. We say that the above Z' is obtained from Z by the curve surgery $(C \subset X) \dashrightarrow (C' \subset X')$.

2.1.4. Examples.

(2.1.4.1) Let V be the total space of the line bundle $\mathcal{O}(-1, -1)$ over $\mathbb{P}^1 \times \mathbb{P}^1$. Both of the projections $\pi_i: \mathbb{P}^1 \times \mathbb{P}^1 \to \mathbb{P}^1$ can be extended to morphisms

$$p_i: (\mathbb{P}^1 \times \mathbb{P}^1 \subset V) \to (C_i \cong \mathbb{P}^1 \subset X_i).$$

It is easy to see that X_i is smooth, and the normal bundle of $C_i \subset X_i$ is $\mathcal{O}(-1) + \mathcal{O}(-1)$. $p_2 \circ p_1^{-1}: X_1 \dashrightarrow X_2$ is bimeromorphic but is not an isomorphism.

This example was first used in a systematic way by Kulikov [62] to study the birational transformations of threefolds that have a basepoint-free pencil of $K3$ surfaces.

(2.1.4.2) Let V be the total space of the line bundle $\mathcal{O}(-1, -1)$ over $E \times E$ for some elliptic curve E. Any of the projections $\pi_{ij}: E \times E \to E$ given by $\pi_{ij}: (z_1, z_2) \mapsto iz_1 + jz_2$, $i, j \in \mathbb{Z}$, can be extended to a morphism

$$p_{ij}: (E \times E \subset V) \to (C_{ij} \cong E \subset X_{ij}).$$

The X_{ij} are singular along C_{ij}, and normal iff $(i, j) = 1$.

Thus there are infinitely many nonisomorphic curve neighborhoods bimeromorphic to each other. (One can produce such examples with rational singularities too.)

We will see several problems about threefolds where the main difficulty turns out be able to understand certain curve surgeries. Frequently the main problem is to show that there are nontrivial curve surgeries. The following definition singles out the curve surgery that "changes X the most":

2.1.5. Definition.

(2.1.5.1) Let $C_1 \subset X_1$ be a three-dimensional curve neighborhood. Let $H_1 \in H^2(X_1, \mathbb{Q})$ be a cohomology class such that $H_1 \cap [C_1^j] < 0$ for every irreducible component C_1^j of C_1. (Choosing H_1 to be negative is just a matter of preference. This choice conforms to the most important special case, the flip, where $H_1 = c_1(K_{X_1})$ is negative.) A three-dimensional curve neighborhood $C_2 \subset X_2$ bimeromorphic to $C_1 \subset X_1$ is called the opposite of $C_1 \subset X_1$ with respect to H_1 if there is a cohomology class $H_2 \in H^2(X_2, \mathbb{Q})$ such that $H_2 \cap [C_2^j] > 0$ for every irreducible component C_2^j of C_2 and

$$(X_2 - C_2, H_2|X_2 - C_2) \cong (X_1 - C_1, H_1|X_1 - C_1).$$

In general, the opposite may not exist and depends on the choice of H_1.

(2.1.5.2) If C_1 is irreducible, then H_1 is unique up to a multiplicative constant since $H^2(X_1, \mathbb{Q}) \cong H^2(C_1, \mathbb{Q}) \cong \mathbb{Q}$. As we will see, there is at most one opposite.

(2.1.5.3) If $C \subset X$ is given, and the choice of H is understood, then the opposite (if it exists) will be denoted by $C^+ \subset X^+$.

The uniqueness of the opposite follows from:

2.1.6. Proposition. *Let $C_i \subset X_i$ $(i = 1, 2)$ be three-dimensional curve neighborhoods. Let $H_i \in H^2(X_i, \mathbb{Q})$ be cohomology classes such that $H_1 \cap [C_1^j] \leq 0$ for every irreducible component C_1^j of C_1, and $H_2 \cap [C_2^j] < 0$*

for every irreducible component C_2^j *of* C_2. *Assume that*

$$(X_1 - C_1, H_1|X_1 - C_1) \cong (X_2 - C_2, H_2|X_2 - C_2).$$

Then the above isomorphism extends to a morphism

$$g: X_1 \to X_2.$$

In particular, if $H_1 \cap [C_1^j] < 0$ *for every irreducible component* C_1^j *of* C_1 *then* $X_1 \cong X_2$.

Proof. Let Z be a desingularization of the graph Γ of the bimeromorphic map $X_1 \dashrightarrow X_2$, and let $q_i: Z \to X_i$ be the projections. Let $E_j \subset Z$ be the exceptional divisors (note that every q_1 exceptional divisor is also q_2 exceptional, and vice versa). By assumption

$$q_1^* H_1 = q_2^* H_2 + \sum d_j [E_j]$$

for some rational numbers d_j. Use the cohomological version of (5.2.5.3) with $L = -H_1$, $M = -q_2^* H_2$, $G = 0$ to conclude that $d_j \le 0$. Interchanging X_1 and X_2 gives that $d_j \ge 0$, thus

$$q_1^* H_1 = q_2^* H_2.$$

If $\Gamma \to X_1$ is not an isomorphism then there is a proper curve $B \subset Z$ such that $q_1(B) =$ point but $q_2(B) \subset C_2$ is one dimensional. Thus

$$0 = q_1^* H_1 \cap [B] = q_2^* H_2 \cap [B] < 0.$$

This is a contradiction. q.e.d.

To get some further results we have to impose some restrictions on $C \subset X$. A weak but useful requirement is the rationality of the singularities of X. The following result collects the basic topological consequences of rationality.

2.1.7. Proposition. *Let* X *be a normal variety, and let* $Z \subset X$ *be a closed subvariety. Then the following hold*:

(2.1.7.1) *The local cohomology sheaves* $\mathcal{H}_Z^0(X, \mathbb{Z})$ *and* $\mathcal{H}_Z^1(X, \mathbb{Z})$ *are zero.*

(2.1.7.2) $H^1(X, \mathbb{Z}) \to H^1(X - Z, \mathbb{Z})$ *is injective.*

(2.1.7.3) *If* X *has rational singularities and* codim$(Z, X) \ge 2$, *then* $\mathcal{H}_Z^2(X, \mathbb{Q}) = 0$.

(2.1.7.4) *If* X *has rational singularities and* codim$(Z, X) \ge 2$, *then* $H^2(X, \mathbb{Q}) \to H^2(X - Z, \mathbb{Q})$ *is injective.*

(2.1.7.5) *If* $x \in X$ *is local, then* $\mathrm{Pic}(X - x) \to H^2(X - x, \mathbb{Z})$ *is an isomorphism.*

(2.1.7.6) *If* $x \in X$ *is local, and* $\dim X \leq 3$, *then* X *is a rational homology manifold iff* X *is* $\mathbb{Q}$*-factorial.*

Proof. (2.1.7.1) and (2.1.7.2) are obvious since normal implies topologically unibranch. To see (2.1.7.3) we may assume that X is local, and $Z = \{0\}$ is the closed point. Using the exact sequence

$$H^1(X - 0, \mathbb{Q}) \to H_0^2(X, \mathbb{Q}) \to H^2(X, \mathbb{Q}) = 0,$$

it is sufficient to see that $H^1(X - 0, \mathbb{Q}) = 0$. Let $f: X' \to X$ be a resolution of singularities and let $E = f^{-1}(0)$. Since X has rational singularities, $R^1 f_* \mathbb{Q} = 0$ (see [108, 2.14]; [60, 12.1.3]). Therefore, $H^1(X - 0, \mathbb{Q}) \cong H^1(X' - E, \mathbb{Q})$ and $H^1(X', \mathbb{Q}) = 0$. By [29, §3] the restriction $0 = H^1(X', \mathbb{Q}) \to H^1(X' - E, \mathbb{Q})$ is surjective. This shows (2.1.7.3) which in turn implies (2.1.7.4). (2.1.7.5) is [21, 6.1], and this implies the last claim.

2.1.8. Lemma. *Let* X *be a variety with rational singularities, and* $Z \subset X$ *be a closed subvariety such that* $\operatorname{codim}(Z, X) \geq 2$. *If* L *is a line bundle on* $X - Z$ *such that its first Chern class is liftable to* $H^2(X, \mathbb{Q})$, *then* L^k *extends to a line bundle over* X *for some* $k > 0$.

Proof. This will be done by induction on $\dim Z$. The $\dim Z = 0$ case follows from (2.1.7.5). Taking a general hyperplane section and using induction we see that L^{k_1} extends across an open dense subset $U \subset Z$ for some $k_1 > 0$. Again by induction $L^{k_2 k_2}$ extends across $Z - U$ for some $k_2 > 0$.

2.1.9. Proposition. *Let* $C \subset X$ *be a three-dimensional curve neighborhood. Assume that* X *has only rational singularities. Let* $H \in H^2(X, \mathbb{Q})$ *be a cohomology class such that* $H \cap [C_j] < 0$ *for every irreducible component* C_j *of* C. *If the opposite* $C^+ \subset X^+$ *exists, then* C *is contractible.*

Proof. Let C_i^+ be the irreducible components of C^+. For every i let D_i^+ and E_i^+ be two-dimensional germs intersecting C_i^+ in a single point and disjoint otherwise. Let $\sum a_i[D_i^+] = \sum a_i[E_i^+] = mH^+$, where $a_i, m \in \mathbb{Z}$ are positive. Let D_i (resp. E_i) be the proper transforms of D_i^+ (resp. E_i^+) in X. Then $D = k\sum a_i D_i$ and $E = k \sum a_i E_i$ extend to Cartier divisors on X by (2.1.8) for some $k > 0$. The ideal sheaf $(\mathcal{O}_X(-D), \mathcal{O}_X(-E)) \subset \mathcal{O}_X$ defines a one-dimensional subscheme $\overline{C}$ of X whose normal bundle $\mathcal{O}_X(D) + \mathcal{O}_X(E)|C$ is negative. Thus, C is contractible by [3, 6.2]; [7, 6.1].

2.1.10. Proposition. *Let* $C_i \subset X_i$ *be three-dimensional curve neighborhoods. Let*

$$g: X_1 \dashrightarrow X_2$$

be a bimeromorphic map. Let $C_1^0 \subset C_1$ *be an irreducible component. Assume that* X_2 *has rational singularities and that* g *is not holomorphic at the generic point of* C_1^0. *Then* C_1^0 *is a rational curve.*

In particular, if X_2 *is the opposite of* X_1, *then every component of* C_1 *is rational.*

Proof. Let Γ be the closure of the graph of g with projections $p_i : \Gamma \to X_i$. By assumption, $p_1^{-1}(C_1^0) \subset \Gamma$ contains a surface E. $p_2(E)$ is positive dimensional. Let $D \subset X_2$ be a general local divisor intersecting $p_2(E)$. Since X_2 has rational singularities, so does D. Therefore, every exceptional curve of $p_2^{-1}(D) \to D$ is rational. One of these exceptional curves maps surjectively onto C_1^0.

2.1.11. Remark. Even if both X_i are smooth, the above C_1^0 can be singular [55, 4.8].

2.1.12. Proposition. *Let* $C_i \subset X_i$ *be bimeromorphic three-dimensional curve neighborhoods. Assume that both have rational* $\mathbb{Q}$*-factorial singularities. Then*

$$\begin{aligned} \#\{\textit{irreducible components of } C_1\} &= \operatorname{rank} H^2(X_1, \mathbb{Q}) \\ &= \operatorname{rank} H^2(X_1 - C_1, \mathbb{Q}) \\ &= \operatorname{rank} H^2(X_2, \mathbb{Q}) \\ &= \#\{\textit{irreducible components of } C_2\}. \end{aligned}$$

The following result is needed in factoring curve surgeries as a succession of "elementary" curve surgeries.

2.1.13. Proposition. *Let* $C_i \subset X_i$ *be three-dimensional curve neighborhoods. Let*

$$g : X_1 \dashrightarrow X_2$$

be a bimeromorphic map. Assume that X_1 *has* $\mathbb{Q}$*-factorial singularities and that* g *is not an isomorphism. Then there is an irreducible component* $C_1^0 \subset C_1$ *which is contractible.*

Proof. As in the proof of (2.1.9) we choose D_i^+ and E_i^+ on X_2. We obtain D and E on X_1. By (2.1.6) there is an irreducible component $C_1^0 \subset C_1$ such that $C_1^0 \cdot D = C_1^0 \cdot E < 0$. Thus, C_1^0 is contractible (cf. [55, 4.10]).

2.1.14. Problem. Prove (2.1.13) without assuming that X_1 has $\mathbb{Q}$-factorial singularities (maybe assuming that it has rational singularities).

In general very little is known about the existence of the opposite of a neighborhood $C \subset X$. A positive answer is known only in the following cases.

2.1.15. Theorem. *Let $C \subset X$ be a three-dimensional contractible curve neighborhood. Then the opposite $C^+ \subset X^+$ exists in any of the following cases:*

(2.1.15.1) *X has terminal singularities, and $K_X|C$ is numerically trivial (H arbitrary)* [98].

(2.1.15.2) *X has canonical singularities, and $K_X|C$ is numerically trivial (H arbitrary)* [47].

(2.1.15.3) *X has canonical singularities, and $-K_X|C$ is ample, $H = K_X$* [82].

(2.1.15.4) *X is toric* [99].

Comments. We will give a proof of (2.1.15.1), which provides a very explicit description of $C^+ \subset X^+$ in the next subsection. See [55] for a shorter proof of (2.1.15.2).

Some important special cases of (2.1.15.3) were done independently by using different methods by Tsunoda [114]; Shokurov [106]; Mori [80]; Kawamata [47]. The proof of the general case is very long and complicated. See [58] for an introduction.

2.2. Flops. The aim of this section is to develop a detailed description of flops in dimension three.

2.2.1. Definition. Let $f\colon C \subset X \to P \in Y$ be a three-dimensional contractible curve neighborhood. Assume that X has terminal or canonical singularities and that $K_X|C = 0$. Let H be a line bundle on X such that $H^{-1}|C$ is ample.

The opposite of $C \subset X$ will be called the flop of $C \subset X$. It will turn out that the flop is independent of the choice of H if X has terminal singularities.

In the terminal singularity case the existence of flops was proved by Reid [98]. The following explicit description is based on an idea of Mori (cf. [55, §2]).

2.2.2. Theorem. *Let $f\colon C \subset X \to P \in Y$ be as above. Assume that X has terminal (resp. canonical) singularities. Then Y has a terminal (resp. canonical) singularity at P. If X has terminal singularities and C is irreducible, then $P \in Y$ and the flop can be described in one of the following ways:*

$$(P \in Y) \cong (0 \in (x^2 + F(y, z, t) = 0) \subset \mathbb{C}^4/\mathbb{Z}_n(a, b, c, d)), \quad (n \geq 1). \tag{2.2.2.1}$$

If $\tau\colon Y \to Y$ is the involution $(x, y, z, t) \mapsto (-x, y, z, t)$, then

$$(C^+ \subset X^+) \cong (C \subset C) \quad \textit{and} \quad f^+ = \tau \circ f,$$

or

(2.2.2.2)
$$(P \in Y) \cong (0 \in (xy - g(z^n, t)h(z^n, t) = 0) \subset \mathbb{C}^4/\mathbb{Z}_n(1, -1, a, 0)) \qquad (n \geq 3).$$

$C \subset X$ can be written explicitly as follows. In

$$(\mathbb{C}^4 \times \mathbb{P}^1)/\mathbb{Z}_n(1, -1, a, 0, -1, 0)$$

with coordinates $(x, y, z, t) \times (p : q)$ the equations of X are

$$xy - gh = 0 \quad \text{and} \quad px - qg = 0.$$

In the affine chart $U = (p \neq 0)$ we let $x_1 = q/p$ and $y_1 = y$. Then

$$(C \subset X) \cap U \cong (x_1\text{-axis} \subset (x_1y_1 - h = 0)) \subset \mathbb{C}^4(x_1, y_1, z, t)/\mathbb{Z}_n(1, -1, a, 0).$$

In the affine chart $V = (q \neq 0)$ we let $y^1 = p/q$ and $x^1 = x$. Then

$$(C \subset X) \cap V \cong (y^1\text{-axis} \subset (x^1y^1 - g = 0)) \subset \mathbb{C}^4(x^1, y^1, z, t)/\mathbb{Z}_n(1, -1, a, 0).$$

To obtain the description of $C^+ \subset X^+$ one only has to interchange g and h in the above formulas.

2.2.3. Remark. If C is reducible, then the statement about the singularities of Y still holds. In case (2.2.2.1) the flop is described the same way. In case (2.2.2.2) the flop will be described in (2.2.8).

2.2.4. Corollary [55, 4.11]. *Notation as above. If X has terminal singularities, then X and X^+ have the same (analytic) singularities.*

Proof. If we have (2.2.2.1), then X and X^+ are even isomorphic. In the second case the singularities are at the origins of the charts U and V, and the isomorphism is clear from the explicit description. The local isomorphism does not take C to C^+, therefore it will not extend to a global isomorphism.

2.2.5. Proof of (2.2.2). Choose $m > 0$ such that $K_X^{[m]}$ is a line bundle. By a suitable analytic version of the Basepoint-Free Theorem (1.4.4) [84, 5.5] we obtain that there is a line bundle L on Y such that $f^*L \cong K_X^{[m]}$. Thus $L \cong K_Y^{[m]}$. A resolution $g: Z \to X$ of X is also a resolution $f \circ g: Z \to Y$ of Y, and $g^*K_X^{[m]} \cong (f \circ g)^*K_Y^{[m]}$. This shows that $P \in Y$ is terminal (resp. canonical). By (1.3.4) we obtain that in the terminal case $P \in Y$ can be described as in (2.2.2.1) or (2.2.2.2).

(2.2.5.1) Let $P \in Y$ be as described in (2.2.2.1). Since the flop of $C \subset X$ is unique (2.1.5.2), it is sufficient to check that $f^+: C^+ \subset X^+ \to Y$

satisfies the conditions of (2.1.5.1). Using the identification $(C^+ \subset X^+) \cong (C \subset X)$ set $H^+ = H^{-1}$. The only requirement that needs verification is that

$$c_1(H^{-1}|X - C) \cong c_1(\tau^*(H|X - C)).$$

Let M be any line bundle on $Y - P \cong X - C$. Observe that $(Y - P)/\tau \cong (\mathbb{C}^3 - 0)/\mathbb{Z}_n$. Therefore, $M \otimes \tau^* M$ is the pullback of a line bundle on $(\mathbb{C}^3 - 0)/\mathbb{Z}_n$, and is torsion. Hence, $c_1(\tau^* M) = c_1(M^{-1})$ as required.

(2.2.5.2) Assume that $P \in Y$ is as described in (2.2.2.2). In (2.2.8) we will give a complete description of all curve neighborhoods $C \subset X$ which contract to $P \in Y$. (2.2.2.2) will be a special case of this more general result.

2.2.6. Proposition. *Assume that* $Y = (xy - F(z, t) = 0) \subset \mathbb{C}^4$ *defines an isolated singularity. Let* $F = f_1 \cdots f_k$ *be the irreducible factors. Set*

$$D_i = (x - f_1 \cdots f_i = f_{i+1} \cdots f_k - y = 0) \subset Y \qquad (i = 1, \cdots, k-1).$$

Then $\operatorname{Pic}(Y - 0)$ *is freely generated by the* $[D_i]$.

We start with the following special:

2.2.6.1. Lemma [27, 1.2]. *Let* $Y = (xy - F(z, t) = 0)$. *Assume that* $F(z, t)$ *is irreducible. Then* $\operatorname{Pic}(Y - 0) \cong 0$. *In fact,* Y *is a topological manifold at the origin.*

Proof. Let $C = (F = 0)$. $C - 0$ is a punctured disc since F is irreducible. Thus by [67, 8.5], if $\Delta_C(t)$ denotes the monodromy on the Milnor fiber, then $\Delta_C(1) = \pm 1$. Also, $\Delta_Y(t) = \Delta_C(t)$ hence $\Delta_Y(1) = \pm 1$ and again by [67, 8.5] we obtain that Y is a topological manifold. Thus by (2.1.7.5) $\operatorname{Pic}(Y - 0) \cong H^2(Y - 0, \mathbb{Z}) \cong 0$.

2.2.6.2. Computations. Let Y_1 be the blow up of Y at $(x = f_1 \cdots f_j = 0)$. In (2.2.2.2) set $g = f_1 \cdots f_j$, $h = f_{j+1} \cdots f_k$, and $C = C_1$.

The proper transform D_i' of D_i on Y_1 is given as follows:

(i) If $i < j$, then D_i' intersects C_1 at the origin of the V-chart, and there it is given by equations

$$x_1 - f_1 \cdots f_i = f_{i+1} \cdots f_j - y^1 = 0.$$

(ii) If $i = j$ then D_i' intersects C_1 transversally at a single point which is not the origin of either charts and is given by equations

$$x_1 - 1 = f_{j+1} \cdots f_k - y_1 = 0 \quad (\text{resp. } x^1 - f_1 \cdots f_j = 1 - y^1 = 0).$$

(iii) If $i > j$, then D_i' intersects C_1 at the origin of the U-chart, and is given there by equations

$$x_1 - f_{j+1} \cdots f_i = f_{i+1} \cdots f_k - y_1 = 0.$$

Proof of (2.2.6). First blow-up $(x = f_1 = 0)$. Then on the U-chart blow-up $(x_1 = f_2 = 0)$ and continue. After $k-1$ steps we obtain a threefold $p: Z \to Y$ with $k-1$ curves $C_1, \cdots, C_{k-1}$ which are p-exceptional. The proper transform $\overline{D}_j$ of D_j intersects $\bigcup C_i$ at a single point of C_j transversally. The singularities of Z are of the form $uv - f_i = 0$. Thus by (2.2.6.1)

$$\operatorname{Pic}(Y - 0) = \operatorname{Pic}\left(Z - \bigcup C_i\right) \cong \operatorname{Pic}(Z) \cong H^2(Z, \mathbb{Z}) \cong H^2\left(\bigcup C_i, \mathbb{Z}\right).$$

Now it is clear that the $\overline{D}_i$ freely generate $\operatorname{Pic} Z$.

(2.2.6.3) For some purposes another basis of $\operatorname{Pic}(Y-0)$ is useful. If we write $F = gh$, then

$$\begin{aligned}(xy - gh, x - g) &= (x - g, y(x-g) - (xy - gh)) \\ &= (x - g, g(h - y)) = (x - g, h - y) \cap (x, g).\end{aligned}$$

Therefore, $[x - g = h - y = 0] = -[x = g = 0] = [x = h = 0]$. Hence,

$$D_i = \sum_{s=i+1}^{k} [x = f_s = 0].$$

Thus, by (2.2.6) $[x = f_i = 0]$, $i = 1, \cdots, k$, generate $\operatorname{Pic}(Y-0)$ and satisfy a single relation

$$\sum [x = f_i = 0] = 0.$$

2.2.7. Proposition [98]; **Mori, Shepherd-Barron, Ue** (unpublished). *Assume that* $Y = (xy - F(z, t) = 0) \subset \mathbb{C}^4/\mathbb{Z}_n(1, -1, a, 0)$ *defines an isolated singularity. Let* $F = f_1 \cdots f_k$, *where* $f_i(0, 0) = 0$, *the* f_i *being* $\mathbb{Z}_n$*-invariants and irreducible among such power series. (Note that there cannot be multiple factors since the singularity is isolated.) Then*

$$\operatorname{Pic}(Y - 0) \cong \mathbb{Z}/n\mathbb{Z} \cdot [K_{Y-0}] + \frac{\mathbb{Z}[x = f_1 = 0] + \cdots + \mathbb{Z}[x = f_k = 0]}{([x = f_1 = 0] + \cdots + [x = f_k = 0])}.$$

Thus, as an abstract group

$$\operatorname{Pic}(Y - 0) \cong \mathbb{Z}/n\mathbb{Z} + \mathbb{Z}^{k-1}.$$

Proof. Let $Y' = (xy - F(z, t) = 0)$ and let $q: Y' \to Y$ be the quotient map. Each f_i decomposes as $f_i = \prod f_{ij}$, where the f_{ij} are irreducible but not necessarily $\mathbb{Z}_n$-invariant.

The kernel of the pullback map

$$q^*: \operatorname{Pic}(Y - 0) \to \operatorname{Pic}(Y' - 0)^{\mathbb{Z}_n}$$

is n-torsion. The torsion in $\operatorname{Pic}(Y-0) \cong H^2(Y-0, \mathbb{Z})$ is dual to the torsion in $H_1(Y-0, \mathbb{Z})$. By [30, X.3.4], $Y'-0$ is simply connected, thus $H_1(Y-0, \mathbb{Z}) \cong \mathbb{Z}/n\mathbb{Z}$. One can easily compute that $[K_{Y-0}]$ has order n in $\operatorname{Pic}(Y-0)$, so that it generates the torsion.

For fixed i the $\mathbb{Z}_n$-action is transitive on the f_{ij} since f_i is irreducible among $\mathbb{Z}_n$-invariant power series. By (2.2.6.3) the classes $[x = f_i = 0]$ generate $\operatorname{Pic}(Y'-0)^{\mathbb{Z}_n}$ and satisfy a single relation

$$\sum [x = f_i = 0] = 0 .$$

The divisors $[x = f_i = 0]$ descend to $Y = 0$, thus q^* is surjective.

2.2.8. Theorem. *Assume that*

$$Y = (xy - F(z, t) = 0) \subset \mathbb{C}^4/\mathbb{Z}_n(1, -1, a, 0)$$

defines an isolated singularity. Let $C \subset X$ be a curve neighborhood with contraction map

$$f \colon C \subset X \to 0 \in Y.$$

(2.2.8.1) *Every such X can be constructed as follows:*

First write $F = f_1 \cdots f_m$, where the f_i are $\mathbb{Z}_n$-invariant but not necessarily irreducible. Let $Y'_0 = (xy - F = 0)$. Next blow up $(x = f_1 = 0) \subset Y'_0$. The resulting Y'_1 has a singularity of the form $x_1y_1 - f_2 \cdots f_m = 0$. Blow up $(x_1 = f_2 = 0) \subset Y'_1$ to obtain Y'_2 and continue. After $m-1$ blow-ups we obtain Y'_{m-1}. Finally let $X = Y'_{m-1}/\mathbb{Z}_n$.

(2.2.8.2) *$C^+ \subset X^+$ is obtained in the same way from the reverse order product $F = f_m \cdots f_1$.*

Proof. Let E be an f-ample line bundle on X. Its restriction to $\operatorname{Pic}(X - C) = \operatorname{Pic}(Y - 0)$ will be denoted again by E. Let $F = g_1 \cdots g_k$, where $g_i(0, 0) = 0$, the g_i being $\mathbb{Z}_n$-invariants and irreducible among such power series. By (2.2.7) we can write

$$[E] = \sum a_i[x = g_i = 0] \quad \text{(modulo torsion)}.$$

By adding a suitable multiple of $0 = \sum[x = g_i = 0]$ and rearranging the g_i we can assume that $0 = a_1 \le \cdots \le a_k$. Set $b_i = a_{i+1} - a_i$, $i = 1, \cdots, k-1$. By (2.2.6.3)

$$[E] = \sum b_i[D_i] \quad \text{(modulo torsion)}.$$

Let $i_1, i_2, \ldots$ be those indices such that $b_i \neq 0$. Let

$$f_j = \prod_{s=i_{j-1}+1}^{i_j} g_s \quad (\text{set } i_{-1} = -1).$$

By the computations (2.2.6.2) if we perform the series of blow-ups in the statement of the theorem, then we obtain a proper modification $\overline{f}\colon Y'_{m-1} \to Y'_0$ such that $E|Y'_0 - 0$ extends to an ample line bundle on Y'_{m-1}. By (2.1.6) this implies that $X \cong Y'_{m-1}/\mathbb{Z}_n$.

The opposite is obtained in a similar way by using $-[E]$;

$$-[E] = (a_k - a_{k-1})[x = g_{k-1} = 0] + \cdots + (a_k - a_1)[x = g_1 = 0],$$

and this implies the second part.

2.2.9. Corollary. *Notation as in* (2.2.2). *Assume that X has terminal singularities and C is irreducible. Then there is an isomorphism $\phi\colon H^2(X, \mathbb{Z}) \to H^2(X^+, \mathbb{Z})$ which makes the following diagram commutative*:

$$\begin{array}{ccc} H^2(X, \mathbb{Z}) & \longrightarrow & H^2(Y - 0, \mathbb{Z}) \\ {\scriptstyle\phi}\downarrow & & \| \\ H^2(X^+, \mathbb{Z}) & \longrightarrow & H^2(Y - 0, \mathbb{Z}) \end{array}$$

(*The horizontal maps are the natural restrictions.*)

Proof. In case (2.2.2.1) $X \cong X^+$ and let $\phi = -\mathrm{id}$. In case (2.2.2.2), $H^2(X, \mathbb{Z})$ is generated by $n[x = g = 0]$, and $H^2(X^+, \mathbb{Z})$ is generated by $n[x = h = 0] = -n[x = g = 0]$. (The n comes in because of the group action.) Let $\phi(n[x = g = 0]) = -n[x = h = 0]$.

2.3. Flips.

2.3.1. Definition. Let $f\colon C \subset X \to P \in Y$ be a three-dimensional contractible curve neighborhood. Assume that X has terminal or canonical singularities and that $-K_X|C$ is ample. $C \subset X$ will also be called an extremal neighborhood (extremal nbd in the terminology of [60]).

The opposite $C \subset X$ with respect to K_X will be called the flip $C \subset X$.

2.3.2 Examples of flips. Examples of flips are not easy to get because X cannot be smooth, in fact X cannot have only hypersurface singularities either. The best hope is to find an example which is globally the quotient of a flop. This is indeed possible.

(2.3.2.1) First we consider the simplest example of flops. $Y = (xy - uv = 0) \subset \mathbb{C}^4$ has an isolated singularity at the origin. Let

$$X = B_{(x, v)}Y \quad \text{and} \quad X^+ = B_{(x, u)}Y.$$

Let $C \subset X$ (resp. $C^+ \subset X^+$) be the exceptional curves of $X \to Y$ (resp. $X^+ \to Y$). Thus, we have the following varieties and maps:

$$(C^+ \subset X^+) \to (0 \in Y) \leftarrow (C \subset X).$$

(2.3.2.2) Consider the action of the cyclic group $\mathbb{Z}_n$: $(x, y, u, v) \mapsto (\zeta x, y, \zeta u, v)$, where ζ is a primitive nth root of unity. This defines an action on all of the above varieties. The corresponding quotients are denoted by a subscript n.

The fixed point set of the action of Y is the 2-plane $(x = u = 0)$. On the projective quadric $(xy - uv = 0) \subset \mathbb{P}^3$ the action has two fixed lines: $(x = u = 0)$ corresponding to the above fixed 2-plane and $(y = v = 0)$ corresponding to the ζ-eigenspace. On X therefore the fixed point set has two components: the proper transform of the $(x = u = 0)$ plane and the image of the $(y = v = 0)$ line. The latter is an isolated fixed point. $(x, v' = vx^{-1}, u)$ give local coordinates at the isolated fixed point. The group action is $(x, v', u) \mapsto (\zeta x, \zeta^{-1}v', \zeta u)$. In particular, the quotient is a terminal singularity (1.3.4).

On X^+ the fixed point set will have only one component and contains the exceptional curve C^+. Thus X_n^+ is smooth.

It is not too hard to compute the intersection numbers of the canonical classes with the exceptional curves. We obtain that

$$C_n \cdot K_{X_n} = -\frac{n-1}{n} \quad \text{and} \quad C_n^+ \cdot K_{X_n^+} = n - 1.$$

Thus, $X_n^+ \to Y_n$ is the flip of $X_n \to Y_n$ for $n \geq 2$.

(2.3.2.3) Before going further let us note two special properties of this example. At the isolated fixed point on X we have coordinates (x, v', u), and the curve C is the v'-axis. A typical local $\mathbb{Z}_n$-invariant section of K_X^{-1} is given by $\sigma = (v'^{n-1} - x)(dx \wedge dv' \wedge du)^{-1}$, which has intersection number $(n-1)$ with C. Since this section is invariant, it descends to a local section σ_n of $K_{X^n}^{-1}$. Let $D_n = (\sigma_n = 0)$. By construction, $D_n \cong \{(v', u) - \text{plane}\}/\mathbb{Z}_n$, which is a DuVal singularity (1.3.3) of type A_{n-1}. Since $C_n \cdot D_n = C_n \cdot K_{X_n}^{-1}$, one can easily see that even globally D_n is a member of $|K_{X_n}^{-1}|$.

Another simple way of getting a surface singularity out of the above construction is to consider the general hyperplane section H_n of Y_n. This is given as the quotient of an invariant section of Y. $v - u^n = 0$ is such a section whose zero set on Y is isomorphic to the singularity $(xy - u^{n+1} = 0)$. This itself is a quotient of $\mathbb{C}^2$ by the group $\mathbb{Z}_{n+1}$. Using this, H_n can be written as a quotient of $\mathbb{C}^2$, and we easily get that H_n is isomorphic to the singularity $\mathbb{C}^2/\mathbb{Z}_{n+1}(1, 1)$.

The first observation made in (2.3.2.3) together with the examination of many other examples leads to the following conjecture:

2.3.3. Reid's Conjecture on General Elephants [100], [60]. The contraction map provides a one-to-one correspondence between the following two sets.

Extremal neighborhoods:

$$EN := \left\{ \begin{array}{l} \text{three-dimensional contractible curve neighborhoods } C \subset X \\ \text{such that } X \text{ has canonical singularities and } -K_X|C \text{ is ample}, \end{array} \right\}$$

and flipping singularities:

$$FS := \left\{ \begin{array}{l} \text{three-dimensional normal singularities } P \in Y \text{ such that } K_Y \text{ is} \\ \text{not } \mathbb{Q}\text{-Cartier, and the general } D \in |-K_Y| \text{ has a DuVal} \\ \text{singularity at } P. \end{array} \right\}$$

Reid's original hope was that this equivalence can be used to obtain a proof of the existence of flips. To do this one needs to produce a "nice" member of $|-K_Y|$ and then to use this member to construct X^+. It is still to be seen whether either of these steps can be done in the spirit envisaged by Reid.

One direction of the conjecture is proved, and the other is also known with some restrictions:

2.3.4. Theorem [60, 3.1]. *For a singularity $P \in Y$ as in* (2.3.3.FS) *there is a curve neighborhood $C \subset X$ as in* (2.3.3.EN) *such that $P \in Y$ is the contraction of $C \subset X$.*

The proof of this result uses the full force of the existence of flips. Thus at the moment it cannot be used to show their existence.

2.3.5. Theorem [82] [60, 1.7]. *Let $f: C \subset X \to P \in Y$ be a three-dimensional contractible curve neighborhood. Assume that X has only terminal singularities, $C \cdot K_X < 0$, and C is irreducible. Then the general member of $|-K_X|$ and the general member of $|-K_Y|$ have only DuVal singularities.*

From the point of view of (2.3.3) the proof is again unsatisfactory. This result appears very close to the end of a nearly complete classification of extremal neighborhoods. At the moment it is easier to prove that flips exist than to show (2.3.5).

The second observation of (2.3.2.3) again gives a general feature of extremal neighborhoods. In this case though there are some exceptions:

2.3.6. Theorem [60, 1.8]. *Let $f: C \subset X \to P \in Y$ be a three-dimensional contractible curve neighborhood. Assume that X has only terminal singularities. $C \cdot K_X < 0$, and C is irreducible. Let $P \in H \subset Y$ be a general hypersurface section of Y through P. Then H is either a cyclic quotient singularity or one of six exceptional singularities listed in* [60, 1.8];

[58, 7.2]; *the exceptional ones are all rational and have multiplicity at most five.*

Comments. We except that the list of exceptions is not much longer if C is allowed to be reducible. However, if X is allowed to have canonical singularities, infinitely many new cases seem to appear. We do not know what to except.

The exceptional hyperplane sections do not correspond in any simple way to extremal neighborhoods $C \subset X$. Several of the cases come from infinitely many different families of extremal neighborhoods. The pull-back f^*H frequently has much worse singularities than H.

The advantage of (2.3.6) is that one can view Y as a family of fairly simple surface singularities. X and X^+ appear as a family of modifications of these surface singularities. This makes it possible to have a reasonably explicit construction of X^+ in terms of H. This description is crucial in understanding more delicate properties of flips. One of the most important applications is to consider flips in families of three-dimensional contractible curve neighborhoods. From this point of view flipping in families is not harder than flipping the individual neighborhoods. This yields:

2.3.7. Theorem [60, 11.7]. *Let $f\colon C_t \subset X_t \to Y_t \colon t \in \Delta$ be a flat family of three-dimensional contractible curve neighborhoods. Assume that X_0 has only terminal singularities, $C_0 \cdot K_{X_0} < 0$, and C_0 is irreducible. Then the flips X_t^+ fit into a flat family over Δ.*

Comments.

(2.3.7.1) It happens frequently that $C_t \colon t \neq 0$ is reducible.

(2.3.7.2) The opposites of a flat family of curve neighborhoods usually do not fit into a flat family. Consider for instance the family of vector bundles

$$0 \to \mathcal{O}_{\mathbb{P}^1}(-3) \to E_t \to \mathcal{O}_{\mathbb{P}^1}(-1) \to 0, \qquad t \in \operatorname{Ext}^1(\mathcal{O}_{\mathbb{P}^1}(-1), \mathcal{O}_{\mathbb{P}^1}(-3)) \cong \mathbb{C}.$$

Then

$$E_t \cong \begin{cases} \mathcal{O}_{\mathbb{P}^1}(-2) + \mathcal{O}_{\mathbb{P}^1}(-2) & \text{if } t \neq 0, \\ \mathcal{O}_{\mathbb{P}^1}(-3) + \mathcal{O}_{\mathbb{P}^1}(-1) & \text{if } t = 0. \end{cases}$$

Let $C_t \cong \mathbb{P}^1 \subset X_t$ be the total space of E_t. Every X_t is toric, and it is not hard to see that

$$X_t^+ \cong \begin{cases} \mathbb{P}^1 \times (x^2 + y^2 + z^2 = 0 \subset \mathbb{C}^2) & \text{if } t \neq 0, \\ \text{a curve neighborhood with an isolated} & \\ \quad \text{singularity of index 3} & \text{if } t = 0. \end{cases}$$

Therefore the X_t^+ do not fit into a flat family. (This example also shows that the so-called "log-flip" is not continuous in families.)

(2.3.7.3) By using (2.3.7) the construction of flips leads to very interesting conjectures concerning deformations of rational singularities. These will be discussed in §6.

3. Applications of minimal models

3.1. Further study of relative Fano models. Let X be an algebraic variety with $\mathbb{Q}$-factorial terminal singularities, which is a relative Fano model (1.4.8.2). Let $F \subset X$ be a general fiber. Then F is an algebraic variety with terminal singularities such that $-K_F$ is ample. Thus, F is a Fano variety, possibly with large Picard number.

In low dimensions the typical examples of Fano varieties are $\mathbb{P}^n$ and the smooth quadric $\mathbb{Q}^n \subset \mathbb{P}^{n+1}$. Thus the basic problem about Fano varieties is:

3.1.1. Problem. How similar are Fano varieties to $\mathbb{P}^n$?

This is a very general question and of not much use without further clarification. Here are some—successively weaker—technical versions of being "similar to $\mathbb{P}^n$."

3.1.2. Definition. Let X be an n-dimensional variety.

(3.1.2.1) X is said to be rational if there is a generically one-to-one map $g: \mathbb{P}^n \dashrightarrow X$.

(3.1.2.2) X is said to be unirational if there is a generically finite map $g: \mathbb{P}^n \dashrightarrow X$.

(3.1.2.3) X is said to be rationally connected if through any two general points $x, y \in X$ there is an irreducible rational curve $C_{x,y} \subset X$.

(3.1.2.4) X is said to be uniruled if there is an $(n-1)$-dimensional variety Y^{n-1} and a generically finite map $g: \mathbb{P}^1 \times Y^{n-1} \dashrightarrow X$.

3.1.3. Theorem.

(3.1.3.1) [12]. *A smooth cubic threefold $X_3 \subset \mathbb{P}^4$ is not rational.*

(3.1.3.2) [41]. *A smooth quartic threefold $X_4 \subset \mathbb{P}^4$ is not rational.*

Methods of proof. The cubic case relies on the observation that the intermediate Jacobian is a birational invariant up to direct factors which are Jacobians of curves. Therefore, one needs to compute the intermediate Jacobian. This method is applicable to several other Fano threefolds and to conic bundles, but it seems to work only in dimension three.

Iskovskikh and Manin prove (along the lines indicated by Fano) that the birational automorphism group of a smooth quartic is finite. Such computations are fairly hard. This method again can be applied more generally, even in higher dimensions [102], [103], [94].

3.1.4. Problems. Is every Fano variety unirational? Is every Fano variety rationally connected?

Every smooth Fano threefold is rationally connected (cf. [38, 4.1]). Miyaoka [72] shows that if the Picard number is one, then there is a connected chain of rational curves through any two points. Unirationality is not known already for smooth quartic threefolds $X_4 \subset \mathbb{P}^4$. The answer is probably negative.

3.1.5. Theorem [73]. *Every Fano variety is uniruled, in fact it is covered by rational curves* C *such that* $-C \cdot K_X \leq 2\dim X$.

For smooth varieties the method of Mori [77] works; this was observed in [52]. In the singular case a more refined version is needed.

3.1.6. Corollary [73]. *Every relative Fano model is uniruled.*

3.1.7. Problems. (3.1.7.1) Are there only finitely many deformation types of Fano varieties with terminal singularities of a given dimension?

(3.1.7.2) Let X be a Fano variety. Is the self-intersection of $-K_X$ bounded by a function depending only on the dimension? Is the index (1.3.2.2) of X bounded by a function depending only on the dimension?

(3.1.7.3) Let X be a Fano variety, and let $x, y \in X$ be sufficiently general points. Is there an irreducible rational curve C_{xy} containing x and y such that

$$C_{xy} \cdot (-K_X) \leq (\text{some function of } \dim X)?$$

Comments. In dimension three there are 104 deformation types of smooth Fano varieties [36], [37], [83], [84].

Kawamata [48] proves (3.1.7.1) in dimension three under the additional assumption that the rank of the Picard group is one.

By [59], [66], (3.1.7.1) is equivalent to (3.1.7.2).

The argument of Iskovskikh [38, 4.1] shows that (3.1.7.3) also implies (3.1.7.1).

3.1.8. In dimension three the most interesting open question about relative Fano models is their birational classification. Given two relative Fano models how do we decide when they are birational? A somewhat simpler question is to decide which ones are rational. This problem is settled for most smooth Fano varieties [5] [38], [39], and a lot of work has been done about conic bundles [5], [102], [103]. See [40] for the conjectured rationality criterion.

3.1.9. Problem. (3.1.9.1) Find "standard" birational models for families of Del Pezzo surfaces over curves. By this we mean that given

$f: X \to C$ whose general fiber is a Del Pezzo surface, find a "standard form" $f': X' \to C$ and a birational map g which fits into a commutative diagram:

$$\begin{array}{ccc} X & \overset{g}{\dashrightarrow} & X' \\ \downarrow f & & \downarrow f' \\ C & = & C \end{array}$$

For example, if the general fiber of f is $\mathbb{P}^2$, then we can choose $X' = C \times \mathbb{P}^2$. By [14] f always has a section. Maybe one can arrange that the section does not pass through any singular points, and every fiber is irreducible and reduced.

(3.1.9.2) Study the birational types of the "standard" birational models. Some results are due to Alekseev [1].

3.2. Further study of minimal models. Minimal models of a given three-fold X are not unique. In some cases there can be infinitely many non-isomorphic minimal models. Fortunately different minimal models are closely related:

3.2.1. Theorem [47], [55]. *Let X_1 and X_2 be three-dimensional minimal models with $\mathbb{Q}$-factorial singularities. Then any birational map $f: X_1 \dashrightarrow X_2$ can be obtained as a composite of flops.*

Proof. First one needs to prove that f is an isomorphism in codimension one and that K_{X_1} is trivial along the locus $\bigcup C_i$, where f is not defined [55, 4.3]. If f is not an isomorphism, then by (2.1.12) one of the C_i can be flopped. By [55, 2.4] after finitely many flops we get X_2. q.e.d.

Since minimal models are not unique, we need to find properties that are invariant under flops. There are surprisingly many such:

3.2.2. Theorem. *Let X be a three-dimensional $\mathbb{Q}$-factorial minimal model. Then the following objects are unchanged under flops, and hence do not depend on the $\mathbb{Q}$-factorial minimal model chosen:*

(3.2.2.1) *The intersection homology groups $IH^i(X, \mathbb{C})$ together with their Hodge structures* [44, 4.12].

(3.2.2.2) *The collection of analytic singularities of X* [55, 4.11].

(3.2.2.3) *The miniversal deformation space* $\operatorname{Def} X$ [60, 12.6].

(3.2.2.4) *The integral cohomology groups $H^i(X, \mathbb{Z})$.*

(3.2.2.5) $\operatorname{Pic} X \subset \operatorname{Weil} X$.

(3.2.2.6) *$h^0(X, \mathscr{O}(D))$ for every Weil divisor D.*

(3.2.2.7) *$h^i(X, \mathscr{O}(mK_X))$ for every i and m.*

Proof. (3.2.2.2) follows from (2.2.4).

If $C \subset X \dashrightarrow C^+ \subset X^+$ is a flop, then the cohomology of X and of X^+ can be computed from a Mayer-Vietoris sequence involving $X - C$, a neighborhood U of C, and $U - C$. U retracts to $C \cong \mathbb{P}^1$, thus the only nontrivial cohomologies of U are H^0 and H^2. Using (2.2.9) we get that the sequences for X and X^+ are isomorphic. (The product structure is not an invariant).

The group of Weil divisors modulo linear equivalence is clearly invariant under flops. The invariance of $\operatorname{Pic} X$ follows from (2.2.9). Any two minimal models are isomorphic in codimension one; this implies (3.2.2.6).

If $X \to Y \leftarrow X^+$ is a flop, then

$$h^i(X, \mathcal{O}(mK_X)) = h^i(Y, \mathcal{O}(mK_Y)) = h^i(X^+, \mathcal{O}(mK_{X^+})).$$

This essentially follows from the fact that K_X is the pullback of K_Y and that Y has rational singularities.

Concerning (3.2.2.1) it is worth pointing out that if X' is any smooth projective variety birational to X, then $IH^i(X, \mathbb{C})$ is naturally a direct summand of $H^i(X', \mathbb{C})$.

3.2.3. The whole point of constructing minimal models was to simplify the global structure of the canonical bundle. The following immediate consequence of (1.4.4) is the first important general result exploiting this:

3.2.4. Theorem [6], [44], [45], [97], [105]. *Let X be a proper algebraic variety with only canonical singularities. Assume that K_X is nef and big. Then $|mK_X|$ has no basepoints for all $m \gg 0$,* $\operatorname{index} X | m$.

3.2.5. Problem. Let X be a proper algebraic variety with only canonical singularities. Assume that K_X is nef and big. Is there a constant N depending only on the dimension such that $|mK_X|$ determines a birational map for $m \geq N$?

Comments. It is certainly not true that one can even get a morphism. This is prevented by the presence of high index singular points.

There has been some positive results in dimension three. If one assumes that $\chi(\mathcal{O}_N) \leq 1$, then one can take $N = 269$ [22]. In general, if $\chi(\mathcal{O}_X) \leq k$, then there is a bound depending on k. $\chi(\mathcal{O}_X)$ can be arbitrarily large for threefolds of general type if the minimal model is sufficiently singular. Compare (3.2.6.2) and (3.2.7).

In dimension three the above problem reduces to finding an N_0 such that the mth-plurigenus is at least 2 for $m \geq N_0$. For this and other purposes it is very useful to have a plurigenus formula. As usual, we can compute only $\chi(\mathcal{O}_X(mK_X))$. Singular Riemann-Roch gives a formula, but an explicit computation of the occurring terms is not easy.

3.2.6. Plurigenus Formula. Let X be a proper threefold with terminal singularities. As was mentioned in (1.3.4.4), and arbitrary terminal singularity $(P_i \in X)$ deforms to a collection S_i of cyclic quotient singularities of the form $\mathbb{C}^3/\mathbb{Z}_r(1, -1, a)$. Let $S = S(X)$ be the collection of all the S_i (with multiplicities counted).

For a pair of integers r, n define $\overline{n}$ by $0 \le \overline{n} < r$ and $n \equiv \overline{n} \pmod r$.

The following is the Barlow-Fletcher-Reid plurigenus formula [100, 10.3]:

$$\chi(\mathscr{O}_X(mK_X)) = \frac{m(m-1)(2m-1)}{12} K_X^{(3)} + (1-2m)\chi(\mathscr{O}_X) + \sum_S \left\{ (m - \overline{m}) \frac{r^2 - 1}{12r} + \sum_{j=0}^{\overline{m}-1} \frac{\overline{aj}(r - \overline{aj})}{2r} \right\} \tag{3.2.6.1}$$

and

$$\chi(\mathscr{O}_X) = -\frac{1}{24} K_X \cdot c_2(X) + \sum_S \frac{r^2 - 1}{24r}. \tag{3.2.6.2}$$

One can rewrite (3.2.6.1) as

$$\chi(\mathscr{O}_X(mK_X)) = \frac{m(m-1)(2m-1)}{12} K_X^{(3)} + \frac{m}{12} K_X \cdot c_2 + \chi(\mathscr{O}_X) + \sum_S \left\{ -\overline{m} \frac{r^2 - 1}{12r} + \sum_{j=0}^{\overline{m}-1} \frac{\overline{aj}(r - \overline{aj})}{2r} \right\}. \tag{3.2.6.3}$$

Comments. Since X has only isolated singularities, c_2 makes sense. Alternatively, one can take c_2 of any resolution of singularities and push it down to X.

The sum in (3.2.6.3) is a periodic function whose period divides index(X). Also, if index$(X)|m$, then the sum is zero, hence we have only the polynomial part.

It is conceivable that (3.2.5) can be answered by understanding the combinatorics of (3.2.6.3). It is not at all clear what the nature of the periodic part is.

The following result helps us understand $K_X \cdot c_2$:

3.2.7. Theorem. *Let X be a projective threefold. Assume that K_X is nef.*

(3.2.7.1) [121], [112] *If X is smooth, then*

$$K_X^{(3)} \le \tfrac{8}{3} K_X \cdot c_2 = -64\chi(\mathscr{O}_X).$$

(3.2.7.2) [64] *If X has isolated singularities (e.g., if X is a minimal model), then*

$$K_X^{(3)} \le 3K_X \cdot c_2.$$

If K_X is not big, then we do not have a vanishing for the higher cohomologies of $\mathscr{O}_X(mK_X)$ on a minimal model. Still the above results can be used to gain some information about the plurigenera:

3.2.8. Theorem [69], [70], [71]. *Let X be a three-dimensional minimal model. Then $h^0(X, \mathscr{O}_X(mK_X)) \geq 1$ for some $m > 0$.*

The above two results of Miyaoka are some of the least understood theorems in threefold theory. I think that it would be worthwhile to investigate them again for simplification or clarification.

The following is one of the basic open problems in the theory of minimal models:

3.2.9. Abundance Conjecture [98, 4.6]. Let X be an n-dimensional minimal model. Then $\mathscr{O}_X(mK_X)$ is generated by global sections for some $m > 0$.

There are some results of Kawamata [46] and Miyaoka [70] about special cases. It seems that even in dimension three new ideas are needed for the proof.

Finally we pose a problem about generalizing the canonical bundle formula of elliptic surfaces to higher dimensional minimal models.

3.2.10. Conjecture (canonical Bundle Formula). Let X be an n-dimensional minimal model. Assume that $\mathscr{O}_X(mK_X)$ is generated by global sections for some $m > 0$. Let $g \colon X \to I(X)$ be the Stein factorization of the corresponding morphism, and $\Delta \subset I(X)$ be the locus of the singular fibers. Then there are

(i) a nef line bundle L on $I(X)$,
(ii) an effective divisor $\sum a_i D_i$ on $I(X)$ such that $\bigcup D_i \subset \Delta$;
(iii) An integer $m > \max a_i$,

such that

$$\mathscr{O}_X(mK_X) \cong g^* \left(L \otimes \mathscr{O}_{I(X)} \left(\sum a_i D_i\right) \otimes \mathscr{O}_{I(X)}(mK_{I(X)})\right) .$$

To be more precise one would like to show that a_i depends only on the degeneration of certain Hodge structures along D_i. See [115], [25], [81, 5.13] for closely related results.

3.3. Applications to threefolds. While the current point of view in three-dimensional algebraic geometry is that minimal models are the basic objects, it is important to note that the theory can be successfully applied to several old problems that seemed completely intractable before. The first such examples are about various birationally invariant properties of threefolds. It is natural to expect that these are easier to study on a birational model whose global structure is comparatively simple.

3.3.1. Theorem [73], [70] ($\kappa = -\infty$ characterization). *Let X be a smooth projective threefold. Then the following three statements are equivalent:*

(3.3.1.1) *There is a rational curve through every point of X.*

(3.3.1.2) *X is uniruled.*

(3.3.1.3) $H^0(X, \omega_X^{\otimes i}) = 0$ *for every $i > 0$.*

Proof. Consider all components of the Hilbert scheme of rational curves on X. There are only countably many such. If there is a rational curve through every point, then one of these components must give a dominant map. If necessary, we can take a lower-dimensional subvariety to get a generically finite dominant map.

If X is uniruled, then there is a rational curve through a general point of X. If we specialize to any $x \in X$, this rational curve will specialize to a collection of rational curves, one of which will pass through x. If $g: \mathbb{P}^1 \times Y \dashrightarrow X$ is generically finite, then

$$h^0(X, \omega_X^{\otimes i}) \leq h^0(\mathbb{P}^1 \times Y, \omega_{\mathbb{P}^1 \times Y}^{\otimes i}) = h^0(Y, \omega_Y^{\otimes i}) \cdot h^0(\mathbb{P}^1, \omega_{\mathbb{P}^1}^{\otimes i}) = 0.$$

Assume (3.3.1.3). Let X' be the model whose existence is guaranteed by (1.4.9). By (3.2.8) it cannot be a minimal model. Therefore, x' is a relative Fano model, thus X' and hence X are uniruled by (3.1.6).

3.3.2. Theorem [44], [6], [25] (Finite generation of the canonical ring). *Let X be a smooth projective threefold. Then the canonical ring*

$$R(X) = \sum_{i=0}^{\infty} H^0(X, \omega_X^{\otimes i})$$

is finitely generated.

Proof. If X is uniruled, then $R(X) \cong \mathbb{C}$. The canonical ring is a birational invariant, therefore it is sufficient to consider the case when X is a minimal model. If K_X is big, then (3.2.4) and [123] imply the result. If $\kappa(X) \leq 1$, then the claim is fairly easy [23]. The remaining case is when X is an elliptic fiber space over a surface S. Here one uses a version of (3.2.10) to reduce the problem to S [25]. The corresponding finite generation problem for surfaces was solved by Fujita [24]. q.e.d.

Minimal models also help in trying to understand the group of birational self-maps: $\operatorname{Bir} X$. This direction was started by Hanamura [31] who later proved several of his results without assuming the existence of minimal models [32]. Here we mention a related result:

3.3.3. Proposition. *Let X be a three-dimensional minimal model. Then there is a natural representation*

$$\operatorname{Bir} X \to \operatorname{Pic} X / \operatorname{Pic}^{\tau} X,$$

whose kernel is a compact group.

Proof. Let $p: X \dashrightarrow X$ be a birational map. For any Cartier divisor D the divisor p^*D is again Cartier by (3.2.2.5). This defines the representation. If H is an ample divisor and p^*H is also ample, then p is an automorphism by (2.1.6) q.e.d.

The existence of flips in families (2.3.7) implies that suitably chosen minimal models of a flat family of smooth threefolds again form a flat family. This way one can prove some results about families of threefolds.

3.3.4. Theorem [60, 1.3] (Deformation invariance of plurigenera). *Let $\{X_t: t \in T\}$ be a flat family of smooth projective threefolds. Assume that T is connected, and that for some $0 \in T$ and some $m > 0$ we have $h^0(X_0, \omega_{X_0}^{\otimes m}) \geq 2$. Then $h^0(X_t, \omega_{X_t}^{\otimes n})$ is independent of $t \in T$ for every n.*

The assumption $h^0(X_0, \omega_{X_0}^{\otimes m}) \geq 2$ is probably not needed.

3.3.5. Theorem [60, 1.4] (Moduli space for threefolds of general type). *Let $\mathcal{M}$ be the functor "families of threefolds of general type modulo birational equivalence," i.e.,*

$$\mathcal{M}(S) = \left\{ \begin{array}{l} \textit{Smooth projective families } X/S \textit{ such that every fiber is a} \\ \textit{threefold of general type. Two families } X^1/S \textit{ and } X^2/S \\ \textit{are equivalent if there is a rational map } f: X^1/S \dashrightarrow X^2/S \\ \textit{which induces a birational equivalence on each fiber.} \end{array} \right.$$

Then there is a separated algebraic space $\mathbf{M}$ *which coarsely represents* $\mathcal{M}$. *Every connected component of* $\mathbf{M}$ *is of finite type.*

3.4. Applications to deformations of surface singularities. Let $S_t: t \in \Delta$ be a one-parameter family of surfaces. Then the total space $X = \bigcup S_t$ is a threefold. Studying birational models of X may lead to various results on families of surfaces. A special case of this problem, when $S_t: t \in \Delta$ is a degeneration of $K3$ surfaces, was the first instance that flops appeared in the literature as a tool [62].

3.4.1. Degenerations of surfaces. Let $S_t: t \in \Delta$ be a one-parameter family of surfaces. We assume that $S_t: t \neq 0$ is smooth but S_0 may be even reducible. By the semistable reduction theorem [50, Chapter 2] after a suitable base change we can take a new degeneration $S'_t: t \in \Delta$ such that S'_0 is reduced with normal crossings only as singularities. Let $X' = \bigcup S'_t$ be the total space of the family.

Due to the special structure of X' the minimal model program for X' is easier than that for arbitrary threefolds. This special case of (1.4.9) was settled earlier by [114], [106], [79], [80], [47] (the latter is the only complete published proof). The end result is the following:

3.4.2. Theorem. *Assume that* $S_t: t \neq 0$ *are not ruled. Then, the relative minimal model of* X'/Δ *is a threefold* $\overline{X}/\Delta$ *satisfying the following properties*:

(3.4.2.1) $\overline{X}$ *has only* $\mathbb{Q}$*-factorial terminal singularities, and* $K_{\overline{X}}$ *is nef on every fiber* $\overline{S}_t$.

(3.4.2.2) $\overline{S}_t$ *for* $t \neq 0$ *is a smooth minimal surface.*

(3.4.2.3) $\overline{S}_0$ *is reduced with only the following types of singularities*:

(i) $(xy = 0) \subset \mathbb{C}^3$ *or* $(xyz = 0) \subset \mathbb{C}^3$;

(ii) $(t = 0)/\mathbb{Z}_n \subset (xy + f(z^n, t) = 0)/\mathbb{Z} \subset \mathbb{C}^4/\mathbb{Z}_n(1, -1, a, 0)$, *where* $(a, n) = 1$.

Comments. If one restricts the birational type of the general fiber S_t, then frequently one can restrict the class of singularities even more. These questions have been worked out in detail for the case where S_t is trivial canonical class.

The above result can serve as a guiding principle to determine which singular surfaces should appear at the boundary of the moduli of surfaces. The best choice seems to be the one given in [61, Chapter 5].

The following question is the only missing ingredient in the construction of a compactification of the moduli of surfaces of general type:

3.4.3. Problem. Find a bound on the order of the group occurring in (3.4.2.3ii) in terms of the general fiber $S_t: t \neq 0$ alone. This is not known even when S_t is a quintic in $\mathbb{P}^3$.

Let $f: X \to Y$ be a flat projective family of surfaces. In general the minimal resolutions of the fibers do not form a flat family. The following result gives a necessary and sufficient condition.

3.4.4. Theorem ([63] for Gorenstein singularities; [61, 2.10] in general). *Let* $f: X \to Y$ *be flat family of projective surfaces with isolated singularities. Assume that* Y *is connected. Then the following are equivalent*:

(3.4.4.1) *The self-intersection* $K_{\overline{X}_y} \cdot K_{\overline{X}_y}$ *of the canonical class of the minimal resolution* $\overline{X}_y$ *of the fiber* X_y *is independent of* $y \in Y$.

(3.4.4.2) *There is a smooth family of projective surfaces* $\overline{f}: \overline{X} \to \overline{Y}$ *and a finite and surjective morphism* $p: \overline{Y} \to Y$ *such that* $(\overline{X})_y \cong \overline{X}_{p(y)}$ *for every* $y \in \overline{Y}$.

3.4.5. Definition.

(3.4.5.1) For a sequence of natural numbers $a_1, \cdots, a_n$ we define the continued fraction $[a_1, \cdots, a_n]$ recursively by

$$[a_n] = a_n, \qquad [a_i, \cdots, a_n] = a_i - \frac{1}{[a_{i+1}, \cdots, a_n]}.$$

(3.4.5.2) Let K_{e-2} be the set of sequences $k_2, \cdots, k_{e-1}$ such that $[k_2, \cdots, k_{e-1}] = 0$. (The indices are chosen to work well for the next definition.) It is known (see, e.g., [110]) that

$$\# K_{e-2} = \frac{1}{e-2}\binom{2e-6}{e-3}.$$

(3.4.5.3) Given a cyclic quotient singularity $\mathbb{C}^2/\mathbb{Z}_n(1, q)$ let

$$\frac{n}{n-q} = [a_2, \cdots, a_{e-1}].$$

(e is the multiplicity of the quotient singularity.)

(3.4.5.4) For a cyclic quotient singularity $\mathbb{C}^2/\mathbb{Z}_n(1, q)$ let

$$K(\mathbb{C}^2/\mathbb{Z}_n(1, q)) = \{[k_2, \cdots, k_{e-1}] \in K_{e-2} | k_i \leq a_i \ \forall i\}.$$

3.4.6. Theorem [61], [11], [110]. *The number of irreducible components of the versal deformation space of* $\mathbb{C}^2/\mathbb{Z}_n(1, q)$ *is exactly* $K(\mathbb{C}^2/\mathbb{Z}_n(1, q))$.

Comments. [61, Chapter 3] establishes a one-to-one correspondence between the components of the versal deformation space and certain partial resolutions of the singularity. The above formula was found by Christophersen by studying explicit equations of the versal deformation space. Some technical details of his proof are still unfinished. Based on the observation of Christophersen, Stevens showed that there is a one-to-one correspondence between the partial resolutions studied in [61] and the above continued fractions.

In§6 we will consider the possibility of extending some of these results to deformations of arbitrary rational surface singularities.

3.5. Applications to the resolution of singularities. Two-dimensional singularities have unique minimal resolutions, and this is very useful in their study. In dimension three there is no unique minimal resolution, and this is one of the reasons why we know much less about three-dimensional singularities.

As a consequence of the Minimal Model Program we obtain two candidates as substitutes for the minimal resolution. Let us formulate their existence as a separate theorem:

3.5.1. Theorem [82]. *Let X be a three-dimensional algebraic variety (not necessarily proper). Then there are two birational modifications*

$$X^{\text{term}} \xrightarrow{t} X^{\text{can}} \xrightarrow{c} X$$

with the following properties:

(3.5.1.1) *X^{term} has only $\mathbb{Q}$-factorial terminal singularities, and X^{can} has only canonical singularities.*

(3.5.1.2) *$K_{X^{\text{can}}}$ is c-ample; $K_{X^{\text{term}}}$ is $(t \circ c)$-nef.*

(3.5.1.3) *X^{can} is unique; X^{term} is unique up to flops.*

(3.5.1.4) *t is crepant, i.e., $t^* K_{X^{\text{can}}} \equiv K_{X^{\text{term}}}$.*

Since we understand three-dimensional canonical and terminal singularities reasonably well, these objects can be viewed as suitable intermediaries between X (which has arbitrary singularities) and a resolution (which is locally nice but globally uncontrollable).

Both (3.4.4) and (3.4.6) can be viewed as applications of this principle. We hope that in the future (3.5.1) will be used even more.

(3.5.2) Here we want to explain how the point of view given by minimal models helps in the resolution process of singularities. This answers an "old dream of many specialists" [28, p. 60]. (To be fair, this is probably not the kind of answer the specialists dreamed about.)

The Jungian method [28] starts with a threefold X and a finite morphism onto a smooth threefold $f: X \to Y$ (e.g., $Y = \mathbb{P}^3$). By embedded resolution of surfaces we can assume that the branch locus $B \subset Y$ is a divisor with normal crossings. This implies that the singularities of X are toroidal, hence it is much easier to resolve them. In dimension three it is however not clear how the local toroidal resolutions can be patched together (especially if we want to preserve projectivity). This is the point where minimal models come in.

Since X^{can} is unique and locally definable, it is sufficient to construct it locally. The local existence of X^{can} follows from the general result, but in the toroidal case it is much easier and was established by [99]. Thus we obtain X^{can}.

In dimension three the structure of canonical singularities is sufficiently understood to proceed with the resolution.

Reid [96], [98] investigates partial resolutions of canonical singularities. He shows how to construct $t: X^{\text{term}} \to X^{\text{can}}$ as a sequence of explicit blow-ups. Most of the time we need to blow up only closed points and curves, but in certain cases some other ideal sheaves are blown up. If the higher-dimensional minimal model program works, then this step should also be possible in higher dimensions.

X^{term} has only isolated, terminal and toroidal singularities. For most applications it is better not to resolve these. Also, the resolution process is now completely local: there are no compatibility questions involved if the local resolution process is via blowing up closed subvarieties that dominate the singular points.

4. How to find extremal rays?

In this section the characteristic of the base field is arbitrary.

The definition of extremal rays is very simple, and in some cases it is easy to exhibit all extremal rays on a threefold. However, by the nature of the definition this requires an overview of all curves on a threefold. This is sometimes very hard. Therefore it is of interest to find other methods that are more local in nature.

Let us review the situation for surfaces. There are three kinds of extremal rays.

If $C \subset S$ is a (-1)-curve, then this fact is shown by the first order infinitesimal neighborhood of C in S. Therefore no global information is required to assert that C is a (-1)-curve, and hence generates an extremal ray.

If $C \subset S$ is a fiber of a $\mathbb{P}^1$-bundle or a line $\mathbb{P}^2$, then local information is not sufficient. Indeed, we can always destroy the extremality of C by blowing up any point away from C. Thus we need some global information. Let $C \subset S$ be a smooth rational curve of self-intersection 0 or 1. Then C generates an extremal ray iff every deformation of C stays irreducible. Thus again a fairly limited amount of global information allows us to decide whether C generates an extremal ray or not.

The aim of this section is to consider the analogous problem for threefolds. Even for smooth threefolds I am unable to give a complete answer at this moment.

4.1. Seemingly extremal rays.

4.1.1. Definition. Let X be an algebraic space.

(4.1.1.1) Let Z_k^+X be the free abelian semigroup generated by closed k-dimensional irreducible and reduced subspaces of X. Thus an element of Z_k^+X is of the form

$$\sum a_i[V_i],$$

where the V_i are k-dimensional closed, irreducible and reduced subspaces, $a_i > 0$ are natural numbers, and the sum is finite.

(4.1.1.2) We define effective algebraic equivalence for elements of Z_1, $Z_2 \in Z_k^+ X$ as follows:

(i) If W is a normal (possibly reducible) pure dimensional variety, and $p: W \to C$ is a flat morphism onto a smooth connected curve, then any fibers are effectively algebraically equivalent.

(ii) If $g: W \to X$ is a proper morphism, and $Z_1, Z_2 \in Z_k^+ W$ are effectively algebraically equivalent, then $g_*[Z_1], g_*[Z_2] \in Z_k^+ X$ are also effectively algebraically equivalent, where g_* is the push-forward of cycles [26, 1.4].

(iii) Finally, take the transitive hull of the relation given by (i) and (ii); i.e., this definition is like the definition of algebraic equivalence but we require all intermediate cycles to be effective.

Effective algebraic equivalence of $Z_1, Z_2 \in Z_k^+ X$ will be denoted by

$$Z_1 \overset{\text{eff}}{\approx} Z_2 .$$

(4.1.1.3) Given a cycle Z the symbol $\mathbb{Q}^+_{\text{eff}}[Z]$ will denote all effective cycles Z_i such that a positive multiple of Z is effectively algebraically equivalent to a positive multiple of Z_i.

One reason for this definition is that effective algebraic equivalence has the following property:

4.1.2. Proposition. *Let $f: X \to Y$ be a proper morphism between algebraic spaces. Let $\sum a_j[V_j]$ and $\sum b_j[U_j]$ be two effective k-cycles which are effectively algebraically equivalent. Assume that every V_j is contained in a fiber of f. Then every U_j is also contained in a fiber of f.*

Proof. Choose morphisms $p_i: W_i \to C_i$, fibers $A_i, B_i \subset W_i$ of p_i, and proper morphisms $g_i: W_i \to X$ which show the effective algebraic equivalence between $\sum a_j[V_j]$ and $\sum b_j[U_j]$; i.e.,

$$\sum a_j[V_j] = g_{1*}[A_1], \qquad g_{i*}[B_i] = g_{i+1*}[A_{i+1}] \quad \text{for } i = 1, \cdots, n-1,$$
$$g_{n*}[B_n] = \sum b_j[U_j].$$

By induction on i we get that $f \circ g_i: W_i \to Y$ maps A_i to a finite point set. Therefore by [13, 1.5] $f \circ g_i$ factors through the Stein factorization of p_i, hence $f \circ g_i(B_i) = f \circ g_{i+1}(A_{i+1})$ is again a finite point set. q.e.d.

Algebraic equivalence does not have the above property if Y is non-projective:

4.1.3. Example. Let $Z \subset \mathbb{P}^4$ be a cubic hypersurface with a single ordinary node locally given by the equation $xy - uv = 0$. Let $p: X \to Z$ be the blow-up of this node. The exceptional divisor of p is $E \cong \mathbb{P}^1 \times \mathbb{P}^1$. Let $L_1, L_2 \subset E$ be two intersecting lines. The normal bundle of E is

$\mathcal{O}_E(-1, -1)$. Therefore $E \subset X$ can be contracted in any of the fiber directions to get a pair of smooth nonprojective threefolds $f_i \colon X \to X_i$. $f_i(L_i) =$ point but $f_i(L_{3-i})$ is one dimensional. This shows that the $L_i \notin \mathbb{Q}^+_{\text{eff}}[L_{3-i}]$.

However, L_1 and L_2 are algebraically equivalent. This can be seen as follows. The projection of Z from the node exhibits X as the blow-up of $\mathbb{P}^3$ along a curve of type (3,3) on a smooth quadric surface Q. Let $F \subset X$ be the exceptional set of the projection $q \colon X \to \mathbb{P}^3$. The restriction $q \colon E \to Q$ is an isomorphism. If a general line $l \subset \mathbb{P}^3$ degenerates to a line on Q, then in the corresponding limit on X we obtain a line on E and three lines on F. Thus, L_1 and L_2 are algebraically equivalent on X.

It is worthwhile to note that if $Z \subset \mathbb{P}^4$ is a generic hypersurface of degree at least five with a single ordinary node, then the corresponding lines L_i are not algebraically equivalent [10].

4.1.4. Definition. Let X be a proper algebraic space (smooth or possibly with canonical singularities).

(4.1.4.1) A curve $C \subset X$ is said to generate a seemingly extremal ray $R = \mathbb{Q}^+_{\text{eff}}[C]$ if the following conditions hold:

(i) $C \cdot K_X < 0$;

(ii) if $Z_1, Z_2 \in Z_1^+ X$ are two 1-cycles such that

$$Z_1 + Z_2 \in \mathbb{Q}^+_{\text{eff}}[C],$$

then Z_1 is numerically equivalent to a multiple of C.

(4.1.4.2) A morphism $f \colon X \to Y$ is called the contraction associated with R if the following conditions hold:

(i) Y is normal and $f_* \mathcal{O}_X = \mathcal{O}_Y$;

(ii) an irreducible curve $D \subset X$ is mapped to a point by f iff there is a cycle Z_2 such that $D + Z_2 \in \mathbb{Q}^+_{\text{eff}}[C]$. R is called contractible if f exists.

4.1.5. Comments.

(4.1.5.1) If C generates an extremal ray, then C also generates a seemingly extremal ray. This is clear from the definition.

(4.1.5.2) It may be more natural to require in (4.1.4.1) that $Z_1 \in \mathbb{Q}^+_{\text{eff}}[C]$. This is a stronger restriction. I do not know if the two versions are equivalent. At the moment I see some advantages of both.

(4.1.5.3) The definition corresponds to what we observed in the surface case: The only information used is about deformations of multiples of C.

Allowing multiples of C and deformations over several irreducible curves is necessary for threefolds.

(4.1.5.4) My hope is that on a projective threefold a seemingly extremal ray looks very much like an extremal ray and that for nonprojective threefolds this is the correct generalization of the notion of extremal rays. There are several results supporting the first hope. Unfortunately, I know very little about the nonprojective case that supports the second hope.

(4.1.5.5) It seems that in positive characteristic there are more deformations of curves than over $\mathbb{C}$. Therefore it is possible that the above definition needs to be modified over $\mathbb{C}$. In the theorems this is reflected by the fact that over $\mathbb{C}$ we need to assume the existence of certain rational curves whose existence can be proved in positive characteristic.

4.1.6. Proposition. *Let X be a smooth projective surface and let Z_1, $Z_2 \in Z_1^+X$ be 1-cycles. Then the following hold:*

(4.1.6.1) *If* $\dim \operatorname{Alb} X = 0$ *or if* Z_1 *is irreducible and* $Z_1 \cdot Z_1 > 0$, *then*

$$Z_1 \equiv Z_2 \quad \textit{iff} \quad \exists m > 0 \textit{ such that } mZ_1 \overset{\text{eff}}{\approx} mZ_2 .$$

(4.1.6.2) *An irreducible curve $C \subset X$ generates an extremal ray iff it generates a seemingly extremal ray. If $D \in Z_1^+X$, then*

$$[D] \in \mathbb{R}^+[C] \Leftrightarrow D \in \mathbb{Q}^+_{\text{eff}}[C] .$$

Every seemingly extremal ray is contractible.

(4.1.6.3) *The usual cone theorem holds for $\overset{\text{eff}}{\approx}$ in the following form: If H is ample on X and $\varepsilon > 0$ are fixed, then for any $Z \in Z_1^+X$ there is an effective cycle*

$$B + \sum a_i E_i \in \mathbb{Q}^+_{\text{eff}}[Z]$$

such that the E_i are extremal rational curves and $\varepsilon B \cdot H \geq B \cdot (-K_X)$.

Proof. $Z_1 \equiv Z_2$ if $\exists m > 0$ such that $[mZ_1] = [mZ_2]$ as elements in $NS(X)$. Thus if $\dim \operatorname{Alb} X = 0$, then mZ_1 and mZ_2 are linearly equivalent. If Z_1 is irreducible, $Z_1 \cdot Z_1 > 0$, and m is sufficiently large, then every linear system numerically equivalent to mZ_1 is nonempty, thus we can deform mZ_1 into mZ_2.

If $C \subset X$ generates an extremal ray, then by definition it also generates a seemingly extremal ray. Conversely, assume that C generates a seemingly extremal ray. We can write

$$C \equiv B + \sum a_i E_i , \tag{4.1.6.4}$$

where $\varepsilon B \cdot H \geq B \cdot (-K_X)$, and the E_i are extremal rational curves. If $C \cdot C > 0$, then (4.1.6.4) converts into an effective algebraic equivalence by (4.1.6.1).Thus $\mathbb{Q}^+_{\text{eff}}[C] = \mathbb{Q}^+_{\text{eff}}[E_1]$. If $C \cdot C \le 0$, then either C is a (-1)-curve, and we are done, or C is a fiber of a not necessarily minimal ruled fibration and for such curves the claim is clear. This shows (4.1.6.2).

It is sufficient to show (4.1.6.3) for irreducible curves. If $C \cdot C > 0$, then the usual cone theorem implies (4.1.6.3). If C is irreducible and $C \cdot C \le 0$, then the previous argument works.

4.1.7. Example. Even for surfaces $\overset{\text{eff}}{\approx}$ may behave unusually. For instance let E be an elliptic curve, and let L be a nontorsion degree zero line bundle. Let $X = \operatorname{Proj}_E(\mathcal{O} + L)$. Then, X has two sections S_1 and S_2 with self-intersection zero. These sections are algebraically equivalent but $S_i \notin \mathbb{Q}^+_{\text{eff}}[S_{3-i}]$.

For threefolds there are only partial results.

4.1.8. Definition. Assume that $C \subset X$ generates a seemingly extremal ray $R = \mathbb{Q}^+_{\text{eff}}[C]$.

(4.1.8.1) We say that R covers X if for every $x \in X$ there is a 1-cycle $Z_x \in \mathbb{Q}^+_{\text{eff}}[C]$ such that $x \in \operatorname{supp} Z_x$.

(4.1.8.2) We say that R rationally covers X if in addition there is a rational component of Z_x containing x. If x is smooth, then in positive characteristic these notions are equivalent (cf. [13, 1.8]).

4.1.9. Theorem. [36, §4] *Let X be a normal threefold with $\mathbb{Q}$-factorial singularities. Assume that $C \subset X$ generates a seemingly extremal ray R which rationally covers X. Then R is contractible, and we have one of the following cases:*

(4.1.9.1) $\dim N_1(X) = 1$; *in particular, X is Fano.*

(4.1.9.2) $\dim N_1(X) = 2$ *and the contraction $g: X \to Y$ is onto a smooth curve Y. The fibers of g are all irreducible.*

(4.1.9.3) *The contraction $g: X \to Y$ is onto a normal projective surface Y. If the characteristic is different from two, then g is generically a $\mathbb{P}^1$ bundle (in the étale topology). In characteristic two the generic fiber can also be a planar double line.*

Proof. Conditions (4.1) and (4.3) of [56] are clearly satisfied, hence the results follow from [ibid. 4.5-6].

4.1.10. Theorem [56, 1.4]. *Let X be a smooth projective threefold whose Kodaira dimension is nonnegative, i.e., $h^0(X, K_X^{\otimes m}) > 0$ for some $m > 0$. Assume that $C \subset X$ generates a seemingly extremal ray R. In characteristic zero assume in addition that C is rational. Then R is contractible. The collection of all curves in R covers a surface E, and we have one of the following situations:*

(4.1.10.1) *E is a smooth minimal ruled surface with typical fiber C and $C \cdot E = -1$. Only the fibers of E are in R.*

(4.1.10.2) $E \cong \mathbb{P}^2$ *and its normal bundle is* $\mathcal{O}(-1)$.

(4.1.10.3) $E \cong \mathbb{P}^2$ *and its normal bundle is* $\mathcal{O}(-2)$.

(4.1.10.4) $E \cong Q$, *where* Q *is a quadric cone in* $\mathbb{P}^3$, *and its normal bundle is* $\mathcal{O}_{\mathbb{P}^3}(-1)|Q$.

The cone theorem for $\overset{\text{eff}}{\approx}$ is not known for smooth threefolds. The proof of [78, 1.4.1] gives the following:

4.1.11. Theorem [78, 1.4.1]. *Let* X *be a smooth projective variety over a field of positive characteristic. If* H *is ample on* X *and* $\varepsilon > 0$ *are fixed, then for any* $Z \in Z_1^+X$ *there are effective cycles.*

$$A + \sum a_i E_i \in \mathbb{Q}^+_{\text{eff}}[Z] \quad \textit{and} \quad A + B + \sum (a_i + b_i) E_i \in \mathbb{Q}^+_{\text{eff}}[Z]$$

such that the E_i *are rational curves satisfying* $0 < E_i \cdot (-K_X) \le 1 + \dim X$, *and* $\varepsilon B \cdot H \ge B \cdot (-K_X)$.

4.1.12. Remarks. (4.1.12.1) This formulation is stronger than [78, 1.4.1]. Of course $mZ \approx B + \sum b_i E$, but again we restrict the kind of algebraic equivalence that we allow.

(4.1.12.2) Essentially nothing is known about characteristic zero versions of this result.

4.2. How to recognize extremal rays? We consider the following question: Given a curve C which generates a seemingly extremal ray, how do we decide whether C generates an extremal ray? We will always assume that the seemingly extremal ray is one of those described in (4.1.9–10).

The above question has a finer version. Namely, if C does generate an extremal ray we may want to know whether

$$[D] \in \mathbb{R}^+[C] \Leftrightarrow D \in \mathbb{Q}^+_{\text{eff}}[C].$$

If this holds, then we will say the seemingly extremal ray generated by C equals the extremal ray generated by C.

4.2.1. Proposition. *Let* X *be a projective variety. Assume that* $C \subset X$ *generates a contractible seemingly extremal ray* R. *Let* $f\colon X \to Y$ *be the associated contraction. If* Y *is projective, then* C *generates an extremal ray. The converse is also true in characteristic zero if* X *has canonical singularities.*

Proof. Assume that Y is projective, and let H be ample on Y. Let $[D] \in \mathbb{R}^+[C]$. Then $f(D) \cdot H = \text{const} \cdot f(C) \cdot H = 0$, hence $f(D)$ is a point.

Conversely, assume that C also generates an extremal ray of $NE(X)$. The contraction of $\mathbb{R}^+[C]$ (1.4.7) has to coincide with f. In positive characteristic the missing ingredient is the vanishing $R^1 f_* \mathcal{O}_X = 0$.

4.2.2. Corollary. *Let X be a normal projective threefold with isolated $\mathbb{Q}$-factorial singularities. Assume that $C \subset X$ generates a seemingly extremal ray R which rationally covers X. Then $C \subset X$ generates an extremal ray.*

Proof. Let $g: X \to Y$ be the morphism constructed in (4.1.9) ($Y =$ point for (4.1.9.1)). If Y is projective, then $C \subset X$ generates an extremal ray. The projectivity of Y is clear except when $\dim Y = 2$. Y is a normal surface, and every point of Y is finitely dominated by a smooth point. Therefore Y has only $\mathbb{Q}$-factorial singularities, hence Y is projective (4.3.4). q.e.d.

If $\dim N(X) > 1$, then it is probably always true that the seemingly extremal ray generated by C equals the extremal ray generated by C. The analogous question for Fano threefolds raises very interesting questions whose answers are known in special cases only.

4.2.3. Problems. Let X be a smooth Fano variety (i.e., $-K_X$ is ample).

(4.2.3.1) Is the vector space $AE_1(X) = \{\text{1-cycles}\}/\{\text{algebraic equivalence}\}$ finitely generated? The answer is yes in positive characteristic by [78, 1.2].

(4.2.3.2) Assume that $\dim X = 3$ and $\operatorname{Pic} X \cong \mathbb{Z}$. Let $C, D \in X$ be two curves. Is $D \in \mathbb{Q}^+_{\text{eff}}[C]$?

The situation turns out to be more complicated for the seemingly extremal rays described in (4.1.10).

4.2.4. Proposition. *Let $D \subset X$ be a Cartier divisor on a smooth projective variety. Assume that $-K_X|D$ is ample and $-D|D$ is nef and numerically nontrivial. Then there is an extremal ray $R \subset \overline{NE}(X)$ such that if $C \subset X$ is an irreducible curve and $[C] \in R$, then $C \subset D$.*

Proof. Observe that the assumptions imply that D is a (possibly singular) Fano variety. Thus the conditions are very restrictive.

Let $H \subset X$ be an ample divisor. Let

$$t = \max\{s | H + sD \text{ in nef}\},$$

and let

$$F = \{z \in \overline{NE}(X) | z \cdot (H + tD) = 0\}.$$

By construction F is an extremal subset of $\overline{NE}(X)$ such that $K_X|(F - 0)$ is negative. Therefore F contains the class of an effective curve, in particular t is rational. Thus F is an extremal face and contains at least one extremal ray.

4.2.5. Corollary. *Let X be a smooth projective threefold, and let $C \subset X$ generate a seemingly extremal ray R which is one of those described in*

(4.1.10). *If* $E \cong \mathbb{P}^2$ *or* $E \cong Q$ *(the quadric cone in* $\mathbb{P}^3$*), then the seemingly extremal ray generated by* C *equals the extremal ray generated by* C.

In the next subsection we will see an example of a seemingly extremal ray of type (4.1.10.1) which is not an extremal ray. In the positive direction we have the following useful result:

4.2.6. Proposition (Pinkham, unpublished). *Let* X *be a smooth projective threefold, and let* $E \subset X$ *be a smooth minimal ruled surface with typical fiber* C *such that* $C \cdot E = -1$. *In* $N_{E|X}^{-1}$ *is not ample, then the seemingly extremal ray generated by* C *equals the extremal ray generated by* C.

Proof. It is clearly sufficient to show that C generates an extremal ray. By (1.2.4) we can write

$$C \equiv [\lim D_i] + \sum a_i E_i,$$

where the E_i generate extremal rays, and $D_i \cdot K_X \geq 0$. $\sum a_i E_i \neq 0$ since $C \cdot K_X = -1$. $\overline{NE}(E)$ is generated by C and by another element $S \in \overline{NE}(E)$ which may not be represented by a cycle with rational coefficients. $N_{E|X}^{-1}$ is not ample iff $S \cdot E \geq 0$.

Split D_i into two parts $D_i = A_i + B_i$, where A_i is made up of those irreducible components that are contained in E, and B_i is made up of those irreducible components that are not contained in E. Thus we obtain

$$C \equiv aC + bS + [\lim B_i] + \sum a_i E_i,$$

where $a, b \geq 0$. If $a \geq 1$, then

$$0 \equiv (a-1)C + bS + [\lim B_i] + \sum a_i E_i,$$

which is impossible since X is projective. Therefore taking intersection with E gives

$$0 > (1-a)C \cdot E = bS \cdot E + [\lim B_i] \cdot E + \sum a_i (E_i \cdot E).$$

On the right-hand side, $S \cdot E \geq 0$ since $N_{E|X}^{-1}$ is not ample and $[\lim B_i] \cdot E \geq 0$ by construction. Thus there is a curve E_1 such that $E_1 \cdot E < 0$, hence $E_1 \subset E$. Of course E_1 has to generate an extremal ray in $NE(E)$ too. $S \notin \mathbb{R}^+[E_1]$ since $S \cdot E \geq 0$ and $E_1 \cdot E < 0$. Thus $[C] \in \mathbb{R}^+[E_1]$.

4.2.7. Corollary. *Notation as in* (4.2.6). *If* C *does not generate an extremal ray in* X, *then* E *is contractible to a point.*

4.2.8. Corollary. *Let* X *be a proper smooth nonprojective threefold. Let* $C \subset X$ *be a smooth rational curve. Assume that* $B_C X$ *is projective.*

Then

either $C \equiv 0$, $N_{C|X} \cong \mathcal{O}(-1) + \mathcal{O}(-1)$, C *is contractible* $f\colon (C \subset X) \to (0 \in Y)$ *and* Y *is projective,*

or $C \not\equiv 0$, *the opposite* $X \dashrightarrow X^-$ *exists and* X^- *is projective.*

Proof. By (4.2.6) C has negative bundle. Therefore C can be contracted, and the inverse flip $X \dashrightarrow X^-$ exists by (4.3.5). Let H' be a very ample divisor on $B_C X$, and let $H \subset X$ be the image of H'.

If $C \cdot H > 0$, then X is projective by (4.1.1). If $C \equiv 0$, then $C \cdot K_X = 0$, and the adjunction formula gives that $N_{C|X} \cong \mathcal{O}(-1) + \mathcal{O}(-1)$. Applying the Basepoint-Free Theorem (1.4.4) to H gives the morphism $f\colon X \to Y$ onto a projective variety Y with a single ordinary node.

Otherwise $C \cdot M < 0$ for some divisor M and $C \cdot H \leq 0$. Hence, for the inverse flip $C^- \subset X^-$ we have

$$C^- \cdot M^- > 0 \quad \text{and} \quad C^- \cdot H^- \geq 0.$$

Therefore, $mH^- + M^-$ is ample on X^- for $1 \ll m$ (5.1.1.).

4.3. Examples of seemingly extremal rays. (4.1.3) shows that there are seemingly extremal rays which are not equal to an extremal ray. It is not too hard to get similar examples of seemingly extremal rays which do not generate an extremal ray. However, in all known constructions the exceptional surface was rational. Moishezon asked whether this was necessarily the case. The following example answers this question.

4.3.1. Example. There is a smooth proper scheme X of dimension 3 and a smooth curve $C \subset X$ such that X is not projective but $B_C X$ is. C can have arbitrarily high genus.

The construction will be done in two steps with auxiliary lemmas collected at the end.

4.3.2. Step 1. Assume that there are projective varieties U and V of dimension three and a morphism $f\colon U \to V$ satisfying the following properties:

(4.3.2.1) U is smooth.

(4.3.2.2) There are two points $P, Q \in V$ such that f^{-1} is an isomorphism over $V - \{P, Q\}$.

(4.3.2.3) $f^{-1}(P) = C$ is a smooth curve of genus g, and $f^{-1}(Q) = L$ is a smooth rational curve.

(4.3.2.4) The normal bundle of C is negative.

(4.3.2.5) The normal bundle of L is the direct sum of two line bundles $\mathcal{O}(a)$ and $\mathcal{O}(b)$ such that $a < 0$, $b < 0$, and $(a, b) = 1$.

(4.3.2.6) $\mathbb{R}^+[C] = \mathbb{R}^+[L] \subset NE(U)$.

Then one can construct an example as in (4.3.1).

Proof. Let $(L^- \subset U^-)$ be the opposite of $L \subset U$ given by (4.3.5), and let $f^-: U^- \to V$ be the corresponding morphism. $U^- - C$ and $U^- - L^-$ are both quasiprojective, but U^- is not projective since $[C]$ and $[L^-]$ generate a line in $NE(U^-)$. We claim that $B_C U^-$ is projective. To see this it is sufficient to show that $B_C U^-/V$ is relatively projective.

Let $H^- \subset U^-$ be a divisor such that $H^- \cdot L^- > 0$. Let $E \subset B_C U^-$ be the exceptional divisor of the blow-up $\pi: B_C U^- \to U^-$. By assumption (4.3.2.4) the normal bundle of E is negative. If m is sufficiently large, then the divisor

$$-mE + \pi^* H^-$$

is $B_C U^-/V$-ample.

Of course U^- has singularities along L^-. By (4.3.5) these singularities are isolated; let $R \subset U^-$ be the singular set. By (4.3.4), $(U^- - R)$ is not quasiprojective. Let $X \to U^-$ be a relatively projective resolution of these singularities such that $(U^- - R) \subset X$. Then X is not projective but $B_C X$ is.

4.3.3. Step 2. Construction of $f: U \to V$ as in (4.3.2).

Construction. Let us start with a surface $S \subset \mathbb{P}^3$ of degree d_2. Assume that S contains a line L and a smooth planar curve C of degree d_1. Assume that $L \cap C = \varnothing$. The normal bundles are

$$N_{L|S} \cong \mathcal{O}(2 - d_2)|L \quad \text{and} \quad N_{C|S} \cong \mathcal{O}(1 + d_1 - d_2)|C.$$

We can apply (4.3.6) to conclude that there is a morphism $\pi: S \to \overline{S}$ onto a projective surface $\overline{S}$ such that L and C are contracted to points, and π is an isomorphism on $S - (L \cup C)$. Let $\overline{G} \subset \overline{S}$ be a smooth ample curve disjoint from $\pi(L \cup C)$, and let $G = \pi^{-1}\overline{G}$.

Embed $S \subset \mathbb{P}^3 \subset \mathbb{P}^4$ and let $V \in \mathbb{P}^4$ be a sufficiently general point. Let $F \subset \mathbb{P}^4$ be the cone over G with vertex V. Thus $S \cap F$ consists of the curve G and a finite point set $P(S, F)$. Finally, let $H \subset \mathbb{P}^4$ be a sufficiently general hypersurface of degree d_3 containing S and F.

4.3.3.1. Lemma. *Assume that d_3 is sufficiently large. Then H has the following properties*:

(Sing H) *the singularities of H are the following*:

(i) *An isolated singularity at V.*

(ii) *Ordinary double points at $P(S, F)$.*

(iii) *Ordinary double points at some other points of S. We may assume that the set of these—$P(S)$—is disjoint from L and C. In general there will be singular points along G.*

(iv) *Ordinary double points at some other points of F. We may assume that the set of these—$P(F)$—is disjoint from G.*

(Div H) *The group of Weil divisors modulo algebraic equivalence is generated by* S, F, *and* $[\mathscr{O}(1)]$ *over* $\mathbb{Q}$.

Proof. The statements about the singularities are easy.

In order to see (Div H) let us take a general hyperplane $W \subset \mathbb{P}^4$. Then $S \cap W$ and $F \cap W$ are two smooth curves intersecting transversally. By the Noether-Lefschetz theorem the Picard group of a general surface of large degree containing these curves has rank three. On the other hand (cf. [18]),

$$\mathrm{Weil}(H) = \mathrm{Pic}(H - \mathrm{Sing}\, H) \hookrightarrow \mathrm{Pic}(W \cap H) \cong \mathbb{Z}^3.$$

This shows (Div H). q.e.d.

We will resolve the singularities of H in three steps. First we blow up the sheaf $\mathscr{O}(S)$ to obtain $p_1 \colon H_1 \to H$, i.e.,

$$H_1 = \mathrm{Proj}_H \sum_{i=0}^{\infty} \mathscr{O}(iS).$$

$\mathscr{O}(S)$ is Cartier outside $P(S) \cup P(S, F)$, thus p_1^{-1} is an isomorphism outside $P(S) \cup P(S, F)$. Let $S_1 \subset H_1$ be the proper transform of S. By construction $\mathscr{O}(S_1)$ is p_1-ample. Also, $S_1 \cong S$ and

$$N_{S_1|H_1} \cong \mathscr{O}(1 + d_2 - d_3)|S.$$

Next we resolve the singularity at V. This gives $p_2 \colon H_2 \to H_1$. The exceptional set is disjoint from S_1. Let $\mathscr{O}(-E)$ be p_2-ample where E is a suitable p_2-exceptional divisor. Let F_2 be the proper transform of F.

Finally we blow up $\mathscr{O}(F_2)$. This gives $p_3 \colon H_3 \to H_2$ which is an isomorphism outside $P(F)$. Let F_3 be the proper transform of F_2. Let $U = H_3$.

Observe that by construction

$$M = \mathscr{O}(aF_3) \otimes p_3^*\mathscr{O}(-bE) \otimes p_3^*p_2^*\mathscr{O}(cS_1) \otimes p_3^*p_2^*p_1^*\mathscr{O}(d)$$

is very ample on U for suitable $a \ll b \ll c \ll d$. We may also assume that $H^1(U, M) = 0$ and that $d = e(d_3 - d_2 - 1)$ for some e.

If $1 \le j \le e - c - 1$ and $d_2 \ll d_3$, then

$$H^1(S_1, M \otimes \mathscr{O}(jS_1)) = H^1(S, \mathscr{O}_S((e - c - j)(d_3 - d_2 - 1)) \otimes \mathscr{O}_S(aG)) = 0,$$

and

$$M \otimes \mathscr{O}((e - c)S_1)|S_1 \cong \mathscr{O}_S(aG) \quad \text{is generated by global sections.}$$

Therefore the conditions of (4.3.6) are satisfied, and we obtain a contraction map $f \colon U \to V$ which is an isomorphism outside S_1. On S_1 it induces the map given by $\mathcal{O}(aF)$. Thus f contracts the two curves L and C and is an isomorphism elsewhere.

The normal bundles of L and C in U can be computed from the sequences

$$0 \to N_{L|S} \to N_{L|U} \to N_{S_1|U}|L \to 0,$$
$$0 \to N_{C|S} \to N_{C|U} \to N_{S_1|U}|C \to 0.$$

In particular they are both negative and $\deg N_{L|U} = 3 - d_3$. We can choose d_3 in such a way that $d_3 - 3$ is a prime number, therefore condition (4.3.2.5) can also be satisfied.

The Picard group of U is generated (over $\mathbb{Q}$) by the classes $\mathcal{O}(S_1)$, $\mathcal{O}(F_3)$, $p_3^* p_2^* p_1^* \mathcal{O}(d)$ and by p_2-exceptional divisors. The p_2-exceptional divisors are all disjoint from L and C and so is F_3.

$$\mathcal{O}(S_1)|S_1 \cong p_3^* p_2^* p_1^* \mathcal{O}(1 + d_2 - d_3)|S_1,$$

thus $\mathbb{R}^+[C] = \mathbb{R}^+[L] \subset NE(U)$. Therefore all the conditions of (4.3.2) are satisfied for a suitable choice of S and H.

4.3.4. Lemma [51, p. 328, Corollary 3]. *Let W be a normal, proper algebraic space with $\mathbb{Q}$-factorial singularities. Let $T \subset W$ be a finite set. If $W - T$ is quasiprojective, then W is projective.*

4.3.5. Lemma. *Let $L \subset U$ be a smooth rational curve in a smooth threefold. Assume that the normal bundle of L is the direct sum of two line bundles $\mathcal{O}(a)$ and $\mathcal{O}(b)$ such that $a < 0$ and $b < 0$. Then the opposite $(L \subset U) \dashrightarrow (L^- \subset U^-)$ exists in the category of algebraic spaces. If $(a, b) = 1$, then U^- has isolated singularities.*

Proof. The problem is local around L. As in [78, 3.33] we obtain that a suitable neighborhood of L in U is analytically equivalent to a neighborhood of L in the total space of the vector bundle $\mathcal{O}_L(-a) + \mathcal{O}_L(-b)$. The latter admits a torus action, and the existence of the opposite becomes an exercise in toric geometry. See [99] for similar computations.

4.3.6. Theorem (Castelnuovo's contractibility criterion). *Let W be a proper algebraic space and let $S \subset W$ be a Cartier divisor. Let M be a line bundle on W generated by global sections. Let $\phi: W \to W'$ be the Stein factorization of the corresponding morphism. Assume the following:*

(4.3.6.1) $H^1(W, M) = 0$,

(4.3.6.2) $H^1(S, \mathcal{O}_S \otimes M(jS)) = 0$ *for* $1 \le j \le k - 1$, *and*

(4.3.6.3) $\mathcal{O}_S \otimes M(kS)$ *is generated by global sections. Let* cont: $S \to \overline{S}$ *be the Stein factorization of the corresponding morphism.*

Then $M(kS)$ is generated by global sections, and the Stein factorization Cont: $W \to \overline{W}$ *of the corresponding morphism has the following properties*:

(4.3.6.4) $\mathrm{Cont}\,|W-S=\phi|W-S$,

(4.3.6.5) $\mathrm{Cont}\,|S=\mathrm{cont}$, *and*

(4.3.6.6) $\overline{W}$ *is projective if* W *is.*

Proof. One can easily check that only the above conditions are used in the proof given in [34, V.5.7] for the classical case.

4.3.7. Remark. The above example is also interesting from the point of view of projectivization by blowing up. Let X be a proper Moishezon threefold. By [75] there is a sequence of smooth blow-ups such that the end result is projective. By (4.3.4) the last necessary blow-up is the blow-up of a curve. We see that this curve need not be rational.

However, the example also shows that there are rational curves around, namely the proper transform of L^-, and we could also blow up the proper transform of L^- to achieve projectivity. Some considerations suggest that this may always be possible.

4.3.8. Conjecture. Let X be a proper Moishezon threefold. Then there is a sequence of blow-ups centered at points or smooth rational curves such that the resulting threefold is projective.

5. Nonprojective threefolds

The aim of this section is to review and simplify some results about nonprojective threefolds and to ask some questions suggested by flips and flops. We will consider only threefolds that are close to being algebraic, namely compact Moishezon threefolds (possibly with some mild singularities). These are the same as proper algebraic spaces. We start by discussing some ampleness criteria.

5.1. Ampleness and projectivity criteria. All results of this subsection are valid over an arbitrary ground field.

The basic ampleness criteria of Nakai-Moishezon, Seshadri and of Kleiman can be formulated for algebraic spaces too. The proofs given in [51], [33] are for schemes, but they apply to algebraic spaces without modification once the conditions are suitably changed.

5.1.1. Theorem (Nakai-Moishezon criterion [86], [74], [51, III.1]). *Let* Z *be a proper algebraic space and let* H *be a line bundle on* Z. *Then* H *is ample on* Z *iff for every irreducible closed subspace* $X\subset Z$ *the* $\dim X$*-fold self-intersection of* $H|X$ *is positive.*

5.1.2. Theorem (Seshadri criterion [33, 1.7]). *Let* Z *be a proper algebraic space, and let* H *be a line bundle on* Z. *Then* H *is ample on* Z *iff there is a positive constant* ε *such that for every irreducible curve* $C\subset Z$

we have

$$H \cdot C \geq \varepsilon \max_{x \in C}\{\text{mult}_x C\}.$$

5.1.3. Theorem (Kleiman criterion [5, IV.2]). *Let Z be a proper algebraic threefold, and let H be a line bundle on Z. Assume that Z has only $\mathbb{Q}$-factorial singularities. Then H is ample on Z iff H induces a strictly positive linear function on*

$$\overline{NE}(Z) - \{0\},$$

and no irreducible curve $C \subset Z$ is numerically equivalent to 0.

In particular, Z is nonprojective iff either $\overline{NE}(Z)$ contains a line (through the origin) or there is an irreducible curve $C \subset Z$ which is numerically trivial.

Proof. If Z is a proper algebraic space of dimension n with $\mathbb{Q}$-factorial singularities, then the quasidivisoriality condition of [51, IV.2] is satisfied by subspaces of dimensions n and $n-1$. In dimension three this is sufficient to make the proof work. I do not know what happens in higher dimensions.

Peternell noted (unpublished) that for smooth Moishezon n-folds in characteristic zero one can prove Kleiman's criterion using projectivization with a sequence of smooth blowups. q.e.d.

For smooth threefolds (5.1.3) was strengthened by [92] using analytic techniques. We will give a simple algebraic proof of his result which also works in the presence of certain singularities.

5.1.4. Theorem (cf. [92]). *Let Z be a proper algebraic threefold, and assume that Z has only normal $\mathbb{Q}$-factorial singularities. Then Z is nonprojective iff there is an irreducible curve $C \subset Z$ such that*

$$-[C] \in \overline{NE}(Z).$$

Proof. The conditions are sufficient by (5.1.3). To get the converse let $C \subset Z$ be an irreducible curve. If C is numerically trivial, then

$$-[C] = [C] \in \overline{NE}(Z),$$

and we are done. So we may assume that there is no such C, thus by (5.1.3) $\overline{NE}(Z)$ contains a line. This means that there are sequences of effective 1-cycles C_i and D_i such that

$$\lim[C_i] + \lim[D_i] = 0 \quad \text{and} \quad \lim[C_i] \neq 0.$$

By the Chow lemma we can find a projective threefold Z' and a proper birational morphism $f \colon Z' \to Z$. There are only finitely many curves $B_j \subset Z$ such that f^{-1} is not an isomorphism generically along B_j. Let

$B_j'' \subset Z'$ be any irreducible curve such that $f\colon B'' \to B_j$ is finite, and let B_j' be a rational multiple of B_j'' such that $f_*[B_j'] = [B_j]$. For any other curve $B \subset Z$ let $B' \subset Z'$ be the proper transform. Thus for any 1-cycle A in Z we have defined a 1-cycle A' in Z' (with possibly rational coefficients) such that if L is a line bundle on Z, then

$$L \cdot A = f^* L \cdot A'.$$

Take an ample divisor H' on Z' and let $H = f(H')$. Since Z is $\mathbb{Q}$-factorial, some multiple of H is Cartier; we may assume that in fact H is Cartier. Then $f^* H = H' + E$, where $E \subset Z'$ is an effective divisor and $f(E) \subset \bigcup B_j$.

$$0 = H \cdot (\lim[C_i] + \lim[D_i]) = H' \cdot (\lim[C_i'] + \lim[D_i']) + E \cdot (\lim[C_i'] + \lim[D_i']).$$

Also, $\lim[C_i] \neq 0$ hence $\lim[C_i'] \neq 0$ and therefore

$$H' \cdot (\lim[C_i'] + \lim[D_i']) \geq H' \cdot (\lim[C_i']) > 0.$$

Hence

$$E \cdot (\lim[C_i'] + \lim[D_i']) < 0.$$

E is an effective divisor, and the B_j' are the only transforms contained in E. Therefore there is at least one curve—call it B_1—and a positive constant ε such that if a_i is the coefficient of B_1 in $C_i + D_i$, then $a_i > \varepsilon$ holds for infinitely many values of i. Now take $C = B_1$ and then

$$-[C] = \varepsilon^{-1} \lim[C_i + D_i - \varepsilon B_1] \in \overline{NE}(Z). \quad \text{q.e.d.}$$

The following result is interesting because it provides a characterization of projectivity without giving a criterion of ampleness.

5.1.5. Corollary. *Let Z be a proper algebraic threefold, and assume that Z has only normal $\mathbb{Q}$-factorial singularities. Then Z is projective iff there is a line bundle L on Z such that $L \cdot C > 0$ for every irreducible curve $C \subset Z$. (L need not be ample.)*

The following is a very interesting open problem:

5.1.6. Problem. Let X be a smooth proper Moishezon threefold. Assume that X is not projective. Can one find an effective 1-cycle $C \subset X$ such that C is numerically trivial?

5.2. Projectivization with flip or flop.

5.1.2 Let us return to a construction used in the proof of (5.1.4). Let Z be a proper normal $\mathbb{Q}$-factorial algebraic space of dimension three. Let $f\colon Z' \to Z$ be a birational projectivization. Let H' be a very ample divisor on Z' and let $H = f(H')$. We may assume that H is Cartier. Let

$B = \bigcup B_j \subset Z$ be the fundamental set of f^{-1}. Then H is very ample on $Z - B$ and has positive intersection with every curve not contained in B. If $H \cdot B_j > 0$, for every j, then Z is projective by (5.1.5). (In fact it is easy to see that H is ample.) If $H \cdot B_j < 0$, then B_j can be contracted (cf. [55, 4.10]), and we can hope that the opposite (2.1.5) exists. If we are lucky, then after finitely many such curve surgeries we obtain a threefold X^+ and a line bundle H^+ such that H^+ is nef and big. In general X^+ may have fairly complicated singularities. We may hope that some multiple of H^+ is basepoint free, thus X^+ dominates a projective variety. I do not know any example where this cannot be done with suitable choice of H. On the other hand, the procedure should involve inverses of flips, which sometimes do not exist. Therefore I do not want to make any conjecture. The following form of the problem is very interesting:

5.2.2. Problem. Let X be a proper algebraic threefold (smooth or with mild singularities). Can one find a (possibly very singular) projective variety X^+ such that X and X^+ are isomorphic in codimension one? This means that there are subsets $B \subset X$ and $B^+ \subset X^+$ and an isomorphism $X - B \cong X^+ - B^+$ such that $\dim B \leq 1$, $\dim B^+ \leq 1$.

This question is especially interesting for threefolds with K_X nef, since extremal ray theory does not give anything for them. For these threefolds the answer is very satisfactory.

5.2.3. Theorem. *Let X be a proper algebraic threefold with $\mathbb{Q}$-factorial canonical singularities. Assume that K_X is nef. Then after finitely many flops one obtains a proper algebraic threefold X^+ with $\mathbb{Q}$-factorial canonical singularities which is a small modification of a projective variety Y. K_Y is nef and Y has only canonical singularities. Furthermore,*

$$IH^i(X, \mathbb{Q}) \cong IH^i(Y, \mathbb{Q}).$$

The isomorphism preserves Hodge structures but it does not preserve products.

If X is smooth, then X^+ is also smooth, and Y has only terminal hypersurface singularities (1.3.4.2).

Proof. Let H be as in (5.2.1) and let $L = H + mK_X$ for a sufficiently large m. Let $C \subset X$ be an irreducible curve such that $L \cdot C < 0$. By [55, 4.10] C is contractible. Since m is sufficiently large, $K_X \cdot C = 0$. Therefore the flop of C exists [47, 6.10] and any sequence of L-flops terminates [55, 6.2]. Hence after finitely many flops we obtain X^+ and L^+ such that X^+ has $\mathbb{Q}$-factorial canonical singularities, K_{X^+} is nef and L^+ is nef and big. By the Basepoint-Free Theorem (1.4.4) some multiple of $L^+ + K_{X^+}$ is basepoint-free. This gives

$$X \overset{q}{\dashrightarrow} X^+ \overset{p}{\longrightarrow} Y,$$

where q is a composition of flops, and p is the morphism induced by a large multiple of $L^+ + K_{X^+} = H^+ + (m+1)K_{X^+}$. There are only finitely many curves $C^+ \subset X^+$ such that $H^+ \cdot C^+ \leq 0$, thus the same holds for $L^+ + K_{X^+}$. In particular, p contracts only finitely many curves. The statement about the intersection homology groups follows easily from general results [55, 4.12].

5.2.4. Corollary. *Let X be a proper algebraic threefold with $\mathbb{Q}$-factorial canonical singularities. Assume that K_X is nef. If X is not projective, then it contains a smooth contractible rational curve.*

One can weaken (5.2.2) by allowing $B \subset X$ or $B^+ \subset X^+$ to have dimension larger than one. If we require only that $\dim B \leq 1$, then any proper birational morphism $f: X^+ \to X$ provides an example. It is more difficult to find X^+ if we require that $\dim B^+ \leq 1$.

5.2.5. Proposition. *Let X be a proper algebraic threefold with terminal singularities. Assume that X is not uniruled. Then there is a projective variety X^+ and closed subsets $B \subset X$ and $B^+ \subset X^+$ such that $\dim B^+ \leq 1$ and $X - B$ is isomorphic to $X^+ - B^+$.*

Proof. By (1.4.9) there is a projective threefold with terminal singularities X^+ such that K_{X^+} is nef, and X^+ is birational to X. We claim that this is the required example. Let Y be a resolution of the graph of a birational equivalence between X and X^+:

$$X \overset{p}{\longleftarrow} Y \overset{q}{\longrightarrow} X^+.$$

We can write

(5.2.5.1)
$$K_Y = p^* K_X + E_1 + F_1,$$
$$K_Y = q^* K_{X^+} + E_2 + G_2,$$

where the E_i, F_1, G_2 are positive linear combinations of divisors satisfying the following conditions: For an irreducible divisor $B \subset Y$,

if B is a component of E_i, then $\dim p(B) \leq 1$ and $\dim q(B) \leq 1$;

if B is a component of F_1, then $\dim p(B) \leq 1$ and $\dim q(B) = 2$;

if B is a component of G_2, then $\dim p(B) = 2$ and $\dim q(B) \leq 1$.

These conditions determine the divisors E_i, F_1, G_2 uniquely. Moreover, since X has terminal singularities, the support of $E_1 + F_1$ contains all p-exceptional divisors. By (5.2.5.1) we have

(5.2.5.2)
$$p^* K_X = q^* K_{X^+} + G_2 + (E_2 - E_1 - F_1).$$

We need the following easy lemma:

5.2.5.3. Lemma. *Let* $g: U \to V$ *be a proper birational morphism between algebraic spaces. Assume that* U *is smooth and projective. Let* $D_i \subset U$ *be the* g*-exceptional divisors,* L *be a line bundle on* V, M *be a* g*-nef line bundle on* U, *and* $G \subset U$ *be an effective divisor such that none of the* D_i *is a component of* G. *Assume that*

$$g^*L \equiv M + G + \sum d_i D_i .$$

Then $d_i \geq 0$ *for every* i.

Proof. Taking general hyperplane sections of U the problem can be reduced to the case when U is a surface. Then G becomes g-nef and hence can be absorbed into M. Let us write

$$g^*L \equiv M + D^+ - D^- ,$$

where the D^+ (resp. D^-) are nonnegative linear combinations of g-exceptional curves without common components. The intersection matrix of the exceptional curves is negative definite, thus if $D^- \neq 0$, then there is an exceptional curve $C \subset \operatorname{supp} D^-$ such that $C \cdot D^- < 0$. Thus

$$0 = C \cdot f^*L = C \cdot M + C \cdot D^+ - C \cdot D^- > 0 .$$

This is a contradiction. q.e.d.

Applying the lemma to (5.2.5.2) we conclude that $F_1 = \emptyset$. Therefore $B = p(E_1 + E_2 + G_2)$ and $B^+ = q(E_1 + E_2)$ satisfy the requirements. q.e.d.

My main interest in this proposition is that in some cases it answers the following question:

5.2.6. Conjecture. Let X be a proper Moishezon space with terminal singularities, and let $S \subset X$ be a proper subset of codimension at least two. Then there is a proper and irreducible curve $C \subset X$ which is disjoint from S. More generally, there should exist such a curve C through any sufficiently general point of X.

To put this into perspective observe that in Zariski's example of a non-projective surface (blow up 12 general points of a plane cubic and contract the proper transform of the cubic) every curve of the surface passes through the unique singular point.

5.2.7. Proposition. *Let* X *be a Moishezon threefold with terminal singularities. Assume that* X *is not uniruled. Let* $S \subset X$ *be a subset of dimension* 1. *Then through any sufficiently general point* $x \in X$ *there is a smooth irreducible curve* $C \subset X$ *which is disjoint from* S.

Proof. We use the notation of (5.2.5). Let $x \in X - (S \cup B)$. Since X^+ is projective, two general hypersurface sections of X^+ through x

give a smooth curve $C^+ \subset X^+$ which is disjoint from B^+, and from the proper transform of S, $p \circ q^{-1}$ is an isomorphism along C^+, hence $C = p \circ q^{-1}(C^+)$ is a curve with the required properties.

For uniruled manifolds a different argument, communicated to me by J.-M. Hwang, settles the conjecture:

5.2.8. Proposition (Hwang, unpublished). *Let X be a smooth, proper uniruled Moishezon manifold of dimension n, and let $S \subset X$ be a subset of codimension at least 2. Then through any sufficiently general point $x \in X$ there is a rational curve $C \subset X$ disjoint from S.*

Proof. Since X is uniruled, there is a (nonproper) variety Y of dimension $n-1$ and a dominant morphism $g: Y \times \mathbb{P}^1 \to X$. For general $y \in Y$ the map

$$\mathscr{O}_{\mathbb{P}^1}(2) + \underbrace{\mathscr{O}_{\mathbb{P}^1} + \cdots + \mathscr{O}_{\mathbb{P}^1}}_{(n-1)-\text{times}} \cong T_{Y\times\mathbb{P}^1}|\{y\} \times \mathbb{P}^1 \to g^* T_X|\{y\} \times \mathbb{P}^1$$

is generically injective. In particular, $g^* T_X|\{y\} \times \mathbb{P}^1$ is generated by global sections. Let $\mathrm{Hom}(\mathbb{P}^1, X)$ be the space parametrizing morphisms of $\mathbb{P}^1$ to X. Let $U \subset \mathrm{Hom}(\mathbb{P}^1, X)$ be an open neighborhood of $g|\{y\}\times\mathbb{P}^1$ such that for every $f \in U$ the pullback $f^* T_X$ is generated by global sections. As in [77, §1] we obtain that U is smooth of dimension $h^0(\mathbb{P}^1, f^* T_X)$ at f.

For $s \in X$ and $p \in \mathbb{P}^1$ let $U_{s,p} \subset U$ be those morphisms f such that $f(p) = s$. Since $f^* T_X$ is generated by global sections, by [ibid, Proposition 3] we obtain that $U_{s,p}$ is smooth of dimension

$$h^0(\mathbb{P}^1, f^* T_X \otimes \mathscr{O}_{\mathbb{P}^1}(-p)) = h^0(\mathbb{P}^1, f^* T_X) - n.$$

Let $U_S \subset U$ be those morphisms whose image intersects S. Then

$$U_S = \bigcup_{p\in\mathbb{P}^1} \bigcup_{s\in S} U_{s,p}.$$

Thus, $\dim U_S \le \dim U - 1$. Therefore, if $f \in U$ is sufficiently general, then the image of f is disjoint from S.

5.3. Moishezon threefolds with $b_2 = 1$.

5.3.1. Notation. For the rest of this section X will be a smooth Moishezon threefold such that the rank of the Néron-Severi group is one. In particular, this condition is satisfied if $b_2(X) = 1$. We fix a generator $\mathscr{O}(1)$ of the Néron-Severi group such that $\mathscr{O}(k)$ is effective for some $k > 0$. The construction of (5.2.1) gives an effective divisor H which has to be numerically equivalent to a positive multiple of $\mathscr{O}(1)$. Therefore, there are only finitely many irreducible curves $C \subset X$ such that $C \cdot \mathscr{O}(1) \leq 0$.

Note that it is possible that the self-intersection $\mathscr{O}(1)^{(3)}$ of $\mathscr{O}(1)$ is negative.

Up to numerical equivalence we can write

$$K_X \equiv \mathscr{O}(m), \qquad m = m_X \in \mathbb{Z}.$$

We try to classify X according to the sign of m.

5.3.2. Theorem. *Notation as above. Assume that* $m > 0$. *Then there is a morphism*

$$f: X \to \overline{X}$$

onto a projective variety $\overline{X}$ *with at most terminal hypersurface singularities* (1.3.4.2), *and there is an ample line bundle* $\mathscr{O}_{\overline{X}}(1)$ *on* $\overline{X}$ *such that*

$$\mathscr{O}_X(1) \cong f^*(\mathscr{O}_{\overline{X}}(1)).$$

f *is small and the self-intersection of* $\mathscr{O}_X(1)$ *is positive. Furthermore, the group of Weil divisors of* $\overline{X}$ *modulo algebraic equivalence has rank one.*

Conversely, if $\overline{X}$ *is a projective variety with at most terminal hypersurface singularities such that the group of Weil divisors modulo algebraic equivalence has rank one, and* $f: X \to \overline{X}$ *is a small resolution, then* X *is a Moishezon threefold with* $\operatorname{rk} NS(X) = 1$ *which is nonprojective if* f *is not an isomorphism.*

Proof. By (5.3.1) there are only finitely many curves $B_j \subset X$ such that $\mathscr{O}(1) \cdot B_j < 0$. Since $m > 0$, we conclude that there are only finitely many curves $B_j \subset X$ such that $K_X \cdot B_j < 0$. If $K_X \cdot B_j < 0$ then by [19, Lemma 5; 58, 5.1] one can deform B_j inside X. This is impossible, thus $\mathscr{O}(1)$ and K_X are nef. By the Basepoint-Free Theorem (1.4.4), some multiple of $\mathscr{O}(1)$ gives the required morphism onto $\overline{X}$.

5.3.2.1. Remark. If X is allowed to have $\mathbb{Q}$-factorial terminal or canonical singularities, then there can be curves which have negative intersection with K_X. These can be flipped, and after finitely many flips some multiple of H^+ becomes basepoint free.

5.3.2.2. Example. Let $\overline{X}$ be a general hypersurface of degree $k \geq 3$ with an ordinary node given locally by the equation $xy - zt = 0$. Let X be obtained by blowing up (locally in the Euclidean topology) the ideal (x, z). Then $\operatorname{Pic} X \cong \mathbb{Z}$, and X is not projective. $K_X = f^*\mathscr{O}(k-5)$.

5.3.2.3. Corollary. *Let* $g: X \to T$ *be a smooth, proper, holomorphic morphism between complex spaces. Assume that* T *is connected and further that for some* $0 \in T$ *the fiber* X_0 *is a smooth projective threefold with*

$b_2(X_0) = 1$ *such that* $m = m_{X_0} > 0$. *Then every fiber of* g *is projective, in fact* g *is projective.*

Proof. We may assume that T is irreducible. Let us consider the line bundle $\omega_{X/T}$. By assumption $\omega^k_{X/T}$ is very ample on X_0 for $k \gg 0$, hence there is a Zariski open set $0 \in U_k \subset Z$ such that $\omega^k_{X_t}$ is very ample on X_t for $t \in U_k$. By upper semicontinuity of h^0 we conclude that $h^0(X_t, \omega^k_{X_t}) < \text{const} \cdot k^3$ for $k \gg 0$ for every $t \in T$. Thus every fiber of g is Moishezon.

From (5.3.2) and (1.4.2) we see that $h^1(X_t, \mathcal{O}_{X_t}(m+s)) = 0$ for $s > 0$. Thus

$$h^0(X_t, \mathcal{O}_{X_t}(m+s))$$

is independent of $t \in T$ for $s > 0$. As in (5.3.2) there are morphisms $\overline{g}: \overline{X} \to T$ and $f: X \to \overline{X}$ such that

$$\mathcal{O}_X(1) \cong f^*\mathcal{O}_{\overline{X}}(1).$$

We claim that f is an isomorphism. If f is not an isomorphism over a point $\tau \in T$, then we take a general disc $h: \Delta \to T$ through τ and consider the family over Δ induced by base change. Thus we have

$$f_\Delta: X_\Delta \to \overline{X}_\Delta.$$

f_Δ is an isomorphism except over finitely many points of $\overline{X}_\Delta$. The central fiber of $\overline{X}_\Delta$ has only isolated hypersurface singularities, hence $\overline{X}_\Delta$ itself has only isolated hypersurface singularities. The claim follows from the next easy result:

5.3.2.4. Lemma. *Let* $0 \in U$ *be a four-dimensional isolated hypersurface singularity. Let* $f: V \to U$ *be a prop morphism. Assume that* f *is an isomorphism over* $U - 0$ *and that* $f^{-1}(0)$ *is at most one-dimensional. Then* f *is an isomorphism.*

Proof. Let $D \subset V$ be a small three-dimensional disc intersecting $f^{-1}(0)$ at a single point. Then $f(D) \subset U$ is a divisor near 0, Cartier on $U - 0$. By [30, XI.3.1.4] this implies that $f(D)$ is Cartier at 0. Therefore f is an isomorphism (cf. (6.1.2)). q.e.d.

If $m = 0$, then (5.2.3) gives the following:

5.3.3. Theorem. *Let* X *be a smooth Moishezon threefold such that* K_X *is trivial. Then after finitely many flops one obtains a smooth Moishezon threefold* X^+ *which is a small resolution of a projective variety* Y. K_Y *is trivial, and* Y *has only terminal hypersurface singularities. Furthermore*

$$H^i(X, \mathbb{Q}) \cong IH^i(Y, \mathbb{Q}).$$

The isomorphism preserves Hodge structures, but it does not preserve products.

The $m < 0$ case seems the hardest, and there are only partial results.

5.3.4. Theorem [87], [88]. *Notation as above. Assume that* $m < 0$. *Then*

(5.3.4.1) $m \geq -4$;

(5.3.4.2) $m = -4$ *iff* $X \cong \mathbb{P}^3$;

(5.3.4.3) $m = -3$ *iff* $X \cong \mathbb{Q}^3$ (*the smooth quadric in* $\mathbb{P}^4$).

5.3.5. Corollary. *Let* X *be a Moishezon threefold which is homeomorphic to* $\mathbb{P}^3$ (*resp.* $\mathbb{Q}^3$). *Then* X *is isomorphic to* $\mathbb{P}^3$ (*resp.* $\mathbb{Q}^3$).

5.3.6. Remark. (5.3.4) and (5.3.5) are claimed [91]. Unfortunately there is a gap in the proof of [ibid, Lemma 3.5]. Therefore (5.3.4) should be attributed to Nakamura, and (5.3.5) seems new. In a letter (March 1990) Peternell informed me that he was preparing an article containing a proof of (5.3.5). Nakamura [87], [88] already observed that (5.3.4) implies (5.3.5) provided $\kappa(X) < 3$.

Proof of (5.3.5). Let L be the generator of $H^2(X, \mathbb{Z})$ with positive self-intersection. As in the case when X is assumed projective [35], [8] one can easily compute that $K_X = -4L$ (resp. $K_X = -3L$). With the notation of (5.3.1) $L = \mathcal{O}(\pm 1)$. Assume that $L = \mathcal{O}(-1)$. Then $K_X = \mathcal{O}(4)$ (resp. $K_X = \mathcal{O}(3)$) hence by (5.3.2) we get that

$$0 < \mathcal{O}(1)^{(3)} = -L^{(3)} < 0,$$

which is impossible. Therefore $K_X = \mathcal{O}(-4)$ (resp. $K_X = \mathcal{O}(-3)$), and the result follows from (5.3.4).

5.3.7. Remarks. (5.3.7.1) The proof presented for (5.3.4) gives useful information about the $m = -2$ case too. I know very little about the case $m = -1$. Recently Nakamura extended his method to the case $m = -2$. It seems that he will be able to go further than (5.3.12).

(5.3.7.2) The idea of the proof is to study the linear system $|\mathcal{O}(1)|$. We have to show that it is very ample. Nakamura [87], [88] studies the base locus very carefully. Here we study the image of the rational map given by $|\mathcal{O}(1)|$. Then only new ingredients are (5.3.8) and (5.3.11). The other lemmas can all be found in [87], [88].

The first step in the proof of (5.3.4) is the following lemma.

5.3.8. Lemma. *Let* X *be a normal proper n-dimensional algebraic space. Let* M *be a Cartier divisor on* X *which is ample in codimension one* (*i.e., there is a codimension two subset* $Z \subset X$ *such that* $M|X - Z$ *is ample*). *Then*

$$h^{n-1}(X, \mathcal{O}(K_X + M)) = 0.$$

Proof. If r is sufficiently large, then the linear system $|rM|$ is very ample when restricted to $X - Z$. Let $f\colon X' \to X$ be a proper, smooth, birational projectivization such that

$$f^*|rM| = |H'| + \sum a_i E_i,$$

where $|H'|$ is basepoint free and big, and $\sum E_i$ is a divisor with normal crossings only. Then

$$f^*M \equiv \frac{1}{r}H' + \sum \frac{a_i}{r}E_i.$$

Let [ ] denote the integral part of a real number. Then by (1.4.2) and [89, 3.6]

$$h^j\left(X', \mathscr{O}\left(K_{X'} + f^*M - \sum \left[\tfrac{a_i}{r}\right] E_i\right)\right) = 0 \quad \text{for } j > 0,$$

and

$$R^j f_* \mathscr{O}\left(K_{X'} + f^*M - \sum \left[\tfrac{a_i}{r}\right] E_i\right) = 0 \quad \text{for } j > 0.$$

Let

$$F = f_* \mathscr{O}\left(K_{X'} + f^*M - \sum \left[\tfrac{a_i}{r}\right] E_i\right).$$

Then $H^j(X, F) = 0$ for $j > 0$, and we have an injection $i\colon F \to \mathscr{O}(K_X + M)$ which is an isomorphism in codimension one. Thus

$$i\colon 0 = h^{n-1}(X, F) \to h^{n-1}(X, \mathscr{O}(K_X + M))$$

is surjective.

5.3.9. Corollary. *Notation as in* (5.3.1). *Assume that* $m < 0$. *Then*

(5.3.9.1) $\operatorname{Pic} X \cong \mathbb{Z}$;

(5.3.9.2.0) $h^0(X, \mathscr{O}(k)) = 0$ *if* $k <$;

(5.3.9.2.1) $h^1(X, \mathscr{O}(k)) = 0$ *if* $k \leq 0$;

(5.3.9.2.2) $h^2(X, \mathscr{O}(k)) = 0$ *if* $k \geq m$;

(5.3.9.2.3) $h^3(X, \mathscr{O}(k)) = 0$ *if* $k > m$;

(5.3.9.3) $\chi(\mathscr{O}_X) = 1$.

Proof. (5.3.9.2.0) and (5.3.9.2.3) are clear. (5.3.9.2.1) and (5.3.9.2.2) are dual. Since $\mathscr{O}(1)$ is ample in codimension one, (5.3.8) implies (5.3.9.2.2) for $k > m$.

The only remaining vanishing is $h^1(X, \mathscr{O}) = 0$. This will be done by studying the Albanese map. Assume that we have a nontrivial albanese map $\operatorname{alb}\colon X \to \operatorname{Alb}(X)$. $\operatorname{Alb}(X)$ is Moishezon and hence projective. Let L be ample on $\operatorname{Alb}(X)$. Then $\operatorname{alb}^* L$ is a nontrivial line bundle on X

which is trivial on the fibers of alb. Since $\operatorname{rk} NS(X) = 1$, this implies that alb is generically finite. Therefore $h^0(X, \mathscr{O}(m)) = h^0(X, \mathscr{O}(K_X)) > 0$ [116, 10.1], which is impossible.

These results imply that $\chi(\mathscr{O}_X) = 1$ and that $\operatorname{Pic} X$ is discrete. If $\operatorname{Pic} X$ contains torsion, then we can construct an étale cover $\widetilde{X} \to X$. (5.3.8) gives that $h^2(\mathscr{O}_{\widetilde{X}}) = h^3(\mathscr{O}_{\widetilde{X}}) = 0$. On the other hand,

$$1 - h^1(\mathscr{O}_{\widetilde{X}}) = \chi(\mathscr{O}_{\widetilde{X}}) = \deg(\widetilde{X}/X)\chi(\mathscr{O}_X) = \deg(\widetilde{X}/X).$$

Therefore $\operatorname{Pic} X$ is torsion free and has rank one.

5.3.10 Corollary. *Notation as above. Then the following hold:*

(5.3.10.1) $m \geq -4$;

(5.3.10.2) *If* $m = -4$, *then*

$$\chi(X, \mathscr{O}(k)) = \tfrac{1}{6}(k+1)(k+2)(k+3) \quad \textit{and} \quad \mathscr{O}(1)^{(3)} = 1.$$

(5.3.10.3) *If* $m = -3$, *then*

$$\chi(X, \mathscr{O}(k)) = \tfrac{1}{6}(k+1)(k+2)(2k+3) \quad \textit{and} \quad \mathscr{O}(1)^{(3)} = 2.$$

(5.3.10.4) *If* $m = -2$, *then*

$$\chi(X, \mathscr{O}(k)) = \mathscr{O}(1)^{(3)}\tfrac{1}{6}k(k+1)(k+2) + k + 1.$$

Proof. (5.3.9) computes $\chi(X, \mathscr{O}(k))$ for $m \leq k \leq 0$. Since $\chi(X, \mathscr{O}(k))$ is a polynomial of degree at most three, knowing it at four places specifies it exactly. The leading coefficient is obtained from Riemann-Roch. q.e.d.

The following is the main step in the proof:

5.3.11 Lemma. *Notation as above. Let* $s = h^0(X, \mathscr{O}(1))$ *and let*

$$h: X \dashrightarrow Y \subset \mathbb{P}^{s-1}$$

be the induced map. If $m \leq -3$ *or if* $m = -2$ *and* $s \geq 4$, *then* h *is generically finite.*

Proof. Step 0. We will need several times the Del Pezzo-Bertini classification of varieties of minimal degree in $\mathbb{P}^n$ (cf. [20]). The following statements will be needed:

If $Y \subset \mathbb{P}^n$ is a nondegenerate k-fold of degree d, then $d \geq n+1-k$. If equality holds and there is a nonempty open subset $U \subset Y$ such that every hyperplane section of U is irreducible, then either $Y = \mathbb{P}^n$ or Y is a quadric hypersurface of rank at least 4.

Since every member of $|\mathscr{O}_X(1)|$ is irreducible, this implies that if $\deg Y = s - \dim Y$, then Y is either $\mathbb{P}^3$ or $\mathbb{Q}^3$ ($\mathbb{P}^1$ and $\mathbb{P}^2$ are excluded since $s \geq 4$).

Step 1. By (5.3.9–10) we get that $s \geq 4$ if $m = -4$ and $s \geq 5$ if $m = -3$. Thus in all cases $s \geq 4$. $\mathcal{O}(1)$ is the effective generator of $\operatorname{Pic} X$, therefore every member of $|\mathcal{O}(1)|$ is irreducible. Since $\dim |\mathcal{O}(1)| \geq 2$, this implies that $|\mathcal{O}(1)|$ is not composed of a pencil. Thus, $\dim Y \geq 2$.

Step 2. Let $H_1, H_2 \in |\mathcal{O}(1)|$ be two different members, and let $C = H_1 \cap H_2$ be the intersection curve (with scheme structure). Then

(i) $H^0(X, \mathcal{O}(1)) \to H^0(C, \mathcal{O}(1)|C)$ is surjective.

(ii) $h^0(\mathcal{O}_C) = 1$.

(iii)

$$h^1(\mathcal{O}_C) = \begin{cases} 0 & \text{if } m \leq -3, \\ 1 & \text{if } m = -2. \end{cases}$$

(iv)

$$h^1(C, \mathcal{O}(-1)|C) = \begin{cases} 0 & \text{if } m = -4, \\ 1 & \text{if } m = -3, \\ s-2 & \text{if } m = -2. \end{cases}$$

All these statements can be obtained from (5.3.9) by standard diagram chasing.

Step 3. Assume that $\dim Y = 2$. Then $Y \subset \mathbb{P}^{s-1}$ is a nondegenerate surface, and hence has degree $d \geq s - 2$. By Step 0 we have in fact $d > s - 2$. f cannot be a morphism since $A_i \cdot \mathcal{O}(1) = 0$ holds for only finitely many curves A_i.

Every section $s \in H^0(X, \mathcal{O}(1))$ gives a map

$$s(-1) \colon \mathcal{O}_X(-1) \to \mathcal{O}_X.$$

Let $I_B \subset \mathcal{O}_X$ be the ideal generated by the images for all $s \in H^0(X, \mathcal{O}(1))$. Then $B = \operatorname{Spec}(\mathcal{O}_X / I_B) \subset X$ is the scheme theoretic base locus of $|\mathcal{O}(1)|$.

For general H_1 and H_2 the intersection curve decomposes as

$$C = F \cup D_1 \cup \cdots \cup D_d,$$

where D_i are the moving components, and $\operatorname{supp} F = \operatorname{supp} B$. The reduced structure of F is independent of the H_i (for general H_i), but the scheme structure along F may depend continuously on the choice of H_i. The moving curves are parametrized by the points of Y, at least generically. For generic $y \in Y$ the arithmetic genus of the curve D_y and the intersection number $\mathcal{O}(1) \cdot D_y$ are independent of y.

Since

$$1 \geq h^1(\mathcal{O}_C) \geq d \cdot h^1(\mathcal{O}_{D_1}),$$

we conclude that $h^1(\mathcal{O}_{D_i}) = 0$. Thus $D_i \cong \mathbb{P}^1$. Also,

$$s - 2 \geq h^1(C, \mathcal{O}_X(-1)|C) \geq d \cdot h^1(D_1, \mathcal{O}_X(-1)|D_1).$$

Since $d > s - 2$, this implies that $h^1(C, \mathcal{O}_X(-1)|D_1) = 0$, so that

$$\deg \mathcal{O}_X(1)|D_i \leq 1.$$

Let $H \in |\mathcal{O}_X(1)|$ be a divisor not containing D_i. H contains B, and D_i intersects B since C is connected by (ii) of Step 2. Therefore, H and D_i have at least one intersection point along F. Thus, their intersection number is 1, and they intersect transversally at a smooth point of H.

Assume that the intersection points of D_y and B do not depend on the choice of $y \in Y$ for generic y. Then $D_1 \cup \cdots \cup D_d$ is connected, hence

$$\begin{aligned} s - 2 &\geq h^1(\mathcal{O}_X(-1)|C) \geq h^1(\mathcal{O}_X(-1)|D_1 \cup \cdots \cup D_d) \\ &= d - \chi(D_1 \cup \cdots \cup D_d) \geq d - 1. \end{aligned}$$

Therefore, $d = s - 1$ and $\chi(D_1 \cup \cdots \cup D_d) = 1$. This implies that the D_i intersect at a single point p, and the embedding dimension of $D_1 \cup \cdots \cup D_d$ at p is $d \geq s-1$. On the other hand, the general $H \in |\mathcal{O}_X(1)|$ is smooth at p, thus $s-1 \leq d \leq 2$. This contradicts $s \geq 4$. Therefore, the intersection points of D_y and F move with y. Thus, $\dim F = 1$, the D_i are disjoint and they intersect the same irreducible component $E \subset F$.

By the above results, the general $H \in |\mathcal{O}_X(1)|$ is generically smooth along E, thus at the generic point of E the scheme B is contained in a smooth surface. In particular, I_B is a local complete intersection at the generic point of E. Therefore, if $s_1, s_2 \in H^0(X, \mathcal{O}(1))$ are sufficiently general, then the inclusion

$$(s_1(-1), s_2(-1)) \subset I_B$$

is an equality at the generic point of E. This means that I_F and I_B agree generically along E.

A general section h of $\mathcal{O}_X(1)$ gives a map $h: \mathcal{O}_C \to \mathcal{O}_C(1)$. Let

$$G = \operatorname{im}[\mathcal{O}_C \xrightarrow{h} \mathcal{O}_C(1)].$$

G is a subsheaf of $\mathcal{O}_C(1)$ which is generated by a global section. Its image is contained in $I_B \otimes \mathcal{O}_C(1)$, hence it is generically zero along E. On each of the curves D_i it is nontrivial with at least one section. Therefore

$$\begin{aligned} s = h^0(X, \mathcal{O}(1)) &= 2 + h^0(C, \mathcal{O}(1)) \geq 2 + h^0(C, G) \\ &\geq 2 + d \geq 2 + s - 1 = s + 1. \end{aligned}$$

This contradiction shows that Y cannot be a surface.

Proof of (5.3.4). The linear system $|\mathcal{O}_X(1)|$ gives a map

$$f\colon X \dashrightarrow Y_d^3 \subset \mathbb{P}^{s-1},$$

where Y is a threefold of degree d. Two general members $H_1, H_2 \in |\mathcal{O}_X(1)|$ intersect in a curve $C = F \cup D$, where F is the fixed part, and D is the irreducible moving curve.

$$\deg \mathcal{O}_X(1)|D = \deg f \cdot \deg Y + b,$$

where $b = 0$ iff f is a morphism. Since $m \le -3$, $D \cong \mathbb{P}^1$ by (iii) of Step 2. By (iv) of Step 2,

$$\deg f \cdot \deg Y + b - 1 = h^1(D, \mathcal{O}_X(-1)|D) \le h^1(C, \mathcal{O}_X(-1)|C) \le \begin{cases} 0 & \text{if } m = -4, \\ 1 & \text{if } m = -3. \end{cases}$$

For $m = -4$ this implies that $b = 0$, $\deg f = \deg Y = 1$. Thus $Y \cong \mathbb{P}^3$, and $f\colon X \to \mathbb{P}^3$ is a birational morphism. Since $\mathbb{P}^3$ is smooth, the exceptional set is a divisor. Since $\operatorname{Pic} X \cong \mathbb{Z}$, there cannot be any exceptional divisor. Thus, f is an isomorphism.

For $m = -3$ we obtained $s \ge 5$, thus $\deg Y \ge 2$. Therefore, again $b = 0$, and f is a birational morphism onto a quadric in $\mathbb{P}^4$. A singular quadric has reducible hyperplane sections, and these give reducible divisors in $|\mathcal{O}_X(1)|$, a contradiction. Thus Y is smooth, and as before $f\colon X \to \mathbb{Q}^4 \subset \mathbb{P}^4$ is an isomorphism.

5.3.12. Theorem. *Notation as above. Assume that* $m = -2$. *Let* $s = h^0(X, \mathcal{O}(1))$ *and let* $f\colon X \dashrightarrow Y$ *be the map given by* $|\mathcal{O}_X(1)|$. *Then the following hold*:

(5.3.12.1) $s \le 7$.

(5.3.12.2) *If* $s \ge 4$, *then either*

(5.3.12.2.1) f *is a morphism with Stein factorization* $X \to \overline{Y} \to Y$, *and* $\overline{Y}$ *is a Fano variety of index two possibly with terminal hypersurface singularities*; *or*

(5.3.12.2.2) $s = 4$, *and* $f\colon X \dashrightarrow \mathbb{P}^3$ *is birational*; *or*

(5.3.12.2.3) $s = 5$, *and* $f\colon X \dashrightarrow \mathbb{Q}^3$ *is birational.*

Remark. The corresponding Fano varieties are the following:

(i) $s = 4$, $\overline{Y}$ is a double cover of $\mathbb{P}^3$ ramified along a quartic.

(ii) $s = 5$, $\overline{Y} = Y$ is a cubic hypersurface in $\mathbb{P}^4$.

(iii) $s = 6$, $\overline{Y} = Y$ is a complete intersection of quadric hypersurfaces in $\mathbb{P}^5$.

(iv) $s = 7$, $\overline{Y} = Y$ is a 3-fold hyperplane section of the Grassmannian $\mathrm{Gr}(1, 4) \subset \mathbb{P}^9$.

A Fano variety of this latter type is called V_5 (see, e.g. [38, II.1]).

5.3.13. Corollary. *Let X be a Moishezon threefold which is homeomorphic to the Fano variety V_5. Then X is isomorphic to the Fano variety V_5.*

Proof. The Betti numbers of V_5 are $b_1 = b_3 = 0$ and $b_2 = 1$ [38, IV.3.5]. Thus, any Moishezon threefold homeomorphic to V_5 has $\chi(\mathcal{O}) = 1$. Let L be the generator of $H^2(X, \mathbb{Z})$ which satisfies $L^{(3)} = 5$. Let $K_X = -xL$. From Riemann-Roch it follows that

$$5x^3 + 4x = 48\chi(\mathcal{O}) = 48.$$

This has $x = 2$ as the only integral solution. As in the proof of (5.3.5), L is the effective generator, thus $K_X = \mathcal{O}_X(-2)$. By (5.3.10.4)

$$s = h^0(X, \mathcal{O}_X(1)) \geq \chi(X, \mathcal{O}_X(1)) = 7.$$

Therefore, by (5.3.12), $|\mathcal{O}_X(1)|$ is base free and maps birationally onto a Fano variety Y of index two and degree 5 in $\mathbb{P}^6$, possibly with terminal hypersurface singularities. If Y indeed has a singular point y, then projecting from y we obtain a threefold $Z \subset \mathbb{P}^5$ of degree 3. Thus Z has minimal degree, and therefore we obtain a contradiction as in Step 0 of (5.3.11). Thus Y is smooth, and $f\colon X \to Y$ has to be an isomorphism.

Proof of (5.3.12). We use the notation of the proof of (5.3.11). If $|\mathcal{O}_X(1)|$ is basepoint free, then $\overline{Y}$ is a Fano variety of index two. Therefore we need to consider the case when there are base points, i.e., $b > 0$. By (ii) of Step 2, $1 = h^1(\mathcal{O}_C) \geq h^1(\mathcal{O}_D)$.

Assume first that $h^1(\mathcal{O}_D) = 1$. Then

$$\begin{aligned}\deg f \cdot \deg Y + 1 &\leq \deg f \cdot \deg Y + b = h^1(D, \mathcal{O}_X(-1)|D) \\ &\leq h^1(C, \mathcal{O}_X(-1)|C) = s - 2.\end{aligned}$$

Since $\deg Y \geq s - 3$, this implies that $\deg f = 1$ and $\deg Y = s - 3$. If $s \geq 6$, then Y contains reducible hyperplane sections, a contradiction. Thus $Y = \mathbb{P}^3$ or $Y = \mathbb{Q}^3$. As before, the singular quadric threefold is excluded.

Now assume that $h^1(\mathcal{O}_D) = 0$. If $b \geq 2$, then the above argument works. If $b = 1$, then as in the proof of (5.3.11) we obtain that $\mathcal{O}_C(1)$ has a subsheaf isomorphic to a line bundle of degree $\deg f \cdot \deg Y$ on D.

Thus

$$s = 2 + h^0(D, \mathcal{O}(1)|D) \geq 2 + h^0(\mathbb{P}^1, \mathcal{O}(\deg f \cdot \deg Y))$$
$$= \deg f \cdot \deg Y + 3.$$

This is the same inequality as before, and leads to the same numerical possibilities.

5.3.14. Examples. There are infinitely many examples of nonprojective threefolds which behave as in (5.3.12.2.2–3). These were independently discovered by Hironaka, Fujiki, and perhaps others.

In $\mathbb{P}^4$ take a smooth quadric Q and blow up a smooth curve of type $(3, n)$ on Q. The proper transform of the quadric has normal bundle $(-1, 2-n)$, thus it can be contracted in one direction to get a smooth Moishezon threefold. For every $n \geq 4$ this gives an example for (5.3.12.2.2).

Similarly, taking a smooth quadric Q in $\mathbb{Q}^3$ and blowing up a curve of type $(2, n)$ produce examples for (5.3.12.2.3).

In all of these cases one can contract the proper transform of Q in the other direction. This gives a singular projective Fano variety satisfying the requirements of (5.2.2).

6. Deformations of rational surface singularities

In this section we want to explore some questions raised again by codimension two modifications, namely various aspects of small resolutions. Applied to deformations of rational surface singularities we obtain several interesting conjectures and results.

6.1. Small modifications of threefold singularities.

6.1.1. Definition. Let $0 \in X$ be a germ of a three-dimensional normal singularity. A small modification of X is a three-dimensional contractible curve neighborhood $C \subset Y$ together with the contraction map $f: C \subset Y \to X$.

The following easy proposition connects small modifications with the divisor class group $\mathrm{Pic}(X - 0)$ [47, 3.1]

6.1.2. Proposition. (6.1.2.1) *Let $f: Y \to X$ be a small modification as above, and let $D \subset Y$ be an f-ample Cartier divisor. Then the following hold*:

(6.1.2.1.1) *$mf(D) \subset X$ is not Cartier if $m > 0$,*

(6.1.2.1.2) *$f_*\mathcal{O}_Y(mD) = \mathcal{O}_X(mf(D))$ for $m \geq 0$, and*

(6.1.2.1.3) *$\sum_{m=0}^{\infty} \mathcal{O}_X(mf(D))$ is a finitely generated $\mathcal{O}_X$ algebra.*

(6.1.2.2) *Conversely, let* $D' \subset X$ *be a divisor such that*
(6.1.2.2.1) *no multiple of* D' *is Cartier, and*
(6.1.2.2.2) $\sum_{m=0}^{\infty} \mathcal{O}_X(mD')$ *is finitely generated* $\mathcal{O}_X$ *algebra.*
Then $Y = \operatorname{Proj}_X \sum_{m=0}^{\infty} \mathcal{O}_X(mD')$ *and the projection map* $f\colon Y \to X$ *give a small modification of* X.

One can always find an ideal $I \subset \mathcal{O}_X$ which is isomorphic to (D') (as a module), and then the mth symbolic power of I is $I^{(m)} \cong \mathcal{O}(mD')$. For this reason the algebra $\sum_{m=0}^{\infty} \mathcal{O}_X(mD')$ is called the symbolic power algebra of D'.

This gives a purely algebraic approach to finding small modifications: we have to compute the local divisor class group and then decide when is the symbolic power algebra finitely generated. There are several results concerning the first problem.

(6.1.3) Let $0 \in X$ be a three-dimensional isolated singularity. Let S be small sphere around 0, and let $Z = X \cap S$ be the link of X. This is a compact 5-dimensional manifold, independent of S (up to diffeomorphism). Then the first Chern class

$$c_1\colon \operatorname{Pic}(X - 0) \to H^2(Z, \mathbb{Z})$$

is an injection, and an isomorphism if $0 \in X$ is rational [21, 6.1]. This result is very useful, but it is frequently very hard to compute Z and its cohomologies. The simplest case should be isolated hypersurface singularities, but even here there are many unsolved problems.

Let $0 \in X \subset \mathbb{C}^4$ be an isolated hypersurface singularity. By [30, X.3.4] its local Picard group is torsion free, and hence a finitely generated free abelian group. There are some easy ways of recognizing nontrivial elements of $\operatorname{Pic}(X - 0)$. Assume for instance that the equation of X can be written as

$$u(\underline{x})v(\underline{x}) - s(\underline{x})t(\underline{x}) = 0.$$

Then

$$D = (u(\underline{x}) = s(\underline{x}) = 0)$$

is a nontrivial element of $\operatorname{Pic}(X - 0)$. Every time a divisor $D \subset X$ is a complete intersection in $\mathbb{C}^4$, we can write the equation of X in the above form.

6.1.4. Example. If X is defined by a homogeneous cubic form $C(\underline{x}) = 0$, then $\operatorname{Pic}(X - 0)$ has rank 6 and is generated by the (cones over the) lines on the corresponding cubic surface in $\mathbb{P}^3$. In analogy with the case of curves on surfaces in $\mathbb{P}^3$ one can expect that $\operatorname{Pic}(X - 0)$ is generated by simple divisors if X itself is simple. See (2.2.7) for another example.

(6.1.5) Let X be a canonical singularity, and let $p\colon Y \to X$ be a resolution of singularities such that the exceptional divisor $E = E_1 \cup \cdots \cup E_k$ has normal crossings only. Then every E_i is rational or ruled [96, 2.14], hence the groups

$$H^2(E_i, \mathbb{Z}) \cong \operatorname{Pic}(E_i)/\operatorname{Pic}^0(E_i)$$

are readily computable. Thus we can also compute $H^2(Y, \mathbb{Z})$. Since

$$\text{(6.1.5.1)} \qquad \operatorname{rank}(\operatorname{Pic}(X - 0)) = \operatorname{rank}(H^2(Y, \mathbb{Z})) - k,$$

we can compute the rank of $\operatorname{Pic}(X - 0)$. However the computations get very messy even in simple examples, unless some shortcuts are used.

Let $X = (f = 0) \subset \mathbb{C}^4$ be an isolated singularity. Let us assign weights to the variables by $w\colon x_i \mapsto w_i \in \mathbb{Z}$. Let $f = f_d + f_{d+1} + \cdots$ be the w-homogeneous decomposition of f. The coordinate ring of the w-tangent cone

$$T(X) := \mathbb{C}[x_1, x_2, x_3, x_4]/(f_d)$$

is w-homogeneous. Let a_i be the dimension of its w-degree i piece.

6.1.6. Theorem [21, 7.5]. *Notation as above. Assume that* $f_d = 0$ *defines an isolated singularity (i.e.,* f *is semiquasihomogeneous in the sense of* [2, 12.1]). *Let* $N = d - \sum w_i$. *Then*

$$\operatorname{rank}(\operatorname{Pic}(X - 0)) \le a_{N+d} - \sum a_{N+w_i},$$

and equality holds if X *is rational.*

Proof. Let $p\colon Y \to X$ be the w-weighted blow up. The fiber over 0 is the weighted hypersurface $E = (f_d = 0) \subset \mathbb{P}(w_i)$. By assumption Y has only quotient singularities, hence the local divisor class group of every point is torsion. By Hodge theory (see, e.g., [21, §8])

$$\operatorname{rank}(H^2(E, \mathbb{C})) = 1 + a_{N+d} - \sum a_{N+w_i}.$$

Now use (6.1.5.1). q.e.d.

At least for terminal singularities this approach should give a complete description of the local divisor class group.

Very little is known about the finite generation of symbolic power algebras of rank-one sheaves: $\sum_{m=0}^{\infty} \mathcal{O}_X(mD')$. There are some old examples due to [95] that show that in general the above symbolic power algebra is not finitely generated. Recently more examples were found by [16]. In the positive direction the best result so far is the following:

6.1.7. Theorem [98, 2.12], [47, 6.1]. *Let* $x \in X$ *be a three-dimensional canonical singularity. Then the symbolic power algebra of any divisor is finitely generated.*

Comments. In general, different divisors $D' \subset X$ may give rise to the same small modification. This defines an equivalence relation on the divisors, or even on the elements of the vector space $\operatorname{Pic}(X - 0) \otimes \mathbb{Q}$. This equivalence relation gives a polyhedral decomposition of the vector space $\operatorname{Pic}(X - 0) \otimes \mathbb{Q}$ [98, §7]; [47, §6]. For terminal singularities one can relate this decomposition to the Weyl chambers of the corresponding Dynkin diagram, but in the canonical case there is no known relation to reflection groups.

It would be very interesting to extend the above results to more general singularities. A reasonable question seems to be the following.

6.1.8. Problem. Let $0 \in X$ be a three-dimensional singularity. Assume that some hyperplane section $0 \in H \subset X$ is a quotient singularity. Is the symbolic power algebra of any divisor finitely generated? What happens if H is any rational surface singularity?

Among the divisors of a singularity X there is a distinguished one: the canonical divisor K_X. Finite generation of its symbolic power algebra is needed for flipping. There is much more known about this special case. One general result is:

6.1.9. Theorem [61, 3.5.b]. *Let $0 \in X$ be a three-dimensional isolated singularity. Assume that some hyperplane section $0 \in H \subset X$ is a quotient singularity. Then the symbolic power algebra of the canonical divisor,*

$$\sum_{n=0}^{\infty} \mathcal{O}_X(nK_X),$$

is a finitely generated $\mathcal{O}_X$-algebra.

6.2. Deformations of rational surface singularities. (6.1.9) was used in [61, §3] to describe the components of the deformation space of a quotient singularity H in terms of certain partial resolutions of H. It seems that a large part of this correspondence can be extended to deformations of any rational surface singularity H. I circulated informal notes about these problems in the past two years and received very useful comments from T. de Jong, J. Stevens, D. van Straten, and J. Wahl. Several of my original conjectures were thus transformed into theorems. The starting point is the following generalization of (6.1.9):

6.2.1. Conjecture. Let $0 \in H$ be a rational surface singularity, and let $0 \in X$ be the total space of a one-parameter smoothing of $0 \in H$. Then the canonical algebra

$$\sum_{n=0}^{\infty} \mathcal{O}_X(nK_X)$$

is a finitely generated $\mathcal{O}_X$-algebra.

6.2.2. If this is true, then the Proj of the above canonical algebra gives $g: Y \to X$, where g is an isomorphism outside the origin, $g^{-1}(0)$ consists of finitely many curves, some multiple of K_Y is Cartier, and K_Y is g-ample. Let $H' = g^{-1}H$; this is a proper modification of H.

Unfortunately, I can prove essentially nothing about the singularities of Y or H'. For instance, is it true that Y has only rational singularities? Is it true that Y is Gorenstein in codimension two? A positive answer to these questions would mean that in codimension one H' has only double normal crossing points. I will see (6.3.3) that H' is not always normal. This makes the situation more complicated, but it also leads to very interesting examples.

For the rest of this section I will pretend that H' satisfies the following condition:

(*) Reduced, Cohen-Macaulay surface with at worst double normal crossing points in codimension one.

Substantial changes are required if H' does not satisfy these conditions.

A positive answer to the above conjecture gives a possible description of the components of versal deformation spaces of rational singularities. The crucial thing to notice is that locally $H' \subset Y$ is very special.

6.2.3. Definition. Let $0 \in S_0$ be a reduced surface singularity such that $S_0 - 0$ is Gorenstein. Let $0 \in Z$ be the total space of a one-parameter smoothing S_t of S_0. We say that the smoothing is $\mathbb{Q}$-Gorenstein (qG for short) if some multiple of the canonical class of Z is Cartier. (This makes sense since Z is normal.)

In fact, in this case the order of K_Z in $\operatorname{Pic}(Z - 0)$ is the same as the order of K_{S_0} in $\operatorname{Pic}(S_0 - 0)$. This follows from:

6.2.4. Lemma. *Let $0 \in Z$ be a three-dimensional singularity with a hyperplane section $0 \in S \subset Z$. Assume that $Z - S$ is smooth (normal would be sufficient). Let $L \in \operatorname{Pic}(Z - 0)$ have finite order. Then $L|S - 0 \in \operatorname{Pic}(S - 0)$ has the same order. Equivalently, the kernel of the restriction map $\operatorname{Pic}(Z - 0) \to \operatorname{Pic}(S - 0)$ is torsion free.*

Proof. Assume that $L \in \operatorname{Pic}(Z - 0)$ has order $k > 0$ but $L|(S - 0) \cong \mathcal{O}_S|(S - 0)$. Using a nowhere zero section of L^k we construct a k-sheeted cover $p: \overline{Z} \to Z$. $p^{-1}(S)$ is connected since $p^{-1}(0)$ is a single point, but $p^{-1}(S) - p^{-1}(0)$ has k connected components isomorphic to $S - 0$. Such a surface singularity is not smoothable by [101, 3.4].

6.2.5. Proposition (Wahl). *Let $0 \in S_0$ be a surface singularity satisfying* (*). *A smoothing $S_t: t \in \Delta$ is qG iff $c_1(K_{S_t}) \in H^2(S_t, \mathbb{Z})$ is torsion. If*

$S'_t: t \in \Delta$ is another smoothing of $S'_0 \cong S_0$ in the same component of $\operatorname{Def} S_0$, *then S'_t is qG iff S_t is qG.*

Proof. Let $0 \in X$ be the total space of the smoothing. Then by the proof of [64, 5.1] $(X - 0, S_t)$ is 2-connected. Hence $H^2(X - 0, \mathbb{Z}) \to H^2(S_t, \mathbb{Z})$ is an injection and $c_1(K_X) \in H^2(X - 0, \mathbb{Z})$ is torsion iff $c_1(K_{S_t}) \in H^2(S_t, \mathbb{Z})$ is.

Since the condition depends only on the topology of S_t, it depends only on the component of the deformation space and not on the particular smoothing.

6.2.6. Definition. A smoothing component of a versal deformation space of a surface singularity $0 \in S$ satisfying $(*)$ is called a qG component if one (or every) smoothing in it is qG.

A singularity will be called a qG singularity if its deformation space has at least one qG component.

6.2.7. Remark. It would be much nicer to have a functorial definition of a qG-family. The reasonable definition is the following:

Let Z/T be flat, Cohen-Macaulay and of relative dimension two. Assume that the locus where the fibers are not Gorenstein is finite over T. Then Z/T is qG iff $(\omega_{Z/T}^{\otimes M})^{**}$ is locally free for some $m > 0$.

At the moment I have some technical difficulties working with this definition.

It is important to note that very few singularities have qG components. One restriction is given by:

6.2.8. Proposition [64, 5.6-9]. *Let $0 \in S$ be a rational singularity with minimal resolution $p: T \to S$ and let $K_{T/S}$ be the relative canonical divisor, written as a $\mathbb{Q}$-linear combination of the exceptional curves. If S has a qG component, then $K_{T/S} \cdot K_{T/S} \in \mathbb{Z}$.*

Proof. Assume for simplicity that there is a compact surface $\overline{S}$ whose only singular point is $0 \in S$ and that $\overline{S}$ admits a qG smoothing $\overline{S}_t$. Let $\overline{T} \to \overline{S}$ be the minimal resolution. Then

$$\begin{aligned} K_{T/S} \cdot K_{T/S} &= K_{\overline{K}/\overline{S}} \cdot K_{\overline{T}/\overline{S}} = K_{\overline{T}} \cdot K_{\overline{T}} - K_{\overline{S}} \cdot K_{\overline{S}} \\ &= K_{\overline{T}} \cdot K_{\overline{T}} - K_{\overline{S}_t} \cdot K_{\overline{S}_t} \in \mathbb{Z}. \end{aligned}$$

In general one can either prove that such a compactification exists or argue as in [64, 5.7].

6.2.9. Corollary.

(6.2.9.1) *A quotient singularity $\mathbb{C}^2/\mathbb{Z}_n(1, q)$ has a qG component if $n | (q+1)^2$* [64, 5.9].

(6.2.9.2) *Let* $0 \in S$ *be a rational singularity such that the dual graph of its minimal resolution has a curve* C *of self-intersection* $(-n)$ *which is intersected transversally by* k *curves of self-intersection* (-2). *If* S *has a qG component, then* $(2n-k)|2(n-2)^2$.

Proof. The first requires careful computation. The second one is easy from (6.2.8). q.e.d.

Extracting two important properties of H' in (6.2.2) we arrive at the following notion which generalizes the notion of P-resolution introduced in [61, 3.8] for quotient singularities.

6.2.10. Definition.

(6.2.10.1) Let H be a rational singularity, and let $g: H' \to H$ be a proper modification. Assume that H' is normal. H' is called in P-modification of H if

(i) $K_{H'}$ is g-ample, and
(ii) every singularity of H' has a qG component.

(6.2.10.2) Let $\mathrm{Def}^{qG} H'$ denote the subset of $\mathrm{Def}\, H'$ consisting of deformations that induce a qG deformation of each singularity of H'. Thus, up to a smooth factor, $\mathrm{Def}^{qG} H'$ is the product of the local $\mathrm{Def}^{qG}(x, H')$ for all $x \in H'$. By (6.2.5–6), $\mathrm{Def}^{qG} H'$ is the union of some components of $\mathrm{Def}\, H'$.

It is less clear what the correct definition is when H' is not normal.

6.2.10′. Definition. (6.2.10′.1) Let H be a rational singularity, and let $g: H' \to H$ be a proper modification. Assume that H' is Gorenstein outside finitely many points. H' is called a P-modification if

(i) $K_{H'}$ is g-ample,
(ii) $R^1 g_* \mathcal{O}_{H'} = 0$, and
(iii) H' has a smoothing which induces a qG smoothing of each singularity of H'.

(6.2.10′.2) Let $\mathrm{Def}^{qG} H'$ denote the closure of the subset of $\mathrm{Def}\, H'$ consisting of smoothings that induce a qG smoothing of each singularity of H'.

6.2.11. Definition. Let H be a rational singularity. A P-modification $g: H' \to H$ is called weakly rigid if it has no positive-dimensional deformations among P-modifications of H, i.e., if $g_t: H'_t \to H$ is a flat deformation of $g: H' \to H$ over Δ, where H'_t is a P-modification for every t, then $H'_t \cong H'$ for every t.

6.2.12. Proposition. *Let* H *be a normal surface singularity, and let* $g: H' \to H$ *be a normal and proper modification which is dominated by* *the minimal resolution of H. Then H' has no positive-dimensional family of deformations among proper modifications of H.*

In particular, a normal P-modification which is dominated by the minimal resolution is weakly rigid.

Proof. Let $H'_t \to H$ be a flat deformation over Δ, and $\overline{H}_t$ be the minimal resolution of H'_t. By [61, 2.10(i)],

$$K_{\overline{H}_t} \cdot K_{\overline{H}_t} \geq K_{\overline{H}_0} \cdot K_{\overline{H}_0}, \qquad t \in \Delta.$$

On the other hand, $\overline{H}_0$ is the minimal resolution of H, and $\overline{H}_t$ is obtained from the minimal resolution of H by blowing up some points (possibly none). Thus

$$K_{\overline{H}_t} \cdot K_{\overline{H}_t} \leq K_{\overline{H}_0} \cdot K_{\overline{H}_0}, \qquad t \in \Delta.$$

Hence the equality holds, and $\overline{H}_t$ is the minimal resolution of H for every t. By [61, 2.10(ii)] the family H'_t can be resolved simultaneously, and the claim is clear. q.e.d.

In order to formulate a general conjecture, we need a definition.

6.2.13. Definition. Let U/S be a flat family of reduced surfaces with rational singularities only. Define a functor $P-\mathrm{mod}(U/S)$ as follows. Given $p\colon S' \to S$ let

$$P\text{-mod}(U/S)(S') = \left\{ \begin{array}{l} \text{Pairs } (Z'/S', g), \text{ where } Z'/S' \text{ is a qG-family} \\ \text{(6.2.7) and } g\colon Z' \to U \times_S S' \text{ is a proper mor-} \\ \text{phism such that for every } s' \in S' \text{ the restriction} \\ g\colon Z' \times_{S'} \{s'\} \to U \times_S \{p(s')\} \text{ is a P-modification.} \end{array} \right.$$

6.2.14. Conjecture. For every U/S as above, the functor $P-\mathrm{mod}(U/S)$ is represented by a separated algebraic space $\mathbf{P}-\mathrm{mod}(U/S)$ which is proper over S.

Comments. One should keep in mind that the tentative definitions of P-modification and qG-family make the conjecture somewhat vague.

The conjecture is a generalization of the results in [4]. Separatedness should be expected since K is assumed to be relatively ample in (6.2.10).

I expect that the proof of [4] can be modified to prove the existence and separatedness of $\mathbf{P}-\mathrm{mod}(U/S)$. The difficult part should be the establishment of properness.

Let $\mathrm{Def}\,H$ be a versal deformation space of a rational singularity, and let $\mathscr{U}/\mathrm{Def}\,H$ be the universal family. Assume that $\mathbf{P}-\mathrm{mod}(\mathscr{U}/\mathrm{Def}\,H)$ exists and is proper. Then for every component $C \subset \mathrm{Def}\,H$ there is a unique component $\mathbf{P}-\mathrm{mod}(C)$ of $\mathbf{P}-\mathrm{mod}(\mathscr{U}/\mathrm{Def}\,H)$ such that $\mathbf{P}-\mathrm{mod}(C)\to C$ is surjective, proper, and birational.

One can view (6.2.14) as a refinement of a conjecture posed in [119, p. 241.]

Even the existence of $\mathbf{P}-\mathrm{mod}(U/S)$ has interesting consequences. Let $g\colon H'\to H$ be a P-modification of H, and H'_t be a sufficiently general qG-smoothing of H'. As in [4] we can assume that H'_t contains no compact curves for $t\neq 0$. By [118, 1.4] and [60, 11.4] the family H'_t can be contracted to a flat family H_t of deformations of H. Thus, $H'_t\to H_t$ can be induced by a morphism $\rho\colon\Delta\to\mathbf{P}-\mathrm{mod}(\mathscr{U}/\mathrm{Def}\,H)$. Let $C_{H'}$ be a component of $\mathbf{P}-\mathrm{mod}(\mathscr{U}/\mathrm{Def}\,H)$ containing $\rho(\Delta)$.

Let $W\subset\mathrm{Def}\,H$ be the open subset parametrizing smooth deformations of H. It is clear that $\mathbf{P}-\mathrm{mod}(\mathscr{U}|W/W)\cong W$. By construction $C_{H'}$ and $\mathbf{P}-\mathrm{mod}(\mathscr{U}|W/W)$ intersect nontrivially. Therefore, there is an irreducible component C_H of $\mathrm{Def}\,H$ such that the induced morphism $C_{H'}\to C_H$ is an isomorphism over an open set. If H' is weakly rigid, then by the valuation criterion of properness $C_{H'}\to C_H$ is even proper. This yields the following:

6.2.15. Conjecture–Corollary. *Let H be a rational surface singularity. There is a natural injective correspondence*

$$\left\{\begin{matrix}\textit{qG-components of deformations of}\\ \textit{weakly-rigid P-resolutions of } H\end{matrix}\right\}\to\{\textit{components of }\mathrm{Def}\,H\}.$$

This result gives an effective procedure to exhibit irreducible components of $\mathrm{Def}\,H$.

6.2.16. Known cases. Conjectures (6.2.1) and (6.2.14) are proved in the following cases:

(6.2.16.1) If C is the Artin component, then $\mathbf{P}-\mathrm{mod}(C)$ exists [4]. The minimal Du Val resolution is the corresponding P-modification. Thus the conjecture is true if the deformation space has only one component. This is for instance the case for rational double and triple points.

(6.2.16.2) quotient singularities [61, Chapter 3];

(6.2.16.3) quotient of simple elliptic and cusp singularities [61, Chapter 5];

(6.2.16.4) quadruple points [111].

(6.2.16.5) [III, Proposition 6] implies 6.2.15 in many cases.

6.3. Examples. (6.2.15) can be used to exhibit examples of components of deformation spaces. It is especially simple to understand those P-modifications that are dominated by the minimal resolution.

Let H be a rational surface singularity with minimal resolution $f\colon \overline{H} \to H$, and $D_i \subset \overline{H}$ be connected subsets of curves, pairwise disjoint. Contracting D_i to a point for every i one obtains a partial resolution

$$\overline{H} \to H' \xrightarrow{g} H .$$

The following is easy:

6.3.1. Lemma. *$K_{H'}$ is g-ample iff every (-2)-curve in $\overline{H}$ is contained in or intersects $\bigcup D_i$.*

Therefore, if every D_i is a resolution of a qG singularity, then H' is a P-modification. The simplest qG singularity is $\mathbb{C}^2/\mathbb{Z}_2(1, 1)$ whose resolution is a single (4)-curve. Thus we obtain:

6.3.2. Example. Let H be a rational surface singularity with minimal resolution $f\colon H \to H$. Every (-4)-curve in $\overline{H}$ gives rise to a component of the versal deformation space of H.

More complicated P-modifications arise in the following examples which were developed jointly with J. Stevens.

6.3.3. Example. Consider any rational singularity with the following dual resolution graph:

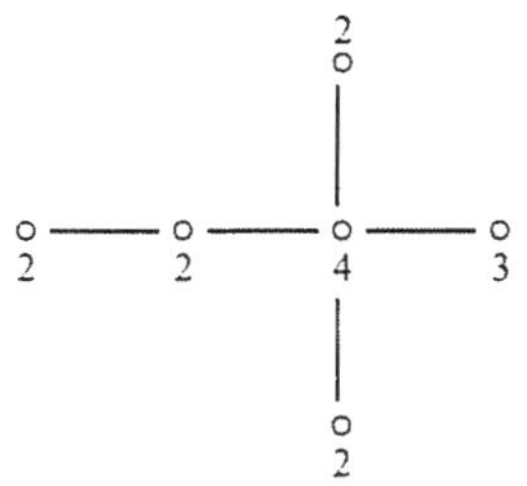

These singularities are rational of multiplicity 5. They have at least one equisingular modulus: the cross ratio of the four curves intersecting the central (-4)-curve.

Two P-modifications are easy to see.

(6.3.3.1) Contract all (-2)-curves. This gives the Artin component.

(6.3.3.2) Contract the (-2) on the left and the (-4)-curve.

There are no other normal P-modifications dominated by the minimal resolution.

(6.3.3.3) There are at least three other normal P-modifications. To obtain these, blow up the intersection point of the (-4)- and the (-3)-curves.

Thus we have

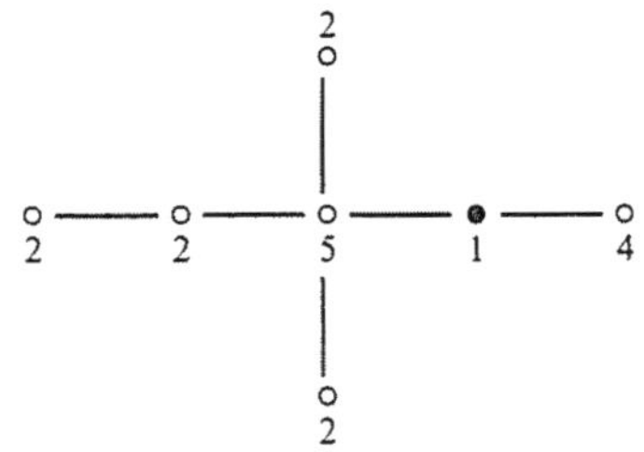

We can contract the (-4)-curve on the right, any of the three

configurations, and the (-2)-curve on the left if it does not intersect any contracted curves. We have to check that after these contractions the canonical class is relatively ample. This is an easy computation.

(6.3.3.4) There is also a nonnormal P-modification. On the central (-4)-curve there are four distinguished points, corresponding to the four intersection points. We denote these intersection points by N, E, S, W corresponding to the directions in the above diagram. There is a unique involution τ of the central (-4)-curve such that $\tau(N) = S$ and $\tau(E) = W$. To get the P-modification, first contract all curves except the central (-4)-curve C. This gives the normal surface $C'' \subset H''$. Then for every $x \in C''$ identify x and $\tau(x)$ (cf. [3, 6.1])) to obtain a nonnormal surface germ $g: C' \subset H' \to H$. Along C' we have generically normal crossings points. There are also two pinch points and two singularities of the form

$$(xy = 0) \subset \mathbb{C}^3/\mathbb{Z}_2(1, -1, 1) \quad \text{and} \quad (xy = 0) \subset \mathbb{C}^3/\mathbb{Z}_3(1, -1, 1).$$

H' has only qG singularities; $xy + t = 0$ is a smoothing at the two non-Gorenstein points. Easy computation gives that

$$C' \cdot K_{H'} = \tfrac{1}{2} C'' \cdot (K_{H''} + C'') = \tfrac{1}{6},$$

hence $K_{H'}$ is g-ample.

We still need to check that H' is qG-smoothable. This is done by using the following technical lemma whose proof we omit:

6.3.4. Lemma. *Let $C' \subset H'$ be the germ of a surface along a smooth curve. Assume that locally along C' the surface H' is one of the following:*

(6.3.4.1) *normal crossing point:* $(xy = 0) \subset \mathbb{C}^3$ *; or*

(6.3.4.2) *pinch point:* $(x^2 - y^2 z = 0) \subset \mathbb{C}^3$ *; or*

(6.3.4.3) *semi-log-terminal point:* $(xy = 0) \subset \mathbb{C}^3/\mathbb{Z}_n(1, -1, a)$, *where* $(a, n) = 1$ *(cf. [61, Chapter 4]). For these singularities the qG deformations are exactly those that can be obtained as* $(xy + tf(z^n, t) = 0) \subset \mathbb{C}^3/\mathbb{Z}_n(1, -1, a, 0)$.

Let the number of pinch points by p, *and* $C'' \subset H''$ *be the normalization of* H''. *Then*

(6.3.4.4) *For* H' *one can define a functor of qG deformations, in particular, the sheaf* $\mathcal{T}^1_{qG}(H') \subset \mathcal{T}^1(H')$. *These sheaves are isomorphic except over a point of type* (6.3.4.3).

(6.3.4.5) $\mathcal{T}^1_{qG}(H')$ *is concentrated along* C', *it has zero dimensional torsion at the pinch points, and the quotient by this torsion is a line bundle* L *on* C'.

(6.3.4.6) $\deg L = C'' \cdot C'' + p$ (*note that* $C'' \cdot C''$ *is automatically an integer by the choice of the singularities* (6.3.4.3)).

Applying the lemma to our situation we get that

$$\mathcal{T}^1_{qG}(H')/(\text{torsion}) \cong \mathcal{O}_{C'}(-4+2+2).$$

Therefore H' has a qG smoothing.

I do not know if these are all the components of $\operatorname{Def} H$ or not.

From the construction it is clear that one can produce similar examples for higher multiplicity rational singularities.

6.3.5. Example. Consider rational singularities of multiplicity 8 such that the minimal resolution has one (-5)-curve C and six (-2)-curves intersecting C transversally.

There are 22 P-modifications dominated by the minimal resolution:

(6.3.5.1) The minimal Du Val resolution; this gives the Artin component.

(6.3.5.2) We can contract any of the six subconfigurations:

This is the dual graph of the quotient singularity $\mathbb{C}^2/\mathbb{Z}_9(1, 2)$.

(6.3.5.3) We can contract any of the 15 subconfigurations:

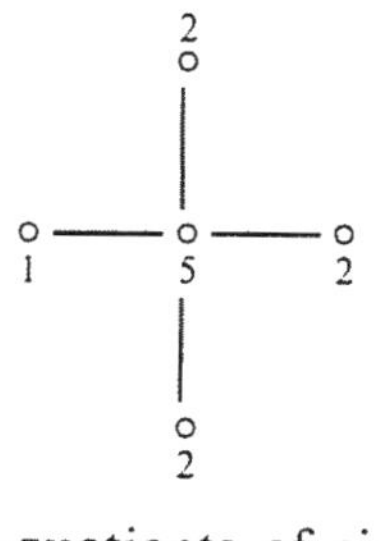

These are dual graphs of $\mathbb{Z}_2$-quotients of simple elliptic singularities of multiplicity 6.

(6.3.5.4) There is no qG component. This follows from (6.2.9.2).

(6.3.5.5) In certain cases there are nonnormal P-modifications. Let $P_1, \cdots, P_6 \in C$ be the six intersection points. Assume that there is an involution $\tau: C \to C$ such that $\tau\{P_1, \cdots, P_6\} = \{P_1, \cdots, P_6\}$, and τ has no fixed points among $\{P_1, \cdots, P_6\}$. Under these assumptions we can repeat the construction in (6.3.3.4) to obtain a nonnormal P-modification.

If we vary the six points generically, then τ will not exist, hence there is no corresponding flat deformation of H'. This shows that the corresponding component does not exist for small generic deformations of the singularity.

Also note that there can be several involutions τ, and these give different P-modifications. To get some examples, choose an isomorphism $C \cong \mathbb{C} \cup \{\infty\}$.

If P_i are the sixth roots of unity, then there are four possible involutions:

$$\tau_0: z \mapsto -z \qquad \tau_i: z \mapsto \zeta^{2i-1}\frac{1}{z} \quad (i = 1, 2, 3),$$

where ζ is a primitive sixth root of unity.

If $\{P_1, \cdots, P_6\} = \{0, \infty, \pm 1, \pm i\}$, then there are six involutions

$$z \mapsto \frac{\pm i}{z}, \quad z \mapsto \eta\frac{z+\eta}{z-\eta}, \quad \text{where } \eta^4 = 1.$$

Again I do not know whether these are all the components of the deformation space.

6.3.6. Example. Consider any rational singularity with the following dual resolution graph:

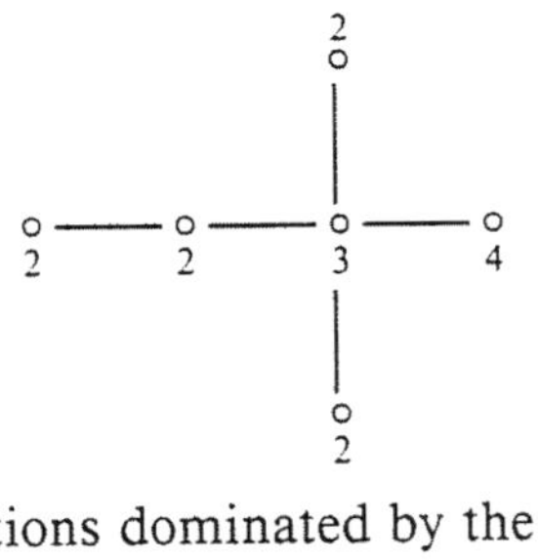

There are five P-modifications dominated by the minimal resolution:

(6.3.6.1) The minimal Du Val resolution.

(6.3.6.2) Contract all (−2)-curves and the (−4)-curve.

(6.3.6.3) Contract any of the three configurations

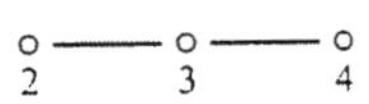

(this is the dual graph of $\mathbb{C}^2/\mathbb{Z}_{18}(1, 5)$) and the (−2)-curve on the left if necessary.

(6.3.6.4) There is also a one-parameter family of P-modifications. To get these, blow up any point x on the (-3)-curve different from the intersection points. Thus we have a (-1)-curve and the following configuration:

(6.3.6.5)

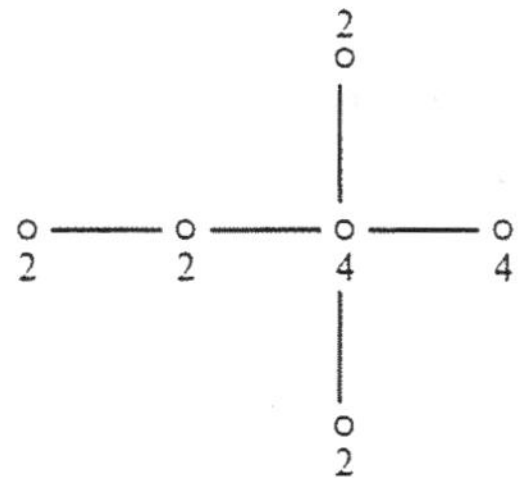

If we contract the above configuration, then we get a proper modification $g: H_x \to H$ with one exceptional curve C_x which is the image of the (-1)-curve. One can compute that

$$K_{H_x} \cdot C_x = \tfrac{1}{5}.$$

The following lemma shows that every H_x is a P-modification.

6.3.7. Lemma. *Any singularity with dual resolution graph as in* (6.3.6.5) *has a qG component.*

Proof. $(x^2z + xy^2 + z^5 + ay^2z^2 = 0) \subset \mathbb{C}^3$ is an elliptic singularity with resolution:

(6.3.7.1)

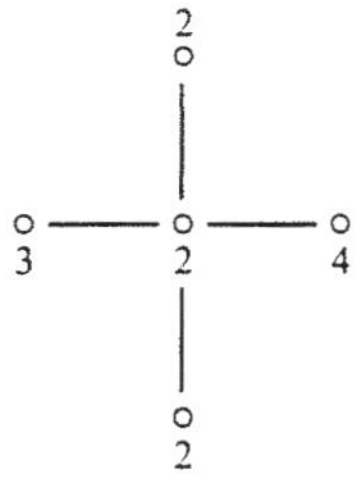

Taking the quotient by the $\mathbb{Z}_5$-action (1,2,3) we obtain a singularity as in (6.3.6.5). Conversely, for any singularity as in (6.3.6.5) one can take the 5-fold cover given by the canonical class, and the cover turns out to be one of the singularities in (6.3.7.1). The required qG smoothing is given by

$$(x^2z + xy^2 + z^5 + ay^2z^2 + t = 0)/\mathbb{Z}_5. \quad \text{q.e.d.}$$

If x degenerates to one of the four intersection points N, E, S or W on the (-3)-curve, the P-modification H_x degenerates into a different type of P-modification. The following are the dual graphs of the minimal resolutions of these four P-modifications; in all cases everything except the

(−1)- curve is to be contracted:

H_N:

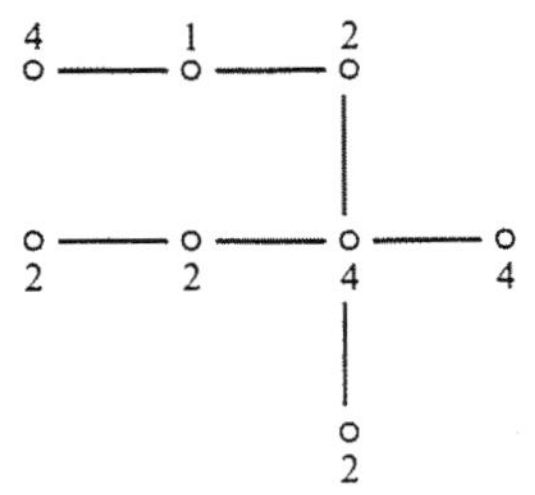

H_S is the same as H_N upside down.

H_W:

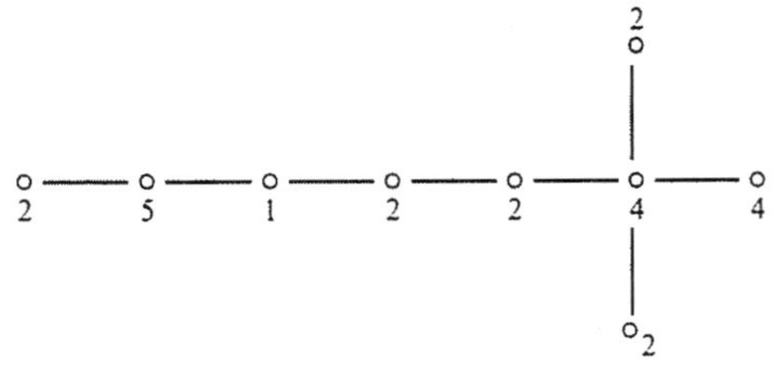

H_E:

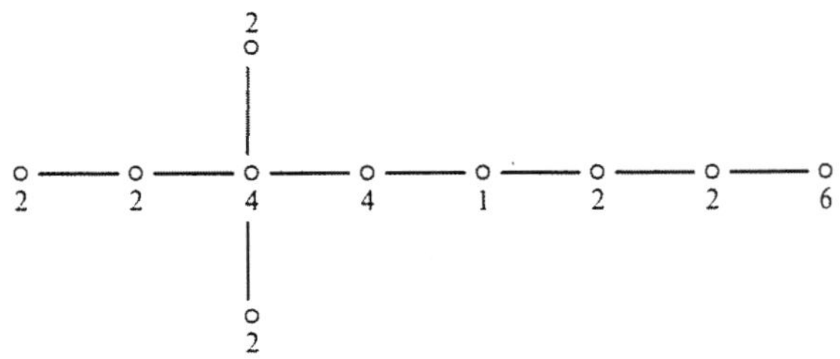

Thus we obtain a flat family of P-modifications parametrized by $\mathbb{P}^1$.

References

[1] V. A. Alekseev, *Rationality condition for three dimensional varieties with a pencil of Del Pezzo surfaces of degree* 4, Math. Notes. **41** (1987) 408–411.

[2] V. I. Arnold, S. M. Gusein-Zade & A. N. Varchenko, *Singularities of differentiable maps*, Birkhäuser, Basel, 1985.

[3] M. Artin, *Algebraisation of formal moduli*. II, Ann. of Math. **91** (1970) 88–135.

[4] ——, *Algebraic construction of Brieskorn resolutions*, J. Algebra **29** (1974) 330–348.

[5] A. Beauville, *Variété de Prym et jacobiennes intermédiaire*, Ann. Sci. École Norm. Sup. **10** (1977) 309–391.

[6] X. Benveniste, *Sur l'anneau canonique de certaines vareétés de dimension* 3, Invent. Math. **73** (1983) 157–164.

[7] J. Bingener, *On the existence of analytic contractions*, Invent Math. **64** (1981) 25–67.

[8] E. Brieskorn, *Ein Satz über die komplexen Quadriken*, Math. Ann. **155** (1964) 184–193.

[9] ——, *Rationale Singularitäten komplexer Flächen*, Invent. Math. **4** (1968) 336–358.

[10] G. Ceresa & A. Collino, *Some remarks on algebraic equivalence of cycles*, Pacific J. Math. **105** (1983) 285–290.

[11] J. Christophersen, *On the components and discriminant of the versal base space of cyclic quotient singularities*, Proc. Warwick Sympos. on Singularity Theory and Appl.

[12] H. Clemens & P. Griffiths, *The intermediate Jacobian of the cubic threefold*, Ann. of Math. **95** (1972) 281–356.

[13] H. Clemens, J. Kollár & S. Mori, *Higher dimensional complex geometry*, Asterisque 166, 1988.

[14] J.-L. Colliot-Thélène, *Arithmétique des variétés rationnelles et problèmes birationnelles*, Proc. Internat. Conf. Math., 1986, 641–653.

[15] S. D. Cutkosky, *Elementary contractions of Gorenstein threefolds*, Math. Ann. **280** (1988) 521–525.

[16] ——, *Weil divisors and symbolic algebras*, Duke Math. J. **57** (1988) 175–183.

[17] V. I. Danilov, *The geometry of toric varieties*, Russian Math. Surveys **33** (1978) 97–154.

[18] P. Deligne, *Le théorème de Noether*, Groupes de Monodromie en Géométrie Algebrique (SGA 7), Lecture Notes in Math. Vol. 340, Springer, Berlin, 1973, 328–340.

[19] L. Ein, *Hilbert scheme of smooth space curves*, Ann. Sci. École Norm. Sup. **19** (1986) 469–478.

[20] D. Eisenbud & J. Harris, *On varieties of minimal degree*, Algebraic Geometry, Bowdoin 1985, Proc. Sympos. Pure Math., Vol. 46, Amer. Math. Soc., Providence, RI, 1987, 3–14.

[21] H. Flenner, *Divisorenclassengruppen quasihomogener Singularitäten*, J. Reine Angew. Math. **328** (1981) 128–160.

[22] A. Fletcher, *Contributions to Riemann-Roch on projective 3-folds with only canonical singularities and applications*, Algebraic Geometry, Bowdoin 1985, Proc. Sympos. Pure Math., Vol. 46, Amer. Math. Soc., Providence, R.I., 1987, 221–232.

[23] T. Fujita, *On L-dimension of coherent sheaves*, J. Fac. Sci. Univ. Tokyo Sect. IA Math. **28** (1981) 215–236; Correction, **29** (1982) 719–720.

[24] ——, *Fractionally logarathmic canonical rings of surfaces*, J. Fac. Sci. Univ. Tokyo **30** (1984), 685–696.

[25] ——, *Zariski decomposition and canonical rings of elliptic threefolds*, J. Math. Soc. Japan **38** (1986) 19–37.

[26] W. Fulton, *Intersection theory*, Springer, Berlin, 1984.

[27] P. J. Giblin, *On the singularities of two related hypersurfaces*, J. London Math. Soc. **7** (1973) 367–375.

[28] J. Giraud, *Desingularization in low dimension*, Resolution of Surface Singularities, Lecture Notes in Math., Vol. 1101, Springer, Berlin, 1984, 51-78.

[29] M. Goresky & R. MacPherson, *On the topology of complex algebraic maps*, Algebraic Geometry, La Rábida, Lecture Notes in Math., Vol. 961, Springer, Berlin, 1982, 119–129.

[30] A. Grothendieck, *Cohomologie local des Faisceaux Cohérant et Théorème de Lefschetz locaux et globaux*, SGA 2, North-Holland, 1968.

[31] M. Hanamura, *On the birational automorphism groups of algebraic varieties*, Compositio. Math. **63** (1987) 123–142.

[32] ——, *Structure of birational automorphism group.* I: *Non-uniruled varieties*, Invent. Math. **93** (1988) 383–403.

[33] R. Hartshorne, *Ample subvarieties of algebraic varieties*, Lecture Notes in Math., Vol. 156, Springer, Berlin, 1970.

[34] ——, *Algebraic geometry*, Springer, Berlin, 1977.

[35] F. Hirzebruch & K. Kodaira, *On the complex projective spaces*, J. Math. Pure. Appl. **36** (1957) 201–216.

[36] V. A. Iskovskikh, *Fano 3-folds*. I, Math. USSR-Izv. **11** (1977) 485–527.
[37] ——, *Fano 3-folds*. II, Math. USSR-Izv. **11** (1978), 469–506.
[38] ——, *Anticanonical models of three-dimensional algebraic varieties*, J. Soviet Math. **13** (1980) 745–814.
[39] ——, *Birational automorphisms of three-dimensional algebraic varieties*, J. Soviet Math. **13** (1980) 815–868.
[40] ——, *On the rationality problem for conic bundles*, Duke Math. J. **54** (1987) 271–294.
[41] V. A. Iskovskikh & Ju. I. Manin, *Three-dimensional quartics and counterexamples to the Lüroth problem*, Math. USSR-Sb. **15** (1971) 141–166.
[42] Y. Kawamata, *A generalisation of Kodaira-Ramanujam's vanishing theorem*, Math. Ann. **261** (1982) 43–46.
[43] ——, *Kodaira dimension of algebraic fiberspaces over curves*, Invent. Math. **66** (1982) 57–71.
[44] ——, *On the finiteness of generators of the pluri-canonical ring for a threefold of general type*, Amer. J. Math. **106** (1984) 1503–1512.
[45] ——, *The cone of curves of algebraic varieties*, Ann. of Math. **119** (1984) 603–633.
[46] ——, *Pluricanonical systems on minimal algebraic varieties*, Invent. Math. **79** (1985) 567–588.
[47] ——, *The crepant blowing-up of 3-dimensional canonical singularities and its application to the degeneration of surfaces*, Ann. of Math. **127** (1988) 93–163.
[48] ——, *Boundedness of $\mathbb{Q}$-Fano threefolds* (to appear).
[49] Y. Kawamata, K. Matsuda & K. Matsuki, *Introduction to the minimal model problem*, Algebraic Geometry, Sendai, Advanced Studies in Pure Math., Vol. 10 (T. Oda, ed.), Kinokuniya & North-Holland, Amsterdam, 1987, 283–360.
[50] G. Kempf, F. Knudsen, D. Mumford & B. Saint-Donat, *Toroidal embeddings*. I, Lecture Notes in Math., Vol. 339, Springer, Berlin, 1973.
[51] S. Kleiman, *Toward a numerical theory of ampleness*, Ann. of Math. **84** (1966) 293–344.
[52] J. Kollár, *Fano varieties of large index*, Vestnik Beloruss Gos. Univ. Ser. I. (1981), 31–34. (Russian)
[53] ——, *The cone theorem*, Ann. of Math. **120** (1984) 1–5.
[54] ——, *The structure of algebraic threefolds—an introduction to Mori's program*, Bull. Amer. Math. Soc. **17** (1987) 211–273.
[55] ——, *Flops*, Nagoya Math. J. **113** (1989) 14–36.
[56] ——, *Extremal rays on smooth threefolds*, Ann. Sci. École Norm. Sup. (to appear).
[57] ——, *Minimal models of algebraic threefolds: Mori's program*, Asterisque **177–178** (1990) 303–326.
[58] ——, *Projectivity of complete moduli*, J. Differential Geometry **32** (1990) 235–268.
[59] J. Kollár & T. Matsusaka, *Riemann-Roch type inequalities*, Amer. J. Math. **105** (1983) 229–252.
[60] J. Kollár & S. Mori (in preparation).
[61] J. Kollár & N. Shepherd-Barron, *Threefolds and deformations of surface singularities*, Invent. Math. **91** (1988) 299–338.
[62] V. Kulikov, *Degenerations of $K3$ surfaces and Enriques surfaces*, Math. USSR-Izv. **11** (1977) 957–989.
[63] H. Laufer, *Weak simultaneous resolution for deformations of Gorenstein surface singularities*, Singularities-Part 2, Proc. Sympos. Pure Math., Vol 40, Amer. Math. Soc., Providence, RI, 1983, 1–27.
[64] E. Looijenga & J. Wahl, *Quadratic functions and smoothings of surface singularities*, Topology **25** (1986) 261–291.
[65] T. Matsusaka, *Polarised varieties with a given Hilbert polynomial*, Amer. J. Math. **94** (1972) 1027–1077.

[66] ——, *On polarized normal varieties.* I, Nagoya Math. J. **104** (1986) 175–211.
[67] J. Milnor, *Singular points of complex hypersurfaces*, Princeton Univ. Press, 1968.
[68] Y. Miyaoka, *Algebraic surfaces with positive indices*, Classification of Algebraic and Analytic Manifolds, Birkhäuser, Basel, 1983, 281–302.
[69] ——, *The Chern classes of Kodaira dimension of a minimal variety*, Algebraic Geometry, Sendai, Advanced Studies in Pure Math., Vol. 10 (I. Oda, ed.), Kinokuniya & North Holland, Amsterdam, 1987, 449–476.
[70] ——, *Abundance conjecture for threefolds: $\nu = 1$ case*, Compositio Math. **68** (1988) 203–220.
[71] ——, *On the Kodaira dimension of minimal threefolds*, Math. Ann. **281** (1988) 325–332.
[72] ——, (in preparation).
[73] Y. Miyaoka & S. Mori, *A numerical criterion for uniruledness*, Ann. of Math. **124** (1986) 65–69.
[74] B. Moishezon, *A criterion for projectivity of complete abstract algebraic varieties*, Amer. Math. Soc. Transl. **63** (1967) 1–50.
[75] ——, *On n-dimensional compact varieties with n algebraically independent meromorphic functions*, Amer. Math. Soc. Transl. **63** (1967), 51–177.
[76] B. Moishezon & M. Teicher, *Simply-connected algebraic surfaces of positive index*, Invent. Math. **89** (1987) 601–643.
[77] S. Mori, *Projective manifolds with ample tangent bundles*, Ann. of Math. **110** (1979) 593–606.
[78] ——, *Threefolds whose canonical bundles are not numerically effective*, Ann. of Math. **116** 91982) 133–176.
[79] ——, *On 3-dimensional terminal singularities*, Nagoya Math. J. **98** (1985) 43–66.
[80] ——, *Minimal models for semistable degenerations of surfaces*, Lectures at Columbia University, 1985 (unpublished).
[81] ——, *Classification of higher-dimensional varieties*, Algebraic Geometry, Bowdoin 1985, Proc. Sympos. Pure Math., Vol. 46, Amer. Math. Soc., Providence, RI, 1987, 269–332.
[82] ——, *Flip theorem and the existence of minimal models for 3-folds*, J. Amer. Math. Soc. **1** (1988) 117–253.
[83] S. Mori & S. Mukai, *Classification of Fano threefolds with $B_2 \geq 2$*, Manuscripta Math. **36** (1981) 147–162.
[84] ——, *On Fano threefolds with $B_2 \geq 2$*, Algebraic Varieties and Analytic Varieties, Advanced Studies in Pure Math., Vol. 1 (S. Iitaka, ed.), Kinokuniya & North-Holland, Amsterdam, 1983, 101–129.
[85] D. Morrison & G. Stevens, *Terminal quotient singularities in dimension three and four*, Proc. Amer. Math. Soc., **90** (1984) 15–20.
[86] Y. Nakai, *A criterion of ample sheaf on a projective scheme*, Amer. J. Math. **85** (1963) 14–26.
[87] I. Nakamura, *Moishezon threefolds homeomorphism to $\mathbb{P}^3$*, J. Math. Soc. Japan **39** (1987) 521–535.
[88] ——, *Threefolds homeomorphic to a hyperquadric in $\mathbb{P}^4$*, Algebraic geometry and commutative algebra in honor of M. Nagata, 1987, 379–404.
[89] N. Nakayama, *The lower semi-continuity of the plurigenera of complex varieties*, Algebraic Geometry, Sendai, Advanced Studies in Pure Math., Vol. 10 (T. Oda, ed.), Kinokuniya, North-Holland, 1987, 551–590.
[90] T. Peternell, *A rigidity theorem for $\mathbb{P}_3(\mathbb{C})$*, Manuscripta Math. **50** (1985) 397–428.
[91] ——, *Algebraic structures on certain 3-folds*, Math. Ann. **274** (1986) 133–156.
[92] ——, *Algebraicity criteria for compact complex manifolds*, Math. Ann. **275** (1986) 653–672.

[93] ——, *Rational curves on a Moishezon 3-fold*, Complex Analysis and Algebraic Geometry, Lecture Notes in Math., Vol. 1194, Springer, Berlin, 1986, 133–144.

[94] V. V. Pukhlikov, *Birational isomorphisms of four dimensional quintics*, Invent. Math. **87** (1987) 303–329.

[95] D. Rees, *On a problem of Zariski*, Illinois J. Math. **2** (1958) 145–149.

[96] M. Reid, *Canonical threefolds*, Géometrie Algébrique Angers (A. Beauville, ed.), Sijthoff & Noordhoof, 1980, 273–310.

[97] ——, *Projective morphisms according to Kawamata*, preprint, Univ. of Warwick, 1983.

[98] ——, *Minimal models of canonical threefolds*, Algebraic Varieties and Analytic Varieties, Advanced Studies in Pure Math., Vol. 1 (S. Iitaka, ed.), Kinokuniya & North-Holland, Amsterdam, 1983, 131–180.

[99] ——, *Decomposition of Toric Morphisms*, Arithmetic and Geometry II, Progress in Math. 36, Birkhäuser, Basel, 1983, 395–418.

[100] ——, *Young person's guide to canonical singularities*, Algebraic Geometry, Bowdoin 1985, Proc. Sympos. Pure Math., Vol. 46, Amer. Math. Soc., Providence, RI, 1987, 345–416.

[101] D. S. Rim, *Torsion differentials and deformations*, Trans. Amer. Math. Soc. **169** (1972) 257–278.

[102] V. G. Sarkisov, *Birational automorphisms of conic bundles*, Math. USSR-Izv. **17** (1981) 177–202.

[103] ——, *On the structure of conic bundles*, Math. USSR-Izv. **120** (1982) 355–390.

[104] N. Shepherd-Barron, *Some questions on singularities in 2 and 3 dimensions*, Thesis, Univ. of Warwick.

[105] V. V. Shokurov, *The nonvanishing theorem*, Izv. Akad. Nauk SSSR Ser. Mat. **49** (1985) 635–651 =Math. USSR-Izv. **19** (1985) 591–604.

[106] ——, Letter to M. Reid, 1985.

[107] Y.-T. Siu & S. T. Yau, *Compact Kähler manifolds of positive bisectional curvature*, Invent. Math. **59** (1980) 189–204.

[108] J. H. M. Steenbrink, *Mixed Hodge structures associated with isolated singularities*, Singularities-Part 2, Proc. Sympos. Pure Math., Vol. 40, Amer. Math. Soc., 1983, 513–536.

[109] J. Stevens, *On canonical singularities as total spaces of deformations*, Abh. Math. Sem. Univ. Hamburg **58** (1988) 275–283.

[110] ——, *On the versal deformations of cyclic quotient singularities*, Proc. Warwick Sympos. Singularity Theory and Appl.

[111] ——, *On rational quadruple points*, Internat. J. Math. (to appear).

[112] H. Tsuji, *Stability of tangent bundle of minimal algebraic varieties*, Topology **27** (1989) 429–442.

[113] ——, *Boundedness of the degree of Fano manifolds with* $b_2 = 1$, preprint, 1990.

[114] S. Tsunoda, *Degenerations of surfaces*, Algebraic Geometry, Sendai, Advanced Studies in Math., Vol. 10 (T. Oda, ed.), Kinokuniya & North-Holland, Amsterdam, 1987, 755–764.

[115] K. Ueno, *Classification of algebraic varieties*. I, Compositio Math. **27** (1973) 277–342.

[116] ——, *Classification theory of algebraic varieties and compact complex spaces*, Lecture Notes in Math., Vol. 439, Springer, Berlin, 1975.

[117] E. Viehweg, *Vanishing theorems*, J. Reine Angew. Math. **335** (1982) 1–8.

[118] J. Wahl, *Equisingular deformations of normal surface singularities*, Ann. of Math. **104** (1976) 325–356.

[119] ——, *Smoothings of normal surface singularities*, Topology **20** (1981) 219–246.

[120] P. M. H. Wilson, *Toward a birational classification of algebraic varieties*, Bull. London Math. Soc. **19** (1987) 1–48.

[121] S. T. Yau, *On the Ricci curvature of a compact Kähler manifold and the complex Monge-Ampère equation.* I, Comm. Pure Appl. Math. **31** (1978) 339–411.

[122] ——, *Problem session*, Seminar on Differential Geometry, Annals. of Math. Studies, No. 102, Princeton Univ. Press, 1982, 669–706.

[123] O. Zariski, *The theorem of Riemann-Roch for high multiples of an effective divisor on an algebraic surface*, Ann. of Math. **76** (1962), 560–615.

UNIVERSITY OF UTAH

SURVEYS IN DIFFERENTIAL GEOMETRY
1 (1991) 201-241

A REPORT ON SOME RECENT PROGRESS ON NONLINEAR PROBLEMS IN GEOMETRY

RICHARD M. SCHOEN

Introduction

In this paper we describe some currently active research areas in the field of geometric partial differential equations. Our initial intention was to write a more comprehensive survey of progress in this general field, but in the end we have chosen a few specific topics of the many more we could have included in a survey paper. The topics which are presented here were chosen primarily because they are of current interest to the author, and in the case of several of them there is very recent progress to be reported. Thus this paper is more of a report on selected currently active research than any attempt at a survey of the field.

In the first section we discuss recent (unpublished) results on conformal deformation of Riemannian metrics to constant (or prescribed) scalar curvature. The major stress here is on large solutions rather than minima for the variational problem. In particular, we include a description of the author's results on the strong Morse inequalities for the Yamabe problem. We also include a brief discussion of a related result (joint with D. Zhang) for the prescribed curvature problem on the 3-sphere S^3 as well as a brief description of recent results on the boundary value problem. Finally we describe progress on complete constant scalar curvature metrics.

In §2 we introduce the Ricci flow equation of R. Hamilton and describe some of the current problems and recent progress on this important nonlinear evolution equation.

In §3 we discuss harmonic maps and some applications. This includes a description of a result of M. Micallef and J. D. Moore on manifolds of positive curvature. We also describe (and slightly amplify) recent work of F. Helein on regularity of weakly harmonic maps into special target manifolds. Finally we give a systematic description of our recent work on the behavior of harmonic maps from manifolds into nonpositively curved Riemannian simplicial complexes. We also briefly describe the application

Received November 19, 1990. Research supported by National Science Foundation Grant DMS89-03076.

of this (joint with M. Gromov) to proving the arithmeticity of lattices in certain rank one Lie groups.

Finally, in §4 we describe the current status of existence and compactness questions for closed and complete constant (nonzero) mean curvature surfaces in three-dimensional Euclidean space. This includes a discussion of recent work of N. Kapouleas (as yet unpublished) which incorporates certain tori of H. Wente into his earlier constructions to produce closed surfaces of constant mean curvature for any genus.

1. Scalar curvature

One of the basic facts about the geometry of surfaces is the uniformization theorem which tells us that every Riemannian metric on a surface is conformally related to a metric of constant curvature. If this metric is geodesically complete and is normalized to have curvature ± 1, 0, it is then unique (in every case except the two-sphere which has a three-parameter family of curvature one metrics in the conformal class of the standard metric). For general manifolds M of dimension three or more there is no known method to construct canonical Riemannian metrics. The most natural determined system of equations for a Riemannian metric is the Einstein system $\mathrm{Ric}(g) = cg$, where $\mathrm{Ric}(g)$ denotes the Ricci tensor of g, and c is a constant necessarily given by $c = n^{-1}R$, $n = \dim M$ and $R =$ scalar curvature of g. The Ricci operator $g \to \mathrm{Ric}(g)$ plays the role of a nonlinear Laplace operator which commutes with the action of the diffeomorphism group on the space of metrics; for a recent account of many aspects of the Einstein equations see [7]. The Einstein equations arise as the Euler-Lagrange equations for a variational integral on the space of metrics. This integral is the Einstein-Hilbert energy which assigns to a Riemannian metric g the value $\mathbf{R}(g) = \int_M R_g \, d\mu_g$, where $d\mu_g$ denotes the Riemannian volume measure of g. The Einstein equations are then the differential equations satisfied by a stationary point of $R(\cdot)$ on the space of unit volume metrics. Analysis of the second variation of $R(\cdot)$ at an Einstein metric g_0 shows that for deformations of g_0 which preserve its conformal class, g_0 is a local minimum modulo a finite-dimensional index space, say of dimension I_c. On the other hand, for deformations which are orthogonal (in the L^2 sense) to both the conformal class and the diffeomorphism group, g_0 is a local maximum modulo a finite-dimensional index space, say of dimension I_T. A theorem of M. Obata (see [45] for detailed discussion) shows that in fact g_0 achieves the minimum of $R(\cdot)$

in its conformal class, and therefore $I_c = 0$. For constant curvature metrics it is easy to show that $I_T = 0$, and it appears that most Einstein metrics which can presently be constructed by analytic means (such as Kähler-Einstein metrics) have $I_T = 0$. A detailed discussion of the total scalar curvature variational problem is given in [52].

The variational problem obtained by restricting $R(\cdot)$ to unit volume Riemannian metrics in a given conformal class is called the Yamabe problem. The Euler-Lagrange equation for the Yamabe problem is the equation $R_g \equiv C$ for a constant $C = \int_M R_g \, d\mu_g$. The first attempt to produce a minimizer for this problem was given by Yamabe [67]. It was pointed out by N. Trudinger [62] that Yamabe's proof contained a serious gap, and this gap was filled for nonpositive conformal classes; i.e., conformal classes containing a metric with everywhere nonpositive scalar curvature. Several years later T. Aubin [4] proved the first result in the positive case. He showed that for $n \geq 6$ and for a nonconformally flat conformal class of metrics the problem has a minimizer. Aubin's method was local, exploiting the local conformal deviation of the given metric from the Euclidean metric. In conformally flat cases the Yamabe problem is clearly a global problem, and in the delicate low-dimensional cases where the equation contains relatively more information the problem is also global. The existence of a minimizer was finally shown in [51]. Explicit examples are given in [52] for which there are arbitrarily many solutions of the Yamabe problem with arbitrarily large energy; i.e., R_g large and unit volume. It is natural to ask whether there is a bound on the energy of solutions in a given conformal class, and whether in any sense the solutions obey a Morse theory as they would if the variational integral satisfied the Palais-Smale condition. We now describe our progress on these questions.

Suppose we fix a background matrix g_0 in our conformal class, and write our unknown metric in the form $g = u^{4/(n-2)} g_0$. It is then well known (see [52]) that the equation of constant scalar curvature becomes

$$Lu + cu^{(n+2)/(n-2)} = 0, \qquad u > 0, \tag{1.1}$$

where $Lu = \Delta_{g_0} u - c(n) R_{g_0} u$, $c(n) = \frac{n-2}{4(n-1)}$. The variational problem is the problem of extremizing $E(u) = -\int_M uLu \, d\mu_{g_0}$ subject to the constraint $\mathrm{Vol}(g) = \int_M u^{2n/(n-2)} \, d\mu_{g_0} = 1$, thus leading to the Euler-Lagrange equation (1.1). We introduce the so-called subcritical regularization of this problem by embedding it in a family of problems indexed by an exponent $p \in [1, \frac{n+2}{n-2}]$. These are the equations

$$Lu + cu^p = 0, \qquad u > 0. \tag{1.1_p}$$

The corresponding variational problem is to extremize $E(u)$ subject to the constraint $V_p(u) \equiv \int_M u^{p+1}\, d\mu_{g_0} = 1$. In this way we connect the eigenvalue problem for L (i.e., $p = 1$) to the Yamabe problem, $p = \frac{n+2}{n-2}$. The constant c in $(1.1)_p$ is dependent on the solution and not specified in advance. It is well known that the variational problem corresponding to $(1.1)_p$ for $p \in [1, \frac{n+2}{n-2})$ does satisfy the Palais-Smale condition, and thus a Morse theory exists for these problems. A consequence of the Morse theory is the strong Morse inequalities which are relations satisfied among the critical points. Rather than attempt to construct a gradient flow for the Yamabe problem, we instead concentrate on convergence of critical points as $p \uparrow \frac{n+2}{n-2}$. Our main analytic result can be stated very simply.

Theorem 1.1. *Assume (M, g_0) is not conformally diffeomorphic to the round n-sphere S^n. There is a constant Λ depending only on the background metric g_0 such that if u is any solution of $(1.1)_p$ for any $p \in [1, \frac{n+2}{n-2}]$, then $\|u\|_{C^2(M)} \le \Lambda$.*

This result implies that all solutions of $(1.1)_p$ have uniformly bounded C^2-norm, and hence there can be no divergence as $p \uparrow \frac{n+2}{n-2}$. The fact that S^n has a noncompact group of conformal transformations implies that the conclusion of Theorem 1.1 is false on S^n. The result of Theorem 1.1 has been announced in [54] together with a detailed proof in the locally conformally flat case.

Note that the constant c in $(1.1)_p$ is the energy of the solution. Thus if one wants to consider large energy solutions it is desirable to multiply u by a constant to normalize the equation. Thus we see that $(1.1)_p$ for $p \in (1, \frac{n+2}{n-2}]$ is equivalent to

$$(1.2)_p \qquad Lu + u^p = 0, \qquad u > 0,$$

where the volume $V_p(u)$ is uncontrolled.

As in the original proof of the existence of a minimizer for the Yamabe problem, there is both a local and a global component to Theorem 1.1. The following result is local in nature and may be viewed as a strong generalization of Aubin's Theorem [4].

Theorem 1.2. *Let g_0 be a smooth Riemannian metric defined in the unit n-ball B_1^n. Suppose there is a sequence of solutions $\{u_j\}$ of $(1.2)_{p_j}$, $p_j \in (1, \frac{n+2}{n-2}]$, such that for any $\varepsilon > 0$ there is a constant $C(\varepsilon)$ so that $\sup_{B_1 \setminus B_\varepsilon} u_j \le C(\varepsilon)$ and $\lim_{j\to\infty}(\sup_{B_1} u_j) = \infty$. It follows that the Weyl tensor $W(g_0)$ satisfies $\|W(g_0)\|(x) \le C|x|^m$ for some integer $m > \frac{n-6}{2}$.*

The previous theorem implies that there cannot be an isolated point of blow-up for a sequence of locally defined solutions of $(1.2)_p$ unless the background conformal structure osculates the flat structure up to high order (for large n). Theorem 1.2 is one of the components required for the proof of Theorem 1.1.

The existence theorem which follows from Theorem 1.1 gives the strong Morse inequalities for the Yamabe problem since these inequalities are well known for $(1.1)_p$ with $p \in [1, \frac{n+2}{n-2})$, and Theorem 1.1 shows that all critical points converge as $p \uparrow \frac{n+2}{n-2}$. In case all solutions of the Yamabe problem are nondegenerate, as will be true for a generic conformal class of Riemannian metrics, the existence result asserts that there are a finite number of solutions of the variational problem. Let C_μ for $\mu = 0, 1, 2, \ldots$ be the number of solutions of Morse index μ. The following system of inequalities holds:

$$(1.3) \qquad (-1)^\lambda \leq \sum_{\mu=0}^{\lambda} (-1)^{\lambda-\mu} C_\mu, \qquad \lambda = 0, 1, 2, \ldots.$$

Since there is a finite number of solutions in all, it follows from (1.3) that

$$(1.4) \qquad 1 = \sum_{\mu=0}^{\infty} (-1)^\mu C_\mu,$$

where the sum on the right is finite. Note that the system (1.3) of Morse inequalities is that satisfied by a proper Morse function on a finite-dimensional Euclidean space. This is quite natural since the space of unit volume metrics in a conformal class is a contractible space. The question arises as to whether higher index solutions of the Yamabe problem exist in typical conformal classes. Explicit examples of this are given in [52] and discussed in detail. A recent result of D. Pollack shows that every Riemannian metric of positive scalar curvature is C^1 close to a smooth metric for which there exists an arbitrarily large number of solutions of the Yamabe problem with arbitrarily high index. To say this more precisely, if g is a metric of positive scalar curvature, and S is any positive integer, then there exists a sequence of smooth metrics g_j converging in C^1 norm to g such that the conformal class of g_j contains at least S solutions of the Yamabe problem for each j. In other words, the set of smooth metrics for which there are arbitrarily many solutions of the Yamabe problem is dense in the C^1 norm on the space of metrics of positive scalar curvature.

It may be useful to give more geometric statements of the analytical results which we have described. The result of Theorem 1.1 for $p = \frac{n+2}{n-2}$ may be described in the following way. Let $\mathscr{S}$ denote the set of all metrics on M with constant scalar curvature equal to 1. Let $\mathscr{M}$ denote the space of all smooth Riemannian metrics on M. The space of smooth functions $C^\infty(M)$ acts on $\mathcal{M}$ by conformal multiplication; that is, if $v \in C^\infty(M)$ and $g \in \mathcal{M}$, then $e^{2v}g \in \mathcal{M}$. The orbit space $\mathcal{C}$ is thus the space of conformal classes of metrics on M. The natural map $\Pi\colon \mathcal{M} \to \mathcal{C}$ is defined, and it is true that at any nondegenerate metric $g \in \mathcal{S}$, the map Π defines a local diffeomorphism from a neighborhood of g in $\mathcal{S}$ onto a neighborhood of $\Pi(g)$ in $\mathcal{C}$ (see [7]). Thus $\mathcal{S}$ is an infinite-dimensional manifold on the complement of the set of degenerate solutions of the Yamabe problem. For more discussion of the singular structure of $\mathcal{S}$ see [7]. Theorem 1.1 then implies

Theorem 1.1$'$. *The projection map $\Pi\colon \mathcal{S} \to \mathcal{C}$ is a proper map with respect to $C^{k,\alpha}$ topologies for k large enough.*

We now describe some recent work on a related variational problem for which blow-up does occur in the subcritical regularization. This is a joint work with Dong Zhang (see [72]). Suppose $K(x) \in C^\infty(S^3)$ is a smooth positive function on S^3. The prescribed curvature problem is then the problem of finding a metric g conformal to the standard metric g_0 on S^3 such that the scalar curvature $R(g)$ is equal to K at each point. By writing $g = u^4 g_0$ for a positive function u on S^3, the desired equation for u is

$$(1.5) \qquad L_0 u + \tfrac{1}{8} K u^5 = 0, \qquad L_0 u = \Delta_{g_0} u - \tfrac{3}{4} u.$$

As above (1.5) is equivalent to the Euler-Lagrange equation for the problem of extremizing $E(u) = -\int_{S^3} u L_0 u \, d\mu_{g_0}$ subject to the "volume" constraint $V_k(u) = \int_{S^3} K u^6 \, d\mu_{g_0} = 1$. We again consider the subcritical regularizations:

$$(1.5)_p \qquad L_0 u + \tfrac{1}{8} K u^p = 0, \qquad p \in [1, 5].$$

For (1.5), work of Kazdan and Warner [35] shows that there may be no solution for certain choices of the function K. As a consequence we see that Theorem 1.1 must fail in this case. Assuming that the critical points of K are nondegenerate and $\Delta K \neq 0$ at each critical point, we let $\mathcal{G} \subseteq S^3$ be the following finite set of points: $\mathcal{G} = \{x \in S^3 : \nabla K(x) = 0, \ \Delta K(x) < 0\}$. For each $x \in \mathcal{G}$, let $i(x) \in \{1, 2, 3\}$ denote the Morse index of K at x. The following result then describes the precise nature of the blow-up occurring in the subcritical regularization.

Theorem 1.3. *Corresponding to each $x \in \mathcal{G}$, there is precisely one solution $u_{x,p}$ of $(1.5)_p$ for $p < 5$ which diverges as $p \uparrow 5$. This solution* *has Morse index* $3 - i(x)$, *and it diverges at the lowest possible energy level by concentration at* x.

It is possible to give a rather precise description of $u_{x,p}$ for $p < 5$ and p near 5. As a consequence of this theorem, we have a version of the strong Morse inequalities for the prescribed curvature problem on S^3. For $\mu = 0, 1, 2$ let $D_\mu = \#\{x \in \mathcal{S}: i(x) = 3 - \mu\}$, and for $\mu = 0, 1, 2, \ldots$ let $C_\mu = \#$ of solutions of (1.5) with Morse index μ. We then have

$$(1.6) \qquad (-1)^\lambda \leq \sum_{\mu=0}^{\lambda} (-1)^{\lambda-\mu}(C_\mu + D_\mu), \qquad \lambda = 0, 1, 2, \ldots.$$

Note that we have assumed here that all critical points of (1.5) are nondegenerate. This holds for K generic. In particular, we see that the equality

$$1 = \sum_{\mu=0}^{\infty}(-1)^\mu C_\mu + \sum_{\mu=0}^{2}(-1)^\mu D_\mu$$

holds, and hence if we let $N(K) = |1 - \sum_{\mu=0}^{2}(-1)^\mu D_\mu|$ we see that the number of solutions of (1.5) is always bounded below by $N(K)$. (One can easily arrange $N(K)$ to be arbitrarily large by choosing K appropriately.) We also remark that without a genericity assumption one can interpret the previous equality as evaluating the degree of a Fredholm map. To put these results in historical context, we note that the first general existence theorem for the prescribed curvature problem on S^2 was due to J. Moser [43] who proved it for K satisfying $K(x) = K(-x)$ on S^2. The higher-dimensional version of Moser's theorem was considered by J. Escobar and the author [21]. Here a solution is constructed by minimization among symmetric metrics, and the result is proved for $n = 3, 4$. The result on S^3 that $N(K) > 0$ implies the existence of a solution is due to A. Bahri and J. M. Coron [5]. The Bahri-Coron result was proven on S^2 by A. Chang and P. Yang [12] and the analogue of our result by Z. C. Han [28]. The main analytic difficulties involved with higher energy blow-up do not occur for $n = 2$, because surfaces of positive curvature are much more restricted metrically than higher-dimensional manifolds of positive scalar curvature. (Certain other difficulties occur for $n = 2$ because the nonlinearity is exponential.) We also remark that N. Korevaar and the author have constructed examples for $n \geq 6$ where infinite energy blow-up occurs in $(1.5)_p$. This suggests that the analogue of our results for higher-dimensional spheres will be difficult to obtain. A partial result for $n = 4$ was obtained by Dong Zhang [72].

As mentioned above, the Yamabe problem has both a local and global component. The global considerations arise in ruling out small energy blow-up on manifolds of dimension 3, 4, and 5 (where Theorem 1.2 gives no information), and at points where the metric is conformally flat to a high order. In the case of conformally flat manifolds these global considerations arise as a special case of a general structure theorem. We now describe our work with S. T. Yau [57]. If (M^n, g_0) is a locally conformally flat Riemannian manifold of dimension $n \geq 3$, then the universal covering $\tilde{M}$ can be mapped conformally into S^n by the so-called developing map $\Phi: \tilde{M} \to S^n$. There is a group homomorphism $\rho: \pi_1(M) \to O(n+1, 1)$, the group of conformal automorphisms of S^n, and the map Φ is equivariant in the sense that $\Phi \circ \gamma = \rho(\gamma) \circ \Phi$. In general, the map Φ may be quite pathological; however, one of the main results of [57] is the following.

Theorem 1.4. *If (M, g_0) is complete, conformally flat with nonnegative scalar curvature, then $\Phi: \tilde{M} \to S^n$ is injective, and the universal covering of M is conformally diffeomorphic to a subdomain of S^n. Moreover, ρ injects onto a discrete subgroup of $O(n+1, 1)$.*

This leads naturally to the question of which domains in S^n have complete conformal metrics of nonnegative scalar curvature. It is shown that a necessary condition on Ω for the existence of such a metric is that the Hausdorff dimension of $S^n \backslash \Omega$ be at most $\frac{1}{2}(n-2)$. There is a more general injectivity criterion derived in [57] which we now describe. Suppose (M, g_0) is a complete Riemannian manifold and $\Phi: M^n \to S^n$ is a conformal immersion into S^n. Let $L_0 u = \Delta_0 u - \frac{n-2}{4(n-1)} R(g_0) u$ be the conformally invariant Laplacian on M. Given a point $x_0 \in M$, it can be shown that L_0 has a minimal positive fundamental solution with pole at x_0, which we denote by G_{x_0}. We then define $p(M, g_0)$ by

$$p(M, g_0) = \inf\{p \in \mathbf{R}^+ : G_{x_0} \in L^p(M \backslash B_1(X_0))\}.$$

Thus $p(M, g_0)$ is a measure of the decay at infinity of G_{x_0}. It is straightforward to see that $p(M, g_0)$ is independent of x_0, and is such that $D(M, g_0) \equiv \frac{n-2}{2} p(M, g_0) \in [0, n]$ for any manifold. Notice also that in case M is the universal covering space of a compact manifold, then $D(M, g_0)$ does not depend on the choice of metric in the quotient, but only the conformal structure; in fact, $D(M, g_0)$ is a quasi-isometry invariant for metrics in the same conformal class. The general injectivity theorem can then be stated.

Theorem 1.5. *Suppose that (M, g_0) is a complete Riemannian manifold, and $\Phi: M \to S^n$ is a conformal immersion, and assume that $|R(g_0)|$*

is bounded for $n = 3, 4$. *If* $D(M, g_0) < \frac{(n-2)^2}{n}$, *then* Φ *is injective and defines a conformal diffeomorphism of* M *with a subdomain of* S^n.

These injectivity theorems may be used to derive a general geometric statement about conformally flat manifolds of constant scalar curvature. The proof of the following result appears in [54].

Theorem 1.6. *If* (M, g_0) *is conformally flat and complete with* $R_{g_0} \equiv 1$, *then any embedded round (i.e., with totally umbilic boundary) ball in* M *is geodesically convex for the metric* g_0.

The proof of this result uses the injectivity theorems together with the Alexandrov reflection method. It is relatively straightforward to prove Theorem 1.1 for conformally flat M from Theorem 1.6, and the proof is discussed in [54].

Finally we discuss the results which have been obtained for the existence of complete conformally flat metrics with constant positive scalar curvature. If we consider simply connected manifolds, then by the injectivity theorem there is no loss of generality in restricting attention to domains Ω in S^n. In case $\Omega = S^n$, the theorem of M. Obata [45] states that all solutions are standard in the sense that they arise by pullback of the standard metric by a conformal transformation of S^n. When $\Omega = S^n - \{x_1\}$ for a point $x_1 \in S^n$, the methods of Gidas, Ni, and Nirenberg [23] imply that there is no such metric on Ω. An old question which has arisen in various contexts is the existence question for $\Omega = S^n - \{x_1, \cdots, x_k\}$ for a prescribed set of k points. When $k = 2$, it was shown by Caffarelli, Gidas, and Spruck [9] that all solutions become rotationally symmetric when the points are put in antipodal position on S^n. (This can be accomplished by a conformal transformation.) It is also shown in [9] that any weak solution of (1.1) on the punctured Euclidean ball is either extendable across the origin or is asymptotic to one of the family of distinct ODE solutions. A general existence theorem for global weak solutions of (1.1), in fact complete conformal metrics of constant scalar curvature, was derived in [53]. This produces solutions on large classes of domains Ω where $S^n - \Omega$ is a totally disconnected set of small Hausdorff dimension. In the case of prescribed singular points we show the following theorem in [53].

Theorem 1.7. *If* $x_1, \cdots, x_k \in S^n$ *with* $k \geq 2$, *then there exists a complete conformal metric with scalar curvature identically one on* $\Omega = S^n - \{x_1, \cdots, x_k\}$.

One of the main difficulties in handling (1.1) analytically is the high degree of instability which solutions tend to have. This difficulty arises already in the Yamabe problem, but it is far more serious when one deals with complete solutions. Here it typically happens that the linearized equation has zero embedded in its continuous spectrum. This problem is overcome in [53] by imposing an infinite number of conditions so that the operator may be inverted, and then varying the geometry of the approximate solution to compensate for these conditions. If one considers domains $\Omega = S^n - \Sigma^k$, where Σ^k is a smooth closed k-dimensional submanifold with $k < \frac{1}{2}(n-2)$, then it is unknown generally whether Ω carries a complete conformal metric with scalar curvature one. For $\Sigma^k = S^k$, a round subsphere, such solutions exist in abundance because $S^n - S^k$ is conformally diffeomorphic to $H^{k+1} \times S^{n-k-1}$, H^{k+1} being the hyperbolic space of dimension $k+1$. It has recently been shown by R. Mazzeo and N. Smale [38] that when Σ^k is a small perturbation of S^k many solutions do exist on $S^n - \Sigma^k$. Their proof involves a delicate linear analysis and perturbation argument. Their solutions have the simplest possible boundary asymptotic behavior. One of the interesting questions in general is to determine the possible boundary asymptotics for global solutions. For example in the case $\Omega = S^n - \{x_1, x_2, \cdots, x_k\}$ for $k \geq 3$ it is not known whether there is a solution on Ω which is asymptotic to the simplest possible ODE solution near one of the points. (The simplest ODE solution near $x = 0$ has the form $c \cdot |x|^{(n-2)/2}$ in stereographic coordinates.)

Finally we mention the Yamabe problem for manifolds with boundary. The most natural homogeneous geometric boundary condition is a mixed Dirichlet-Neuman condition which states that the mean curvature of the boundary in the new metric vanishes. This boundary condition is of the form

$$\frac{\partial u}{\partial \nu} + \frac{n-2}{2} H_{g_0} u = 0 \quad \text{on } \partial M,$$

where (M, g_0) is a Riemannian manifold with smooth compact boundary, H_{g_0} denotes the mean curvature of ∂M with respect to g_0, and ν denotes the outward unit (with respect to g_0) normal. Thus one constructs metrics (M, g) satisfying $R(g) \equiv 1$ on M and $H_g \equiv 0$ on ∂M. This problem has been solved by J. Escobar [20] in the sense that he has produced a minimizing solution of the variational problem. A variant on this problem might be to prescribe H_g to be constant on ∂M and R_g to be constant in M. Escobar has considered this in [19] in case $R_g \equiv 0$ in M and $H_g \equiv 1$ on ∂M. This amounts, for $\Omega \subseteq \mathbf{R}^n$, to finding positive harmonic functions in Ω satisfying a critical nonlinear boundary condition.

2. Evolution of Riemannian metrics

One of the most striking developments in geometric partial differential equations during the last decade has been the successful application of heat equation methods to discuss the evolution of Riemannian metrics on smooth manifolds. The central evolution equation which arises in this context is the Ricci flow equation introduced by R. Hamilton [25]. The idea here is that the Ricci curvature may be thought of as a second-order differential operator from the space of metrics to the symmetric quadratic tensors. It is well known that if one introduces harmonic coordinates locally on M for a metric g, then the Ricci curvature takes the form

$$R_{ij} = -\tfrac{1}{2}\Delta_g(g_{ij}) + Q_{ij}(g, \partial g),$$

where Q_{ij} is an expression which is a quadratic polynomial in the partial derivatives ∂g. Thus minus twice the Ricci operator becomes locally a quasilinear elliptic operator with diagonal leading order term which is the Laplacian in the unknown metric g. This suggests that the equation

$$\frac{\partial g}{\partial t} = -2\,\mathrm{Ric}(g) \tag{2.1}$$

should be an essentially parabolic equation for the evolving metric g. This is the Ricci flow equation. In order to globalize the above discussion concerning harmonic coordinates, we may introduce a fixed background metric $\hat{g}$ on M. The difference D of the Levi-Civita connections may then be expressed as a global first-order differential operator applied to the metric g. Precisely we have

$$D^i_{jk} = \Gamma^i_{jk} - \hat{\Gamma}^i_{jk} = \frac{1}{2}\sum_l g^{il}(g_{jl;k} + g_{kl;j} - g_{jk;l}), \tag{2.2}$$

where we have chosen a local basis, and "semicolon" is used to denote covariant differentiation with respect to the background metric $\hat{g}$. Similarly, the difference of the Ricci tensors of g and $\hat{g}$ is given by

$$R_{ij} - \hat{R}_{ij} = \sum_k \left\{ D^k_{ij}; k - D^k_{ki;j} + \sum_l (D^k_{kl}D^l_{ji} - D^k_{jl}D^l_{ki}) \right\}. \tag{2.3}$$

By direct calculation from (2.2) we find that the second-order terms of the right-hand side of (2.3) may be written

$$-\frac{1}{2}\sum_{k,l} g^{kl} g_{ij;kl} + \frac{1}{2}\sum_{k,l} g^{kl}(g_{il;jk} + g_{jl;ik} - g_{kl;ij}).$$

If we introduce the vector field $X = X(g)$ given by

$$X^p(g) = -\sum_{i,k,l} g^{pi} g^{kl} \left(g_{ik;l} - \frac{1}{2} g_{kl;i} \right),$$

we see that (2.3) may be written

$$\text{Ric}(g) = \text{Ric}(\hat{g}) - \tfrac{1}{2} F(g) - \tfrac{1}{2} \mathcal{L}_X g, \tag{2.4}$$

where $\mathcal{L}_X g$ denotes the Lie derivative of g with respect to X, and $F(g)$ is an elliptic operator of the form

$$F_{ij}(g) = \sum_{k,l} g^{kl} g_{ij;kl} + Q_{ij}(g, \nabla g),$$

where Q is a quadratic polynomial in the covariant derivatives ∇g of g (taken with respect to $\hat{g}$) with coefficients depending on g. To solve the initial value problem for (2.1) we first solve the quasilinear parabolic equation

$$\begin{aligned} \frac{\partial \overline{g}}{\partial t} &= F(\overline{g}) - 2\,\text{Ric}(\hat{g}), \\ \overline{g}(0) &= g_0 \end{aligned} \tag{2.5}$$

for a given initial metric g_0. This has a local solution by standard parabolic theory. The corresponding time dependent vector field $X(t) = X(\overline{g}(t))$ is then given, so we may integrate the system of ordinary differential equations to determine a family of maps $\Phi_t \colon M \to M$:

$$\begin{aligned} \frac{d\Phi}{dt}(t, x) &= X(t, x), \\ \Phi(0, x) &= x. \end{aligned} \tag{2.6}$$

We then let $\Phi_t(x) = \Phi(t, x)$, and we have $\Phi_0 =$ Identity. Now define $g(t) = \Phi_t^* \overline{g}(t)$, and compute

$$\frac{\partial g}{\partial t} = \Phi_t^* \left(\frac{\partial \overline{g}}{\partial t} \right) + \mathcal{L}_{X(t)} g(t).$$

By (2.4) and (2.5) we obtain

$$\Phi_t^* \left(\frac{\partial \overline{g}}{\partial t} \right) = \Phi_t^* (-2\,\text{Ric}(\overline{g}) - \mathcal{L}_X \overline{g}) = -2\,\text{Ric}(g) - \mathcal{L}_X g.$$

Therefore $g(t)$ is a solution of the initial value problem for the Ricci flow equation (2.1) with initial metric $g(0) = g$. The local existence theorem was originally proven by R. Hamilton [25] using a less transparent argument. The proof given here is essentially that of D. DeTurck [16]. The philosophy of this argument is quite clear in that (2.1) is parabolic except for the difficulty arising from the diffeomorphism invariance of the Ricci operator. The time dependent family of diffeomorphisms Φ_t has the effect of fixing this gauge freedom.

The main geometric interest in (2.1) lies in the large time behavior of solutions, although the local existence theorem has also proven to be useful as we will see later. The simplest large time behavior which one might expect for a parabolic equation is that the solutions approach a time independent solution. The time independent solutions of (2.1) are Ricci flat metrics; however this class of solutions can be enlarged by additionally imposing a volume constraint on the solution. This leads to an equivalent equation, called the normalized Ricci flow,

$$\frac{\partial g}{\partial t} = -2\left(\mathrm{Ric}(g) - \frac{r}{n}g\right), \tag{2.7}$$

where $r = \mathrm{Vol}(g)^{-1} \int_M R(g)\, d\mu_g$. Equation (2.7) may be derived directly from (2.1) by setting $\overline{g} = (V_0/\mathrm{Vol}(g))^{2/n} g$, where $g(t)$ satisfies (2.1) and one makes an appropriate change of the time variable $\tau = \tau(t)$. One sees that $\overline{g}(\tau)$ is then a solution of (2.7), and $\mathrm{Vol}(\overline{g}(\tau)) \equiv V_0$ for all t. The time independent solutions of (2.7) are, of course, the Einstein metrics. Thus studying the large time behavior of (2.7) presents an analytic approach to solving the elliptic Einstein equation.

The first global result obtained via the Ricci flow was done by R. Hamilton [25] for compact (M, g_0) with $\dim M = 3$ and $\mathrm{Ric}(g_0) > 0$. This result was later simplified and extended to four dimensions by Hamilton [26] in which case the assumption on g_0 was that g_0 have positive curvature operator. We give a brief summary of the main ideas involved here. The main estimate of both papers is a strong pinching estimate on the eigenvalues of the curvature operator of the form

$$\max\{|\lambda_i - \lambda_j|\} \leq CR^{1-\varepsilon_0}$$

for a constant C depending only on the initial metric, and a fixed small positive number ε_0. This estimate implies that at points where the curvature is large (note that the scalar curvature provides an upper bound on the full curvature tensor) the eigenvalues are relatively very close together. Thus the crucial high curvature regions of the solution are becoming round. This pinching estimate together with the result that the flow preserves the positivity assumption made on g_0 are derived through the analysis of a quadratic first-order autonomous system of ordinary differential equations on the vector space of linear endomorphisms of the Lie algebra of $\mathrm{SO}(n)$.

This vector space may be considered as the space of curvature operators, and the dynamical system on this space is the ODE part of the parabolic evolution equation satisfied by the curvature matrix of g. The idea then is to look for a *convex* invariant set for this ODE system so that the parabolic system will preserve the same set. This is reasonable to expect because the heat equation should preserve convex constraints. Hamilton then constructs convex invariant sets by finding Lyapunov functions for the flow. A pinching estimate of the type described above then implies convergence of the normalized flow to a constant curvature metric. In the final part of this scheme, it is necessary to go from pointwise pinched curvature to globally pinched curvature. This can be done formally if $\dim M \geq 3$ by the twice contracted second Bianchi identity which states that

$$\frac{n-2}{2n} R_{ji} = \sum_j T_{ij;j}$$

in an orthonormal basis, where T denotes the trace-free Ricci tensor. Thus if T is small, one can expect R to be nearly constant so that it might be reasonable to derive global curvature pinching from pointwise pinching.

It is presently unknown whether the Ricci flow converges if $\dim M \geq 5$ and g_0 has positive curvature operator. While the positivity is preserved under the flow, the necessary pinching estimates have not been derived. The corresponding ODE system is more difficult to analyze because the Lie algebra is higher-dimensional and more complicated. The Micallef-Moore theorem which we will discuss in the next section does imply that a manifold of positive curvature operator is covered by a topological sphere.

The general method outlined above, because it relies on the twice contracted second Bianchi identity, also does not work on surfaces. Recently the Ricci flow has been understood on compact two-manifolds by somewhat different methods. This problem was discussed in [27] by R. Hamilton. The normalized Ricci flow on surfaces is the equation $\frac{\partial g}{\partial t} = (r-R)g$, where r denotes the average value of R, and R is twice the Gaussian curvature. Thus $r = (\mathrm{Area}(g))^{-1} 4\pi\chi(M)$ by the Gauss-Bonnet theorem, and since the flow is normalized the area is constant in time. For surfaces of positive genus it is shown in [27] that the flow exists for all time and converges to the unique constant curvature metric in the conformal class of g_0. As one would expect, the most subtle case is $M = S^2$. In this case an interesting notion which arises and plays an important role in the arguments is the notion of a soliton (actually travelling wave) solution. This means a solution of the normalized Ricci flow which has the form $g_t = F_t^* g_0$ for some smooth curve of diffeomorphisms $F_t\colon M \to M$, $t \in [0, \infty)$, with $F_0 = \mathrm{id}$. Thus the geometry for a soliton solution remains constant; in fact the solution at every time is isometric to the solution at time zero. On the other hand, if the path F_t is divergent, then the solution g_t will diverge in the space of metrics. In [27] it is shown that if $R(g_0) > 0$ on S^2, then the Ricci flow converges to a constant curvature metric. In detail the proof is quite different from the higher-dimensional case and relies on a Harnack-type inequality and the decrease of an entropy integral in order to bound the geometry of g_t for all time. Finally to prove convergence of g_t it is shown that divergence would imply the existence of a soliton solution on S^2. An explicit argument however excludes this possibility. The Ricci flow analysis for smooth metrics on S^2 was later completed by B. Chow [13] who removed the assumption of positivity on the initial metric g_0. Thus even a dumbbell metric on S^2 will converge under the Ricci flow to a constant curvature metric. The Ricci flow on positively curved two-dimensional orbifolds has been studied by B. Chow and L. F. Wu [14]. The interesting feature of this problem is that constant curvature metrics need not exist. It is shown in this case that the solution of the Ricci flow converges asymptotically to a soliton solution.

The Ricci flow on complete manifolds has been studied by W. X. Shi [58]. He has established short time existence for complete initial metrics of bounded curvature, and in certain cases has obtained long time results. In particular, he has recently announced [59] the result that a complete noncompact Kähler manifold with positive bounded holomorphic sectional curvature, maximal volume growth, and weak decay on the average total scalar curvature on balls is biholomorphic to $\mathbf{C}^n$. The proof involves showing that the Ricci flow exists for all time and converges to a flat metric. In the Kähler case, it was shown by H. D. Cao [10] that for compact manifolds with $c_1 = 0$ or $c_1 < 0$ the Ricci flow converges to a Kähler-Einstein metric.

The Ricci flow has also been applied recently by M. Min-Oo [41] and R. Ye [70] to construct Einstein metrics on Ricci-pinched manifolds of negative scalar curvature. A consequence of the results of R. Ye [71] is the existence of an abundance of Einstein metrics on higher-dimensional manifolds $n \geq 5$ whose sectional curvatures are pinched near a negative constant. Finally, the Ricci flow has been used by D. Yang [68], [69] to convert integral to pointwise bounds on curvature, and hence to establish certain results of Riemannian geometry under integral rather than pointwise bounds on curvature.

3. Harmonic maps and applications

For a map $u: M^n \to X^k$, we can assign to u various energies if M and X have Riemannian metrics g, h respectively. This can be done by considering the pullback metric u^*h, and diagonalizing this with respect to g. The eigenvalues $\lambda_1, \cdots, \lambda_n$ of u^*h are then the squares of the "principal stretches" of u, and the symmetric scalar functions of $\lambda_1, \cdots, \lambda_n$ are well defined functions on M, independent of basis or coordinate choice. The simplest of these scalar functions is the trace, $\mathrm{Tr}_g(u^*h) = \sum_{i=1}^n \lambda_i$, which we refer to as the energy density of u, and denote $e(u)$. The total energy of u is then given by $E(u) = \int_M e(u)\, d\mu_M$. The critical points of $E(u)$ on the space of maps are referred to as harmonic maps. The theory of harmonic maps includes the theory of geodesics (with constant speed parametrization) when $n = 1$, and the theory of harmonic functions when $X = \mathbf{R}$.

The theory for $n = 2$ has played an important role in minimal surface theory since the work of C. B. Morrey in the 1940's. It has only been relatively recently, beginning with the work of Sacks and Uhlenbeck, that the higher critical point theory for $n = 2$ has been developed. For $M = S^2$, the result which one can expect is a partial Morse theory for the energy functional. We state one result of this type which is a slight refinement of [48] due to Micallef-Moore [40].

Theorem 3.1. *Suppose $\pi_{p+2}(X) \neq 0$ for some integer $p \geq 0$. There exists a nonconstant harmonic map $u: S^2 \to X$ whose Morse index (for the energy functional) is at most p.*

This theorem has found a striking geometric application to the study of manifolds of nonnegative curvature in [40]. To describe these results, we consider the complexification of the tangent space $T_xX \otimes \mathbf{C}$ at a point x, and extend the curvature tensor to complex vectors by linearity. The complex sectional curvature of a two-dimensional subspace π of $T_xX \otimes \mathbf{C}$ is then defined by $K(\pi) = \langle R(v, w)v, w\rangle$, where $\{v, w\}$ is any unitary basis of π, and $\langle\cdot, \cdot\rangle$ denotes the Hermitian inner product on $T_xX \otimes \mathbf{C}$. A subspace π is said to be isotropic if every vector $v \in \pi$ has square zero; that is, $(v, v) = 0$, where $(\cdot, \cdot)$ denotes the complex linear extension of the real inner product to $T_xX \otimes \mathbf{C}$. We say that X has *positive isotropic sectional curvature* if $K(\pi) > 0$ for every isotropic two-plane π. If the dimension $k = \dim X$ is less than four, then every X has positive isotropic curvature because there are no isotropic two-dimensional subspaces π in this case. In general, the classical conditions of (pointwise) $\frac{1}{4}$-pinching and positive curvature operator are easily seen (see [40]) to imply positive isotropic sectional curvature. The main harmonic map theorem of [40] can then be stated.

Theorem 3.2. *Assume X has positive isotropic sectional curvature. Then any nonconstant harmonic map $u\colon S^2 \to X^k$ has Morse index at least $[[\frac{k}{2}]] - 1$, where $[[\cdot]]$ denotes the integer part of a number.*

Combining Theorems 3.1 and 3.2 we conclude that $\pi_p(X) = 0$ for $p = 2, 3, \cdots, [[\frac{k}{2}]]$ $(k \geq 4)$. An immediate consequence is that X is a homotopy sphere if $\pi_1(X) = 0$. Using known results from topology one concludes

Theorem 3.3. *A compact simply connected manifold with positive isotropic sectional curvature is homeomorphic to the sphere.*

It should be observed that the condition of positive isotropic sectional curvature does not imply that $\pi_1(X)$ is finite (in particular, it does not imply positive Ricci curvature). For example, in dimension four any locally conformally flat manifold with positive scalar curvature has positive isotropic sectional curvature. It is well known that the fundamental group of such a manifold can be a free group.

The present state of affairs with the regularity properties for harmonic maps is slightly better than at the time of our survey [50]. Recall that a regularity theory for minimizing maps exists [55]. There has recently been progress on the structure of the singular set for minima by R. Hardt, F. H. Lin, and L. Simon [29], [60]. For example, they show that the singular set for a minimizing map from $\Omega \subset \mathbf{R}^4$ to S^2 consists of curves and isolated points. Recall that a map $u \in H^1(M, X)$ is a weak solution if it satisfies the Euler-Lagrange equation in the distributional sense. A map u is *stationary* if it is a weak solution, and in addition satisfies the condition that its energy is critical with respect to variations of the type $u \circ F_t$, where $F_t\colon M \to M$ is a smooth path of diffeomorphisms of M fixing the boundary. It can be shown (see [50] for a more detailed discussion) that stationary maps satisfy the monotonicity property for the scale invariant energy in balls. This property plays an important role in the regularity theory for minimizing maps [55]. For $n = \dim M = 2$, the stationary property imlies the holomorphicity of the Hopf differential. It was shown by M. Grüter [24] that weakly conformal weak solutions $(n = 2)$ are regular. This was extended to stationary maps in [50]. For $n \geq 3$, no regularity theorem exists for stationary maps, although such a theory should be true. In the past few years it has been shown that the stationarity condition is much more restrictive than the weak solution property. For example, in [47] many weak solutions $(n = 3)$ which are not stationary are constructed. Note that smooth weak solutions are clearly stationary, so the distinction exists only for maps with singularities.

While the regularity properties of weak solution are still not understood, there has been a recent development by F. Helein [30], who has shown that for maps from a surface M^2 to the standard sphere S^k, weak solutions are in fact regular. This result exploits the special form of the equation for maps to the round sphere. We make some remarks concerning this phenomenon. These observations arose in a discussion with Rugang Ye. First there is a result about linear equations in two variables due to H. Wente [64]. In general, a solution u of $\Delta u = f$ with $f \in L^1_{\text{loc}}(\Omega)$, $\Omega \subset \mathbf{R}^2$, fails to be continuous. We may consider a special class of L^1 functions generated in the following way: Let $v: \Omega \to \mathbf{R}^{2q}$ be a map which is locally in H^1 and let $\omega = \sum_{j=1}^q dx_j \wedge dy_j$ be the standard symplectic form on $\mathbf{R}^{2q}$. We may then write $v^* w = f dx \wedge dy$ with $f \in L^1_{\text{loc}}(\Omega)$. Let $\mathcal{F}$ denote the space of functions f which arise in this way for some q; that is, the L^1_{loc} functions which can be expressed as sums of Jacobians of H^1 maps. Wente's theorem can be stated:

Theorem 3.4. *Let $\Omega \subset \mathbf{R}^2$ be a domain, and let u be a solution of $\Delta u = f$ in Ω. If $f \in \mathcal{F}$, then u is continuous on any compact subdomain of Ω.*

Now, let us consider maps $u: \Omega \to S^k$. The Euler-Lagrange equation for harmonic maps is

$$\Delta u^j = -|\nabla u|^2 u^j, \qquad j = 1, \cdots, k+1,$$

where

$$S^k = \{u \in \mathbf{R}^{k+1} : |u|^2 = 1\}, \qquad |\nabla u|^2 = \sum_{\alpha=1}^{2} \sum_{i=1}^{k+1} \left(\frac{\partial u^i}{\partial x^\alpha}\right)^2.$$

This equation can be rewritten as

$$\Delta u^j = \sum_{i=1}^{k+1} \sum_{\alpha=1}^{2} \frac{\partial u^i}{\partial x^\alpha} \left(\frac{\partial u^j}{\partial x^\alpha} u^i - \frac{\partial u^i}{\partial x^\alpha} u^j\right), \qquad j = 1, \cdots, k+1,$$

since

$$\sum_{i+1}^{k+1} \frac{\partial u^i}{\partial x^\alpha} u^i = \frac{1}{2} \frac{\partial}{\partial x^\alpha}(|u|^2) = 0.$$

Finally observe that for any i, j we have

$$\sum_{\alpha=1}^{2} \frac{\partial}{\partial x^\alpha} \left(\frac{\partial u^j}{\partial x^\alpha} u^i - \frac{\partial u^i}{\partial x^\alpha} u^j\right) = 0,$$

so that the 1-form β_{ij} on Ω given by

$$\beta_{ij} = \sum_{\alpha=1}^{2} \left(\frac{\partial u^j}{\partial x^\alpha} u^i - \frac{\partial u^i}{\partial x^\alpha} u^j \right) dx^\alpha$$

is co-closed. Thus $*\beta_{ij} = -dh_{ij}$ for a function $h_{ij} \in H^1_{\mathrm{loc}}(\Omega)$ (we may assume Ω is simply connected since this theory is local). Thus we may rewrite the harmonic map equation as

$$\Delta u^j = \sum_{i=1}^{k+1} *(du^i \wedge *\beta_{ij}) = \sum_{i=1}^{k+1} *(du^i \wedge dh_{ij}).$$

In particular, $\Delta u^j \in \mathcal{F}$ for $j = 1, \cdots, k+1$, and hence u is continuous. It then follows that u is smooth. Thus we get the following result.

Theorem 3.5 (F. Helein). *Weakly harmonic maps from surfaces into round spheres are regular.*

It is unknown whether the regularity theorem holds for arbitrary target spaces. We now make some observations which clarify the geometry of the situation above, and show that Theorem 3.5 does hold whenever X is a compact homogeneous Riemannian manifold. Recently F. Helein has extended his method to cover the case in which X is homogeneous. First we associate to any Killing vector field of X a conjugate function $v \in H^1_{\mathrm{loc}}(\Omega)$ which satisfies an equation of the form $\Delta v = f$ with $f \in \mathcal{F}$. Thus v is continuous by Theorem 3.4. We then show that if the Killing vector fields span the tangent space of X at each point, then the map u is itself regular.

Let K be a Killing vector field on X, and let $K^\#$ denote the associated 1-form. We observe that $u^*(K^\#)$ is an $L^2_{\mathrm{loc}}(\Omega)$ 1-form which is weakly co-closed whenever $u\colon \Omega \to X$ is a weakly harmonic map. To see this, observe that

$$u^*(K^\#) = \sum_{\alpha=1}^{2} \left\langle u_* \left(\frac{\partial}{\partial x\alpha} \right), K \right\rangle dx^\alpha,$$

so that if $\varphi \in C_0^\infty(\Omega)$, then

$$\int_\Omega \langle u^*(K^\#), d\varphi \rangle \, dx^1 dx^2 = \int_\Omega \sum_{\alpha=1}^{2} \left\langle u_* \left(\frac{\partial}{\partial x^\alpha} \right), K \right\rangle \frac{\partial \varphi}{\partial x_\alpha} \, dx^1 dx^2.$$

First note that if u is *any* map in $H^1_{\mathrm{loc}}(\Omega, X)$, then u can be approximated locally in H^1 norm by smooth maps. Let $\{u_p\}$ be a smooth approximating sequence, and observe that because K is a Killing vector field we have

$$\int_\Omega \langle u_p^*(K^\#), d\varphi\rangle\, dx^1\, dx^2 = \int_\Omega \sum_{\alpha=1}^2 \left\langle u_{p*}\left(\frac{\partial}{\partial x^\alpha}\right), \nabla_{\partial/\partial x^\alpha}(\varphi K)\right\rangle dx^1\, dx^2,$$

where ∇ denotes the connection on the pullback of the tangent bundle of X by u_p. This is true because in an orthonormal basis we have

$$\sum_{\alpha=1}^2 \left\langle u_{p*}\left(\frac{\partial}{\partial x^\alpha}\right), \nabla_{\partial/\partial x^\alpha} K\right\rangle = \sum_{\alpha=1}^2 \sum_{i,j=1}^k K_{i;j} \frac{\partial u_p^i}{\partial x^\alpha}\frac{\partial u_p^j}{\partial x^\alpha} = 0$$

since $K_{i;j}$ is antisymmetric. Thus we may let $p \to \infty$, and deduce

$$\int_\Omega \langle u^*(K^\#), d\varphi\rangle\, dx^1\, dx^2 = \int_\Omega \sum_{\partial=1}^2 \left\langle u_*\left(\frac{\partial}{\partial x^\alpha}\right), \nabla_{\partial/\partial x^\alpha}(\varphi K)\right\rangle dx^1\, dx^1.$$

Since φK is an H_0^1 section of $u^*(TX)$, the fact that u is weakly harmonic implies that this vanishes for each $\varphi \in C_0^\infty(\Omega)$. Thus $u^*(K^\#)$ is weakly co-closed, and we may write $u^*(K^\#) = *dv$ for a function v which is unique up to an additive constant. Moreover, $v \in H^1_{\mathrm{loc}}(\Omega)$ and may be thought of as a conjugate function to u in the direction of the Killing field K. We now compute $\Delta v = *d * dv = *d(u^*(K^\#))$. We observe that if u is a smooth (not necessarily harmonic) map, then we have (in an orthonormal basis)

$$\begin{aligned} d(u^*(K^\#)) &= \sum_{i=1}^k \left[\frac{\partial}{\partial x^1}\left(\frac{\partial u^i}{\partial x^2}K^i\right) - \frac{\partial}{\partial x^2}\left(\frac{\partial u^i}{\partial x^1}K^i\right)\right] dx^1 \wedge dx^2 \\ &= \left[\left\langle \frac{\partial u}{\partial x^2}, \nabla_{\partial/\partial x^1} K\right\rangle - \left\langle \frac{\partial}{\partial x^1}, \nabla_{\partial/\partial x^2} K\right\rangle\right] dx^1 \wedge dx^2. \end{aligned}$$

If $X \subset \mathbf{R}^Q$ isometrically, then K may be extended as a smooth vector field on $\mathbf{R}^Q$, and

$$d(u^*(K^\#)) = -\sum_{i=1}^Q du^i \wedge d(K^i \circ u).$$

By approximation as above this holds weakly for any H^1 map u, and therefore $\Delta v \in \mathcal{F}$. Hence by Theorem 3.4, $v \in C^0(\Omega, X)$.

Now let $\{K_1, \cdots, K_s\}$ be a basis for the vector space of Killing vector fields on X. For each K_l we construct a conjugate function v_l as above, and these v_l satisfy equations $\Delta v_l = f_l \in \mathcal{F}$, $l = 1, \cdots, s$. Let $0 \in \Omega$ be any point, and $\sigma > 0$ with $B_\sigma(0) \Subset \Omega$. We then have the equation

$$\int_{B_\sigma(0)} |\nabla v_l|^2 \, dx^1 \, dx^2 = -\int_{B_\sigma(0)} (v_l - \overline{v}_l) f_l \, dx^1 \, dx^2 + \int_{\partial B_\sigma(0)} (v_l - \overline{v}_l) \frac{\partial v_l}{\partial r} \, ds$$

for any constant $\overline{v}_l$. Taking $\overline{v}_l$ to be the average of v_l on $\partial B_\sigma(0)$ and applying the Poincaré inequality on $\partial B_\sigma(0)$ as well as the Schwarz inequality we obtain

$$\int_{B_\sigma(0)} |\nabla v_l|^2 \, dx^1 \, dx^2 \leq \int_{B_\sigma(0)} |v_l - \overline{v}_l| |f_l| \, dx^1 \, dx^2 + \sigma \int_{\partial B_\sigma(0)} |\nabla v_l|^2 \, ds.$$

Since v_l is continuous, if σ is chosen small enough, $\sup_{B_\sigma(0)} |v_l - \overline{v}_l|$ can be made smaller than any preassigned number. On the other hand, $|f_l| \leq c e(u)$ for a constant c depending only on X. Thus for sufficiently small σ we have

$$\int_{B_\sigma(0)} |v_l - \overline{v}_l| |f_l| \, dx^1 \, dx^2 \leq \varepsilon_0 \int_{B_\sigma(0)} e(u) \, dx^1 \, dx^2$$

for any $\varepsilon_0 > 0$. Summing on l then gives

$$\sum_{l=1}^{s} \int_{B_\sigma(0)} |\nabla v_l|^2 \, dx^1 \, dx^2 \leq \varepsilon_0 s \int_{B_\sigma(0)} e(u) \, dx^1 \, dx^2 + c\sigma \int_{\partial B_\sigma(0)} e(u) \, ds,$$

where we have used the obvious inequality $\sum_{l=1}^{s} |\nabla v_l|^2 \leq c e(u)$. Now if we assume that $\{K_1, \cdots, K_s\}$ span the tangent space of X at each point, then clearly for any vector $w \in T_x X$ and any $x \in X$ there is a constant $c > 0$ so that $\|w\|^2 \leq c \sum_{l=1}^{s} \langle w, K_l \rangle^2$. Since $|\nabla v_l|^2 = \sum_{\alpha=1}^{2} \langle \frac{\partial u}{\partial x^\alpha}, K_l \rangle^2$, we see that

$$\int_{B_\sigma(0)} e(u) \, dx^1 \, dx^2 \leq \varepsilon_0 c \int_{B_\sigma(0)} e(u) \, dx^1 \, dx^2 + c\sigma \int_{\partial B_\sigma(0)} e(u) \, ds,$$

and hence if $\varepsilon_0 c \leq \frac{1}{2}$, then

$$\int_{B_\sigma(0)} e(u) \, dx^1 \, dx^2 \leq 2c\sigma \frac{d}{d\sigma} \int_{B_\sigma(0)} e(u) \, ds.$$

This Morrey-type inequality then implies that $\int_{B_\sigma(0)} e(u) \, dx^1 \, dx^2$ decays like a power of r, and hence by Morrey's Lemma [42] the map u is Hölder continuous and therefore smooth. Hence we have established the following result of Helein.

Theorem 3.6. *If the Killing vector fields of X span the tangent space at each point (e.g., if X is homogeneous), then any weakly harmonic map from a surface to X is regular.*

There has been a good deal of activity on harmonic maps and related variational problems generated by the theory of liquid crystals. We will not discuss this direction here, but instead refer the interested reader to the survey paper of F. H. Lin [37] and the references therein.

One of the most interesting geometric applications of harmonic mapping theory has been to the study of rigidity questions for nonpositively curved manifolds. The Mostow rigidity theorem [44] was proven for Hermitian locally symmetric spaces by Y. T. Siu [61] by showing that certain harmonic maps are holomorphic. The existence theory of Eells and Sampson [17] asserts that maps into manifolds of nonpositive curvature can be deformed to harmonic maps. In certain cases the Bochner method can be used to show that harmonic maps satisfy more equations (such as the Cauchy-Riemann equations in Siu's case). The Bochner argument was improved in a useful way by J. Sampson [49] who showed that harmonic maps from a Kähler manifold to a manifold of negative curvature operator have rank at most two. These results were systematically analyzed by J. Carlson and D. Toledo [11] who studied the possible ranks of harmonic maps from Kähler manifolds into locally symmetric spaces. The Kähler assumption on the domain was removed by K. Corlette [15] who realized that one gets conditions on the second derivatives of a harmonic map whenever the domain manifold possesses a parallel form w of any degree. In case the stabilizer of the form $w(x) \in \bigwedge^p T_x^*M$ in $\mathrm{GL}(T_xM)$ is a compact group, it follows that the map is totally geodesic; that is, has vanishing second covariant derivatives. Corlette then used this to establish super-rigidity results for homomorphisms of lattices in the rank-1 groups $\mathrm{Sp}(n, 1)$ and F_4^{-20} into Lie groups (so-called Archimedian super-rigidity). Recall that super-rigidity had been conjectured by A. Selberg and proved by G. Margulis for semisimple algebraic groups of real rank at least two (see [73]). On the other hand it is known to fail for lattices in the rank-1 groups $\mathrm{SO}(n, 1)$ for all n and $\mathrm{SU}(n, 1)$ for $n = 2, 3, 4$. (See [73] for references to this literature.)

It was pointed out to the author by M. Gromov that the applications of harmonic maps to rigidity problems could be substantially amplified if one could allow the image space X to be a singular space such as a simplicial complex. In particular, one can study p-adic representations of a lattice by constructing maps into Bruhat-Tits buildings associated with simple algebraic p-adic groups. (See K. Brown [8] for a readable treatment of Euclidean buildings.) This program has been successful, and it can be shown that p-adic super-rigidity holds for lattices in $\mathrm{Sp}(n, 1)$ and F_4^{-20}. We now describe our results in this direction. Complete details will appear in a forthcoming paper with M. Gromov. To make plausible the idea that there might be a good theory of harmonic maps into certain types of singular spaces, recall that if X is a nonpositively curved manifold and $u: \Omega \to X$ is a harmonic map from an open subset $\Omega \subseteq M$, then there is a uniform Lipschitz estimate for any $\Omega_1 \Subset \Omega$ of the form

$$\sup_{\Omega_1} e(u) \leq c(M, \Omega_1) \int_{\Omega} e(u)\, d\mu_g,$$

where the constant c is independent of both u and the target space X. The traditional proof of this inequality is based on the Bochner formula

$$\tfrac{1}{2}\Delta e(u) = |\nabla du|^2 - \sum_{i,j=1}^{n} \langle \mathscr{R}^X(u_*(e_i \wedge e_j)), u_*(e_i \wedge e_j)\rangle + \mathrm{Ric}^m(du, du),$$

where $e_1, \cdots, e_n$ is a local orthonormal tangent frame on M and $\mathscr{R}^X$ denotes the curvature operator of X. Since the vectors $u_*(e_i \wedge e_j)$ are simple, we see that if X has nonpositive sectional curvature, then $e(u)$ is a subsolution of a linear operator of the form $\Delta_M + c$, where c depends only on a lower bound for the Ricci curvature of M. The pointwise bound on $e(u)$ then follows from standard mean-value type inequalities. The Lipschitz estimate suggests that there may be an existence theory for Lipschitz harmonic maps into singular spaces of nonpositive curvature. We now proceed to formulate such a result.

Let X be a Riemannian simplicial complex; that is, a (locally compact) simplicial complex the faces of which are endowed with Riemannian metrics extending smoothly to the closure such that lower-dimensional faces have the metric induced from inclusion in the closure of higher-dimensional faces. For example, a one-dimensional Riemannian simplicial complex is a graph with a length assigned to each edge. We assume for simplicity that X is embedded in a Euclidean space $\mathbf{R}^Q$ in such a way that each face has the induced Riemannian metric from $\mathbf{R}^Q$.

Assuming that X is connected, it is clear that any two points $x_0, x_1 \in X$ can be joined by a path $\gamma: [0, 1] \to X$ which is Lipschitz as a map to $\mathbf{R}^Q$. We can then define the Riemannian distance function $d(x_0, x_1)$ by

$$d(x_0, x_1) = \inf\{L(\gamma): \gamma \text{ a Lipschitz path from } x_0 \text{ to } x_1\}.$$

It is easy to see that $d(\cdot, \cdot)$ is a metric, and that (X, d) is a complete metric space. Moreover, the infimum is attained by a Lipschitz path. To see this, we introduce the energy of a path $E(\gamma)$, and observe that standard analytic methods produce an energy minimizing path γ from x_0 to x_1. Of course, as a map $\gamma: [0, 1] \to \mathbf{R}^Q$, γ is Hölder continuous with the

Hölder exponent $\frac{1}{2}$ by the fundamental theorem of calculus. We now claim that the function $\frac{d\gamma}{dt}$, the distributional derivative of γ, has constant length almost everywhere on $[0, 1]$. To see this, let $\zeta(t)$ be a smooth real-valued function with compact support in $(0, 1)$, and consider the path of maps $\gamma_s(t)$ given by $\gamma_s(t) = \gamma(t + s\zeta(t))$. By the minimizing property of γ we see directly that $E(\gamma) \leq E(\gamma_s)$, and hence the function $s \mapsto E(\gamma_s)$ has a minimum at $s = 0$. We examine this function more carefully:

$$E(\gamma_s) = \int_0^1 \left\| \frac{d\gamma_s}{dt} \right\|^2 dt = \int_0^1 \left\| \frac{d\gamma_s}{dt} \right\|^2 (1 + s\zeta'(t))^{-1}\, d\tau,$$

where $\tau = t + s\zeta(t)$, s being a *fixed* small number. Now $\frac{d\gamma_s}{dt}(t) = \frac{d\gamma_s}{d\tau} \cdot (1+s\zeta'(t))$ by the chain rule, and hence $E(\gamma_s) = \int_0^1 \|\frac{d\gamma}{d\tau}\|^2 (1+s'(t))\, d\tau$. Since $\frac{d}{ds}E(\gamma_s) = 0$ at $s = 0$, we conclude

$$\int_0^1 \left\| \frac{d\gamma}{dt} \right\|^2 \zeta'(t)\, dt = 0$$

for every smooth function $\zeta(t)$ with compact support in $(0, 1)$. This implies that $\|\frac{d\gamma}{dt}\|$ is equal to a constant almost everywhere on $[0, 1]$. We must then have $\|\frac{d\gamma}{dt}\| = L(\gamma)$ a.e., and hence $E(\gamma) = L(\gamma)^2$. Now if γ_1 is *any* Lipschitz path from x_0 to x_1, we can parametrize γ_1 proportionate to arclength so that $\|\frac{d\gamma_1}{dt}\| = L(\gamma_1)$ a.e. Thus

$$L^2(\gamma_1) = E(\gamma_1) \geq E(\gamma) = L^2(\gamma),$$

and we see that γ realizes the distance between x_0 and x_1. In general, γ will be no smoother than Lipschitz; however, any open segment of γ which lies in a face of X is a curve of least length in that face, and hence a smooth geodesic.

In order to expect a theory of continuous harmonic mappings from higher-dimensional manifolds to X it will be necessary to assume that X has nonpositive curvature in a suitable sense. (Without this assumption harmonic maps will sometimes be discontinuous.) Let $y, x_0, x_1 \in \mathbf{R}^n$, and let $x(s)$ be the unit speed geodesic from x_0 to x_1 parametrized on $[0, l]$, $l = |x_0 - x_1|$. Notice that the function $f_0(s) = |x(s) - y|^2$ is determined by the conditions $f_0''(s) = 2$, $f_0(0) = |x_0 - y|^2$, $f_0(l) = |x_1 - y|^2$. We say that X has *nonpositive curvature* if for any three points $y, x_0, x_1 \in X$, the function $f(s) = d^2(x(s), y)$ satisfies $f(s) \leq f_0(s)$, where $x(s)$, $s \in [0, l]$, $l = d(x_0, x_1)$, denotes a unit speed geodesic from x_0 to x_1, and f_0 is the function satisfying $f_0'' = 2$, $f_0(0) = d^2(x_0, y)$,

$f_0(l) = d^2(x_1, y)$. Thus the condition states that points of the side of a geodesic triangle opposite to y in the space X are at least as close to y as they would be in a Euclidean triangle with the same side lengths. The reader can consult [8] for an elementary discussion of nonpositively curved metric spaces. The following properties can be derived from the definition of nonpositive curvature. First, any two points in X can be joined by precisely one geodesic path. Second, if x_0, x_1 and y_0, y_1 are two pairs of points in X, and we parametrize the geodesic paths from x_0 to x_1 and from y_0 to y_1 by $x(t), y(t)$ for $t \in [0, 1]$, where t is proportional to arclength along each of the paths, then the function $g(t) = d(x(t), y(t))$ is a convex function of t. Note that this second property implies that geodesics from a point spread more quickly than Euclidean geodesics since we may take $x_0 = y_0$ to conclude that $d(x(t), y(t)) \geq (t/t')d(x(t'), y(t'))$ for $0 \leq t' \leq t \leq 1$. Finally, we observe that for any $\lambda \in [0, 1]$ and $y \in X$ we can define a map $R_{\lambda, y}: X \to X$ by setting

$$R_{\lambda, y}(x) = x(\lambda),$$

where $x(t)$, $t \in [0, 1]$, denotes the constant speed geodesic from y to x. By our previous discussion we see that each $R_{\lambda, y}$ is a Lipschitz map; in fact

$$d(R_{\lambda, y}(x_0), R_{\lambda, y}(x_1)) \leq \lambda d(x_0, x_1),$$

so the Lipschitz constant is at most λ. Moreover, the family of maps $R_{\lambda, y}$, $\lambda \in [0, 1]$, defines a deformation retraction of X to the point y, so that X is necessarily contractible.

The direct method enables us to construct an energy minimizing map $u: \Omega \to X$ with a given map φ specified on $\partial\Omega$ (say, for example, that φ is Lipschitz). Here Ω is a compact manifold with boundary. This map will be a priori in the space $H^1(\Omega, X)$ defined by

$$H^1(\Omega, X) = \{v \in H^1(\Omega, \mathbf{R}^Q): v(x) \in X \text{ a.e. } x \in \Omega\}.$$

We now discuss some important properties of energy minimizing maps. Since our discussion is local on the domain manifold, we will assume the domain metric is Euclidean. Generally, certain allowable error terms appear in the calculations. We will merely sketch the arguments here (full details will appear elsewhere), so we omit this complication. The first observation is that the usual monotonicity inequality for harmonic maps holds. This is derived by considering a variation of the map u of the following type: Let $\zeta(x)$ by a smooth function with compact support in Ω, and for $|\tau|$ small consider the diffeomorphism of Ω given by $f_\tau(x) = (1+\tau\zeta(x))x$. Consider the maps $u_\tau = u \circ F_\tau$. Then the function $\tau \mapsto E(u_\tau)$

has a minimum at $\tau = 0$, because $u = u_0$ is energy minimizing. To analyze this condition we perform a change of variables as we did in the geodesic case above; that is, let $y = F_\tau(x)$ and observe

$$E(u_\tau) = \int_\Omega \sum_{i,j,k} \frac{\partial y_j}{\partial x_i}\frac{\partial y_k}{\partial x_i}\left(\frac{\partial u}{\partial y_j}\cdot\frac{\partial u}{\partial y_k}\right)\det\left(\frac{\partial x_p}{\partial y_q}\right) dy.$$

In particular, $E(u_\tau)$ is a differentiable function of τ, and by direct calculation we have

$$\frac{d}{d\tau}E(u_\tau)|_{\tau=0}$$
$$= \int_\Omega \left[|\nabla u|^2(2-n)\zeta - |\nabla u|^2\sum_i x_i\frac{\partial \zeta}{\partial x_i} + 2\sum_{i,j}\frac{\partial \zeta}{\partial x_i}x_j\frac{\partial u}{\partial x_j}\cdot\frac{\partial u}{\partial x_i}\right] dx.$$

Taking ζ to be an approximation to the characteristic function of a ball, say $B_\sigma(0) \subset \mathbf{R}^n$, we get (since u minimizes)

$$0 = (2-n)\int_{B_\sigma}|\nabla u|^2\,dx + \sigma\int_{\partial B_\sigma}|\nabla u|^2\,d\Sigma - 2\sigma\int_{\partial B_\sigma}\left|\frac{\partial u}{\partial r}\right|^2 d\Sigma. \tag{3.1}$$

We can gain additional information by using the fact that X has nonpositive curvature. Let $\zeta(x) \geq 0$ be a smooth nonnegative function with compact support in Ω, and for small $\tau \geq 0$ consider the variation of u given by $u_\tau = F_{1-\tau\zeta(x),y} \circ u$, where $F_{\lambda,y}: X \to X$ was constructed above, and $y \in X$ is any chosen point. The following calculation can be justified; we omit the details here but operate formally:

$$\frac{\partial u_\tau}{\partial x_i} = dF_{1-\tau\zeta,y}\left(\frac{\partial u}{\partial x_i}\right) - \tau\frac{\partial \zeta}{\partial x_i}\cdot\frac{\partial F_{1-\tau\zeta,y}}{\partial \lambda}.$$

From the contracting property of $F_{\lambda,y}$ it follows that

$$E(u_\tau) \leq \int_\Omega [(1-\tau\zeta)^2|\nabla u|^2 - \tau\nabla\zeta\cdot\nabla(d^2(u(x),y))]\,dx + o(\tau).$$

Using the fact that $E(u) \leq E(u_\tau)$ we obtain

$$0 \leq -\tau\int_\Omega [2\zeta|\nabla u|^2 + \nabla\cdot\nabla(d^2(u(x),y))]\,dx + o(\tau).$$

Dividing by τ and letting τ go to zero we have

$$\int_\Omega [2\zeta|\nabla u|^2 + \nabla\zeta\cdot\nabla(d^2(u(x),y))]\,dx \leq 0$$

for any $\zeta \geq 0$ with compact support. Choosing ζ to approximate the characteristic function of $B_\sigma(0)$ yields

$$2\int_{B_\sigma} |\nabla u|^2\, dx \leq \int_{\partial B_\sigma} \frac{\partial}{\partial r}(d^2(u(x), y))\, d\Sigma. \tag{3.2}$$

If we let $E(\sigma) = \int_{B_\sigma} |\nabla u|^2\, dx$ and $I(\sigma) = \int_{\partial B_\sigma} d^2(u(x), y)\, d\Sigma$, we can compute logarithmic derivatives

$$\frac{I'(\sigma)}{I(\sigma)} = \frac{n-1}{\sigma} + (I(\sigma))^{-1} \int_{\partial B_\sigma} \frac{\partial}{\partial r}(d^2(u, y))\, d\Sigma,$$

whereas from (3.1) we obtain

$$\frac{E'(\sigma)}{E(\sigma)} = \frac{n-2}{\sigma} + 2(E(\sigma))^{-1} \int_{\partial B_\sigma} \left|\frac{\partial u}{\partial r}\right|^2 d\Sigma.$$

Therefore

$$\frac{I'(\sigma)}{I(\sigma)} - \frac{E'(\sigma)}{E(\sigma)} = \frac{1}{\sigma} + (E(\sigma)I(\sigma))^{-1} \left[E(\sigma) \int_{\partial B_\sigma} \frac{\partial}{\partial r}(d^2(u, y))\, d\Sigma \right.$$
$$\left. -2I(\sigma) \int_{\partial B_\sigma} \left|\frac{\partial u}{\partial r}\right|^2 d\Sigma\right],$$

which together with (3.2) implies the inequality

$$\frac{d}{d\sigma} \log\left(\frac{I(\sigma)}{\sigma E_\sigma}\right) \leq (E(\sigma)I(\sigma))^{-1} \left[\left(\int_{\partial B_\sigma} d(u, y) \frac{\partial}{\partial r} d(u, y)\right)^2\right.$$
$$\left. - \left(\int_{\partial B_\sigma} d^2(u, y)\right)\left(\int_{\partial B_\sigma} \left|\frac{\partial u}{\partial r}\right|^2\right)\right]. \tag{3.3}$$

Since $|\frac{\partial}{\partial r} d(u, y)| \leq |\frac{\partial u}{\partial r}|$, by the Schwarz inequality we see that

$$\frac{d}{d\sigma}\left(\frac{\sigma E(\sigma)}{I(\sigma)}\right) \geq 0. \tag{3.4}$$

For any $x \in \Omega$, $\sigma > 0$, $y \in X$ we define $\mathrm{Ord}(x, \sigma, y)$ by

$$\mathrm{Ord}(x, \sigma, y) = \frac{\sigma \int_{B_\sigma(x)} |\nabla u|^2}{\int_{\partial B_\sigma(x)} d^2(u, y)\, d\Sigma}.$$

The reason for this notation is that for a harmonic function u,

$$\lim_{\sigma \downarrow 0} \mathrm{Ord}(x, \sigma, u(x)) = \mathrm{Order}_x(u - u(x));$$

that is, the order with which u attains its value $u(x)$ at x. In particular, for harmonic functions (or harmonic maps into smooth manifolds of nonpositive curvature) this limit is a positive integer.

Generally if $x \in \Omega$ and $\sigma > 0$, then the function

$$y \mapsto \int_{\partial B_\sigma(x)} d^2(u, y)\, d\Sigma$$

is a proper convex function on X and hence has a unique minimum point $y_{x,\sigma} \in X$. The function $y \mapsto \operatorname{Ord}(x, \sigma, y)$ thus has a unique maximum point at $y_{x,\sigma}$. We now define $\operatorname{Ord}(x)$ by

$$\operatorname{Ord}(x) = \lim_{\sigma \to 0} \operatorname{Ord}(x, \sigma, y_{x,\sigma})$$

observing that $\sigma \mapsto \operatorname{Ord}(x, \sigma, y_{x,\sigma})$ is a monotone increasing function of σ so that the limit exists. Moreover, for a fixed $\sigma > 0$, the function $x \mapsto \operatorname{Ord}(x, \sigma, y_{x,\sigma})$ is a continuous function, and hence it follows that the function $x \mapsto \operatorname{Ord}(x)$ is upper semicontinuous since it is the decreasing limit of a family of continuous functions. It can be seen that $\operatorname{Ord}(x) \geq 1$ for almost every $x \in \Omega$ and hence for *all* $x \in \Omega$ by upper semicontinuity. Suppose we are at a point $x_0 \in \Omega$ where $\alpha = \operatorname{Ord}(x_0)$. Fix $\sigma_0 > 0$ so that $B_{\sigma_0}(x_0) \Subset \Omega$, and let $\sigma_1 \in (0, \sigma_0)$. Because of the monotonicity of the ratio we have

$$\sigma \int_{B_\sigma(x_0)} |\nabla u|^2 \geq \alpha \int_{\partial B_\sigma} d^2(u, y_1)\, d\Sigma$$

for all $\sigma \in [\sigma_1, \sigma_0)$, where $y_1 = y_{x_0, \sigma_1}$. Combining this with (3.2) yields

$$\begin{aligned} \alpha I(\sigma) &\leq \frac{1}{2}\sigma \int_{\partial B_\sigma(x_0)} \frac{\partial}{\partial r}(d^2(u(x), y_1))\, d\Sigma(x) \\ &= \frac{1}{2}(\sigma I'(\sigma) - (n-1) I(\sigma)). \end{aligned}$$

Therefore $\frac{d}{d\sigma}(\sigma^{-2\alpha-(n-1)} I(\sigma)) \geq 0$ for $\sigma \in [\sigma_1, \sigma_0]$, and hence it follows that $\sigma_1^{-(n-1)} I(\sigma_1) \leq c\sigma_1^{2\alpha}$ (we have now fixed σ_0). Since the function $d^2(u(x), y_1)$ is a subharmonic function (as we saw above), by the mean-value inequality we obtain

$$\sup_{x \in B_{\sigma_{1/2}}(x_0)} d^2(u(x), y_1) \leq c\sigma_1^{2\alpha}.$$

In particular, it follows that $d(u(x), u(x_0)) \leq c|x - x_0|^\alpha$ for $x \in B_{\sigma_0}(x_0)$. As a consequence we conclude that u is Lipschitz in the interior of Ω.

Proposition 3.7. *A minimizing harmonic map from* Ω *to* X *is Lipschitz in the interior provided that* X *has nonpositive curvature.*

It is possible to make the transition from the Dirichlet problem to the homotopy problem (see [50]) to prove, for example, an extension of the Eells-Sampson theorem with singular image space. For the purpose of the statement assume that X is a Riemannian complex whose universal covering space $\tilde{X}$ has nonpositive curvature. Let $\mathscr{M}$ be a smooth compact Riemannian manifold, and $\varphi: M \to X$ be a given continuous map.

Proposition 3.8. *There is a Lipschitz harmonic map* $u: M \to X$ *which is freely homotopic to* φ. *Moreover,* u *minimizes energy among all Lipschitz maps homotopic to* φ.

In order to obtain more information than a Lipschitz bound on a harmonic mapping u, we make the observation that the ratio $\mathrm{Ord}(x, \sigma, y)$ is invariant under both rescalings of the domain and the range. To say this precisely, observe that if $\lambda, \mu > 0$, then we can consider a rescaled map $u_{\lambda,\mu}(x) = \mu u(\lambda x)$ which is a new harmonic map from $\lambda\Omega = \{\lambda x: x \in \Omega\}$ to the complex $\mu X = \{\mu v: v \in X\}$. Of course the centers of dilation in both Ω and $\mathbf{R}^Q$ are arbitrary. Notice that the complex μX again has nonpositive curvature since distances are multiplied by a constant factor. It is easily checked that $\mathrm{Ord}^u(x, \sigma, y)$ is equal to $\mathrm{Ord}^{u_{\lambda,\mu}}(\lambda x, \lambda\sigma, \mu y)$. In particular, by translation of coordinates assume that $0 \in \Omega$, $u(0) = 0 \in X \subset \mathbf{R}^Q$. For any small $\lambda > 0$, let μ be defined by

$$\mu = \max\{d(u(x), 0): |x| \leq \lambda\}.$$

The map $u_{\lambda,\mu}$ then has the property that it maps the unit ball in $\mathbf{R}^n$ into the unit ball (but no smaller ball) centered at 0 in μX. Now let $\{\lambda_i\}$ be a sequence tending to zero, and let $u_i = u_{\lambda_i,\mu_i}$ be defined as above. The sequence $\{u_i\}$ has a uniform Lipschitz bound on compact subsets of $\mathbf{R}^n$, and since $d^2(u_i, 0)$ is subharmonic we have

$$\max\{d^2(u_i(x), 0): |x| = 1\} = 1$$

for every i. (Note that distance is computed here in $\mu_i X$.) Therefore a subsequence again denoted $\{u_i\}$ converges to a map $u_0: \mathbf{R}^n \to X_0$ which is a harmonic map from $\mathbf{R}^n$ to the tangent cone of X at 0 which we denote X_0. Note that X_0 is a geometric cone in $\mathbf{R}^Q$ so that the distance from a point $v \in X_0$ to 0 is simply $|v|$, the Euclidean norm of v. Because of the scale invariance of the order function one can deduce that $\mathrm{Ord}^{u_0}(0, \sigma, 0) = \mathrm{Ord}^u(0) \equiv \alpha$ for every $\sigma > 0$. Now since $R_{\lambda,0}(v) = \lambda v$, it can be seen that equality holds in (3.2), and hence we have

$$E(\sigma) = \frac{1}{2}\sigma^{n-1}\frac{d}{d\sigma}(\sigma^{1-n}I(\sigma)).$$

Using the fact that $E(\sigma) = \sigma^{-1}\alpha I(\sigma)$, we then conclude that $\sigma^{1-n}I(\sigma) = \sigma^{2\alpha}I(1)$. Since the right-hand side of (3.3) must be zero, it follows that $\frac{\partial u}{\partial r} = h(r)u$ for some function $h(r)$, so that $u_0(r, \xi) = g(r)u_0(1, \xi)$ for some Lipschitz function $g(r)$ with $g(1) = 1$. From the fact that $\sigma^{1-n}I(\sigma) = \sigma^{2\alpha}I(1)$ we deduce that $g(r) = r^\alpha$, and hence u_0 is homogeneous of degree $\alpha = \mathrm{Ord}(0)$.

In general, homogeneous harmonic maps can be quite complicated, however, there is one particularly simple class of such maps which we now describe. Suppose there is an isometric embedding $E\colon \mathbf{R}^m \to X_0$ which has totally geodesic image. Given a harmonic map $w\colon \mathbf{R}^n \to \mathbf{R}^m$, we can then construct a harmonic map $E \circ w\colon \mathbf{R}^n \to X_0$. If $E(0) = 0$, and w is a homogeneous map (hence given by spherical harmonics of some degree), then $E \circ w$ is also a homogeneous map. The following important lemma tells us that homogeneous maps of degree 1 are all described in this way.

Lemma 3.9. *Suppose $u_0\colon \mathbf{R}^n \to X_0$ is homogeneous of degree 1. There exists an isometric embedding $E\colon \mathbf{R}^m \to X_0$ for some $1 \le m \le n$ and a linear map $L\colon \mathbf{R}^n \to \mathbf{R}^m$ so that $u_0 = E \circ L$.*

The essential reason for the truth of this lemma is that the map u_0 has least possible order at 0, and we must have $\mathrm{Ord}(x) = 1$ for every $x \in \mathbf{R}^n$ since the function $\mathrm{Ord}(x)$ is upper semicontinuous. This tells us that the map u_0 is essentially homogeneous of degree 1 about every point, and in particular is totally geodesic. It can then be shown that the map u_0 is of the above type.

In order to use the Bochner method for harmonic maps into singular spaces it seems to be necessary to discuss higher differentiability properties of harmonic maps. We discuss now a result of this type. We first define what we mean by a smooth point of a harmonic map u. The key observation here is that if there is an isometric totally geodesic embedding E from a neighborhood U of a point p_0 in a smooth Riemannian manifold into X, then a harmonic map $w\colon B_\varepsilon(x_0) \to U$ gives rise to a harmonic map $u = E \circ w\colon B_\varepsilon(x_0) \to X$. Such a map deserves to be considered smooth. Thus for a general harmonic $u\colon \Omega \to X$ we make the following definition.

Definition 3.10. A point $x_0 \in \Omega$ is a regular point of u if u factors locally through an isometric totally geodesic (local) embedding of a smooth manifold into X.

Thus a point x_0 is regular if $u = E \circ w$ as above in a neighborhood of x_0. Let $\mathbf{R}(u)$ denote the regular set, and observe that $\mathbf{R}(u)$ is an open subset of the interior of Ω. Let $\mathscr{S}(u) = \Omega \backslash \mathbf{R}(u)$ denote the singular set of u.

Without further hypothesis on X there are unlikely to be many regular points. We now impose a further hypothesis on X which will imply that the set of regular points is very large. This hypothesis is satisfied in important examples such as buildings [8].

Hypothesis (H). Any two adjacent simplices of X are contained in the image of a totally geodesic isometric embedding of a Euclidean space into X.

One may think of Hypothesis (H) as providing us with sufficiently many totally geodesic submanifolds of X. Without some hypothesis there are likely to be very few of these, and as a result very few regular points of a harmonic map u into X.

We now state the first result which provides a local description of order one points for a harmonic map into a complex X satisfying (H).

Theorem 3.11. *Assume X satisfies* (H), *and $u\colon \Omega \to X$ is harmonic. There exists a positive number ε_0 depending only on n and X such that for any $x_0 \in \Omega$ we have* $\operatorname{Ord}^u(x_0) \geq 1 + \varepsilon_0$ *or* $\operatorname{Ord}(x_0) = 1$. *If* $\operatorname{Ord}(x_0) = 1$, *then there is a ball $B_{r_0}(x_0)$ for some $r_0 > 0$ such that $u(B_{r_0}(x_0))$ is contained in a totally geodesic subcomplex X_1 of X such that X is isometric to $\mathbf{R}^d \times X_2^{k-d}$ for some integer d with $1 \leq d \leq k$ and for a $(k-d)$-dimensional complex X_2 which also satisfies* (H). *Thus the map u may be written as $u = (u_0, u_1)$, where $u_0\colon B_{r_0}(x_0) \to \mathbf{R}^d$ is a smooth rank d harmonic map, and $u_1\colon B_{r_0}(x_0) \to X_1$ satisfies* $\operatorname{Ord}^{u_1}(x_0) \geq 1 + \varepsilon_0$.

We will refer to the integer d given in Theorem 3.11 as the rank of u at x_0. If $\operatorname{Ord}^u(x_0) > 1$, we will say that u has rank 0 at x_0. Note also that the upper semicontinuity together with the gap $(1, 1 + \varepsilon_0)$ omitted in the values of $\operatorname{Ord}(x)$ implies that the set $\mathscr{S}_1$ defined by

$$\mathscr{S}_1 = \{x \in \Omega\colon \operatorname{Ord}(x) > 1\}$$

is a relatively closed subset of Ω. Moreover, the set $\mathscr{S}_1$ can be shown to have Hausdorff dimension at most $n-2$ by an argument which goes back to H. Federer [22]. To illustrate the argument we show that the set $\mathscr{S}_1$ is discrete for $n = 2$, and we omit the remaining details for $n \geq 3$ as they will appear elsewhere. Assume on the contrary that $0 \in \mathscr{S}_1 \subset \Omega \subset \mathbf{R}^2$ is an accumulation point of $\mathscr{S}_1$. Let $\{x_j\} \subset \mathscr{S}_1$ be a sequence with $x_j \neq 0$ and $0 = \lim x_j$. Let $\lambda_j = |x_j|$, and construct the dilated sequence

$\{u_j\}$ as above. The u_j are then harmonic maps, a subsequence of which converges to a homogeneous map $u_0: \mathbf{R}^2 \to X_0$ where we assume as above that $0 \in X$, and X_0 denotes the tangent cone to X at 0. Notice that by construction each of the u_j has a point on the unit circle $|x_j|^{-1}x_j$ at which the order is greater than $1+\varepsilon_0$. It then follows that u_0 has at least one point $x_0 \in S^1$ at which $\operatorname{Ord}^{u_0}(x_0) > 1$. Since u_0 is homogeneous, we have $\operatorname{Ord}^{u_0}(\lambda x_0) = \operatorname{Ord}^{u_0}(x_0)$ for $\lambda > 0$, and hence $\operatorname{Ord}^{u_0}(0) \geq \operatorname{Ord}^{u_0}(x_0)$. Further, since $\frac{d}{d\lambda}u(\lambda x_0) = 0$ for $\lambda \geq 0$ it follows that the ray $\{\lambda x_0 : \lambda \geq 0\}$ is mapped to 0 by u_0. We now rescale at the point x_0, and consider the maps u_{0i} given by

$$u_{0i}(x) = \mu_i u_0(x_0 + \lambda_i x),$$

where μ_i is chosen as above. A subsequence of u_{0i} then converges to a homogeneous harmonic map $u_1: \mathbf{R}^2 \to X_0$. Since the map

$$u_{0i}\left(\lambda x + \frac{\lambda - 1}{\lambda i}x_0\right) = \lambda^\alpha u_{0i}(x)$$

for each i and every $\lambda > 0$, we may differentiate in λ and set $\lambda = 1$ to obtain $\nabla u_{0i} \cdot x_0 = 0(\lambda_i)$. Hence the limit u_1 satisfies $\nabla u_1 \cdot x_0 = 0$. If we take $x_0 = (1, 0)$, then u_1 is independent of the first coordinate x_1, and hence $u_1(s) = u_1(0, s)$ is a constant speed geodesic. But clearly it then follows that $\operatorname{Ord}^{u_1}(0) = 1$, a contradiction. This shows that $\mathscr{S}_1$ is a discrete set of points. Note that by Theorem 3.11 for $k = \dim X = 1$ we have $\mathscr{S} \subseteq \mathscr{S}_1$, and hence the Hausdorff dimension of $\mathscr{S}$ is at most $n-2$. Note also that for a regular point $x_0 \in \Omega$, there is a neighborhood of x_0 in which u may be represented by a smooth harmonic map into a Euclidean space. In particular, it follows that $\Delta e(u) = 2|\nabla\nabla u|^2$, where $|\nabla\nabla u|^2$ represents the sum of squares of the second derivatives of u. The following result will be important for the application of the Bochner method for maps into complexes.

Proposition 3.12. *The Hausdorff dimension of $\mathscr{S}$ is a most $n-2$. For any $\Omega_1 \Subset \Omega$, there is a sequence of Lipschitz functions $\{\psi_i\}$ with $\psi_i \equiv 0$ in a neighborhood of $\mathscr{S}$, $0 \leq \psi_i \leq 1$, and $\psi_i(x) \to 1$ for all $x \in \Omega_1 - \mathscr{S}$ such that*

$$\lim_{i\to\infty} \int_{\Omega_1} |\nabla\nabla u||\nabla\psi_i|\,dx = 0.$$

This result states that in an average sense the second derivatives of u blow-up more slowly than the reciprocal of the distance to $\mathscr{S}$. We outline the proof of this result. It is proven by induction on $k = \dim X$. For $k = 1$ we have already seen that $\mathscr{S} \subseteq \mathscr{S}_1$, and hence the first statement follows. To prove the second, let $\varepsilon > 0$ and $d > (n-2)$. Let Ω_2 be a compact subdomain of Ω with $\Omega_1 \Subset \Omega_2$, and choose a finite covering $\{B_{r_j}(x_j): j = 1, \cdots, l\}$ of the compact set $\mathscr{S} \cap \overline{\Omega}_2$ satisfying $\sum_{j=1}^{l} r_j^d \leq \varepsilon$. Let φ_j be a Lipschitz function which is zero on $B_{r_j}(x_j)$ and identically one on $\mathbf{R}^n - B_{2r_j}(x_j)$ such that $|\nabla \varphi_j| \leq 2r_j^{-1}$. We assume also that $x_j \in \mathscr{S} \cap \overline{\Omega}_2$. Let φ be defined by $\varphi = \min\{\varphi_j : j = 1, \cdots, l\}$, and observe that φ vanishes in a neighborhood of $\mathscr{S} \cap \overline{\Omega}_2$, and $\varphi \equiv 1$ on $\mathbf{R}^n - \bigcup_{j=1}^{l} B_{2r_j}(x_j)$. Now let $\psi = 1 - \varphi^2(1-\varphi)^2$, and observe

$$\int_{\Omega_1} |\nabla\nabla u||\nabla \psi|\, dx \leq 2 \left(\int_{\Omega_1} |\nabla\nabla u|^2 |\nabla u|^{-1} \varphi^2 (1-\varphi)^2\, dx \right)^{1/2} \times \left(\int_{\Omega_1} |\nabla u||\nabla \varphi(1-\varphi)|^2\, dx \right)^{1/2}$$

by the Schwarz inequality. On the other hand, an elementary result for harmonic maps (see [56]) implies that on the regular set we have for a positive ε_n depending only on n

$$\Delta|\nabla u| \geq \varepsilon_n |\nabla\nabla u|^2 |\nabla u|^{-1}.$$

Let ζ be a smooth function with support in Ω_2 and $\zeta \equiv 1$ on Ω_1, and observe

$$\int_{\Omega_1} |\nabla\nabla u|^2 |\nabla u|^{-1} \varphi^2 (1-\varphi)^2\, dx \leq \int_{\Omega} |\nabla\nabla u|^2 |\nabla u|^{-1} \varphi^2 (1-\varphi)^2 \zeta^2\, dx.$$

Then using the above differential inequality we obtain

$$\int_{\Omega} |\nabla\nabla u|^2 |\nabla u|^{-1} \varphi^2 (1-\varphi)^2 \zeta^2\, dx \leq 4 \int_{\Omega} |\nabla u| \varphi(1-\varphi)\zeta \nabla|\nabla u| \cdot \nabla(\varphi(1-\varphi)\zeta),$$

which easily implies

$$\int_{\Omega} |\nabla\nabla u|^2 |\nabla u|^{-1} \varphi^2 (1-\varphi)^2 \zeta^2\, dx \leq c \int_{\Omega} |\nabla u||\nabla(\varphi(1-\varphi)\zeta)|^2\, dx.$$

Combining this with our previous bound yields

$$\int_{\Omega_1} |\nabla\nabla u||\nabla \psi|\, dx \leq c \int_{\Omega} |\nabla u| (\zeta^2 |\nabla \varphi(1-\varphi)|^2 + \varphi^2(1-\varphi)^2 |\nabla \zeta|^2)\, dx.$$

Therefore

$$\int_{\Omega_1} |\nabla\nabla u||\nabla \psi|\, dx \leq c \sum_{j=1}^{l} r_j^{-2} \int_{B_{2r_j}(x_j)} |\nabla u|\, dx.$$

Recall that we can estimate the decay of the energy on small balls in terms of the value of $\mathrm{Ord}(\cdot)$ at the center; in particular, since $\mathrm{Ord}(x_j) \geq 1 + \varepsilon_0$ we have

$$\int_{B_{2r_j}(x_j)} |\nabla u|^2 \, dx \leq c r_j^{n+2\varepsilon_0}.$$

Therefore

$$\int_{\Omega_1} |\nabla\nabla u||\nabla \psi| \, dx \leq c \sum_{j=1}^{l} r_j^{n-2+\varepsilon_0} \leq c\varepsilon$$

provided $d \leq n - 2 + \varepsilon_0$. The desired result now follows by choosing a sequence $\varepsilon_i \to 0$ and a corresponding sequence $\{\psi_i\}$ of functions. This completes the proof for $k = 1$. To verify the inductive step, first we cover the set $(S \cap \mathscr{S}_1) \cap \overline{\Omega}_2$ with balls and construct a function ψ_0 as above. Next we cover the set $(\mathscr{S} - \bigcup_{j=1}^{l} B_{r_j}(x_j)) \cap \overline{\Omega}_2$ with balls $\{B_{r_p}(y_p) : p = 1, \cdots, q\}$ in which the map u can be written $u = (u_0, u_1)$ as in the conclusion of Theorem 3.11. We then use the inductive assumption to construct a function ψ_p vanishing in a neighborhood of $\mathscr{S} \cap \overline{B_{r_p}(y_p)}$ and identically one outside a slightly larger neighborhood with

$$\int_{B_{r_p}(y_p)} |\nabla\nabla u||\nabla \psi_p| \, dx \leq \varepsilon 2^{-p}.$$

We finally set $\psi = \min\{\psi_0, \psi_1, \cdots, \psi_q\}$, and conclude

$$\int_{\Omega_1} |\nabla\nabla u||\nabla \psi| \, dx \leq c\varepsilon.$$

Choosing a sequence of ε's and ψ's hence completes the argument.

The main application of these results is to apply the Bochner method for harmonic maps into complexes which satisfy (H). To illustrate this point we extend a vanishing theorem of K. Corlette [15] to this setting. Suppose that M is a quarternionic Kähler manifold which is compact without boundary, and that $u : M \to X$ is a harmonic map where the universal cover $\tilde{X}$ of X is a complex of nonpositive curvature satisfying (H). Let ω be the quaternionic Kähler 4-form on M, and notice that at a regular point of u we can form the exterior product $\omega \wedge du$ which is locally a 5-form with values in the pullback of the tangent bundle of a smooth flat manifold containing the image of u locally. The main vanishing result then comes from the conclusion that $\omega \wedge du$ is co-closed with respect to the exterior derivative arising from the pullback connection ∇. The local computation of [15] shows that

$$d_\nabla \delta_\nabla (\omega \wedge du) = 0$$

since in our case the image manifold is flat. Let ψ be a function constructed from Proposition 3.12 which vanishes in a neighborhood of $\mathcal{S}$ and is identically one outside a slightly larger neighborhood such that

$$\int_M |\nabla\nabla u||\nabla\psi|\, d\mu_M \leq \varepsilon$$

for any given $\varepsilon > 0$. We then form the invariant scalar function $\psi\langle \omega \wedge du, d_\nabla \delta_\nabla(\omega \wedge du)\rangle$ on M. Integrating by parts yields

$$\int_M \psi \|\delta_\nabla(\omega \wedge du)\|^2\, d\mu \leq c \int_M |\nabla u||\nabla\nabla u||\nabla\psi|\, d\mu.$$

Since $|\nabla u|$ is bounded, we have

$$\int_M \psi \|\delta_\nabla(\omega \wedge du)\|^2\, d\mu \leq c\varepsilon,$$

and since $\varepsilon > 0$ is arbitrary we conclude that $\delta_\nabla(\omega \wedge du) \equiv 0$ on the regular set of u. As in [15] this implies that u is totally geodesic, and since the image is flat we would have a local parallel one-form on M if $du \not\equiv 0$. Therefore we conclude that $du \equiv 0$, so that u is constant. It is possible to apply this to study p-adic representations of lattices in $\mathrm{Sp}(n, 1)$ and F_4^{-20} with the image complex taken to be a Euclidean building (see [8]), and prove the arithmeticity of such lattices. This is carried out in a forthcoming joint work with M. Gromov.

We close this section with some historical remarks concerning certain aspects of the proof we have outlined. The monotone ratio which plays a central role originates for harmonic functions with the Three Circles Theorem of J. Hadamard. The statement for a harmonic function $u(x)$ is that the log of the average of u^2 on ∂B_r is a convex function of $\log r$. It is easily seen that the first logarithmic derivative of this function is given by

$$\left(\int_{\partial B_r} u^2\, d\Sigma\right)^{-1} \left(r \int_{B_r} |\nabla u|^2\, dx\right).$$

The fact that this is monotone increasing then expresses the convexity statement. This kind of ratio was used by F. J. Almgren [3] in his study of multivalued harmonic mappings, and also by F. H. Lin [37] for harmonic maps into the cone over a manifold (a problem which arose in the theory of liquid crystals). Also we remark that the more standard monotonicity formula (3.1), the monotonicity of $r^{2-n}E_r(u)$, plays an important role in the regularity theory for minimizing harmonic maps given by the author and K. Uhlenbeck [55]. As we have seen above the monotonicity of the ratio is obtained by combining this with an inequality (3.2) expressing the strong convexity of the distance function in a space of nonpositive curvature.

4. Complete constant mean curvature surfaces

In this section we summarize the striking progress which has been made in the past several years on the global question concerned with constant mean curvature surfaces in the Euclidean space $\mathbf{R}^3$. Let $\Sigma^2 \hookrightarrow \mathbf{R}^3$ be an immersed two-dimensional surface. If g, h denote the first and second fundamental forms of Σ, we then have the mean curvature H given by

$$H = \sum_{i,j=1}^{2} g^{ij} h_{ij}.$$

Thus H is the trace of the second fundamental form with respect to g. The equation of nonzero constant mean curvature may be written $H = 1$ since the sign of H is reversed by reversing the choice of unit normal, and the magnitude of H may be normalized by a homothety. The only obvious closed surface Σ of constant mean curvature is a sphere of the appropriate radius. Notice that a sphere lies in a nontrivial family of solutions with three parameters which may be taken as the coordinates of the center. The constant mean curvature equation is the Euler-Lagrange equation for the variational problem associated with the isoperimetric inequality; that is, a surface Σ has constant mean curvature if and only if Area(Σ) is extremal for variations of Σ, which preserve the enclosed three-dimensional volume. Thus the physical model described by constant mean curvature surfaces is that of a soap bubble where gravity is neglected. It is then a natural question to ask whether a mathematical soap bubble is necessarily round. This question was popularized by H. Hopf [31], and was sometimes referred to as a Hopf conjecture, although it is presently an unsolved problem to determine if Hopf actually conjectured an affirmative answer to this question or whether he simply proposed it as an interesting geometric problem. In any case the question has generated a wealth of deep and important mathematics.

The first result on this problem was given in 1853 by J. H. Jellet [32] who showed that a closed surface of constant mean curvature which is star-shaped with respect to some point is a round sphere. This result was greatly improved by H. Hopf [31] who showed that any surface Σ of constant mean curvature which is homeomorphic to S^2 is necessarily a round sphere. Hopf's method of proof has been important for a variety of problems, so we recall it. The main observation is that the second fundamental form h is a symmetric divergence free tensor on Σ since the Codazzi equations imply

$$\sum_{i,j=1}^{2} g^{ij} h_{ki;j} = H_{;k} = 0.$$

If we choose isothermal coordinates (x^1, x^2) locally on Σ, and set $z = x^1 + \sqrt{-1}x^2$, then the locally defined function $\varphi(z) = (h_{11} - h_{22}) - 2\sqrt{-1}h_{12}$ becomes holomorphic. The global quadratic differential $\Phi = \varphi(z)\,dz^2$ is thus a regular holomorphic quadratic differential on Σ. It is an elementary matter to show that the Riemann sphere S^2 cannot support a nonzero holomorphic quadratic differential. Thus if Σ is simply connected, then $\Phi \equiv 0$ and hence Σ is umbilic. Therefore Σ is a round sphere. Generally Φ is referred to as the Hopf differential of Σ.

The next chapter in the story is the result of A. D. Alexandrov [2] who proved that a closed embedded surface Σ of arbitrary genus with $H \equiv 1$ is a round sphere. In this case the method developed to prove this result has proved to be extremely useful and powerful. This is the Alexandrov method of plane reflection, a symmetry argument relying on the strong maximum principle.

From a variational point of view, the classical isoperimetric inequality implies that the round sphere is the unique least area surface bounding a given volume. That the round sphere is the only locally minimizing surface for the variational problem was shown by L. Barbosa and M. do Carmo [6].

We have now presented all of the evidence known before 1983 concerning the question of whether closed constant mean curvature surfaces are round. Since this evidence is all affirmative, it came as quite a surprise when H. Wente [65], [66] constructed a constant mean curvature torus. We first remark that the equation $H = 1$ can locally be reduced to either the sinh-Gordon equation $\Delta w + \sinh 2w = 0$ or to the harmonic map equation into the standard S^2. The reduction to the sinh-Gordon equation involves working in isothermal coordinates in which the Hopf differential is given by dz^2. This can be achieved globally for tori, and thus the problem can be reduced to studying periodic solutions. Wente's argument involves studying sufficiently interesting families of doubly periodic solutions and showing that some of these give rise to closed surfaces. After Wente's paper, the construction was made more explicit by U. Abresch [1] and R. Walter [63]. These ideas were extended, and a classification result proved by U. Pinkall and I. Sterling [46] for constant mean curvature tori. This makes contact with earlier work on the sinh-Gordon equation by N. Ercolani and G. Forest [18]. There has been further work done by Bobenko.

While there are useful general methods to deal with the construction of constant mean curvature tori, these use heavily the special structure of the torus and are difficult to generalize to higher genus. In 1987 N. Kapouleas [33], [34] introduced a new approach to the construction of complete and closed surfaces of constant mean curvature. To understand this approach one needs to understand slightly more complicated explicit surfaces of constant mean curvature. There is an important classical family of such surfaces introduced by Delaunay in 1841. These are the surfaces of revolution satisfying $H \equiv 1$. This family includes the cylinder as well as singly periodic surfaces which approximate in a singular limit infinite strings of tangent spheres. The idea is to use Delaunay surfaces as building blocks to construct more complicated surfaces. The proper way to think of the nearly singular Delaunay surfaces is that they are made up of positively curved pieces which are spherical, negatively curved which approximate catenoids with small necks, and flat regions which join the two. It turns out that the positively and negatively curved regions play more or less equivalent roles in the constructions. Kapouleas understood in a very precise way the kinds of configurations which lead to constant mean curvature surfaces and was able to construct closed surfaces of any genus greater than two. He also constructed larger classes of complete properly embedded surfaces of constant mean curvature. It turned out to be impossible using the Delaunay surfaces as building blocks to construct surfaces of genus 1 or 2. An analysis of Wente's tori suggests that they are made up of spherical positively curved regions, negatively curved regions which approximate scaled-down Enneper surfaces, and flat regions joining the two. Notice that the Enneper surface and the catenoid are closely related minimal surfaces, and are in fact the only two complete minimal surfaces whose Gauss map is injective. In addition certain of the Wente surfaces can be described rather explicitly, so Kapouleas has very recently succeeded in understanding Wente tori (or pieces of them) as building blocks along with the Delaunay surfaces. This has made it possible for him to construct closed constant mean curvature surfaces of arbitrary genus as well as many other previously unknown types of complete surfaces.

Since the work of Alexandrov gave us a method of controlling embedded constant mean curvature surfaces, it is reasonable to expect that complete noncompact embedded $H = 1$ surfaces may be more rigid than immersed surfaces. In fact a result of N. Korevaar, R. Kusner, and B. Solomon [36] shows that an annular end of such a surface is strongly asymptotic to a Delaunay surface. An earlier work of W. Meeks [39] should also be mentioned here. Korevaar and Kusner have also announced a compactness theorem which yields uniform control on an embedded constant mean curvature surface provided certain weak geometric quantities are controlled.

References

[1] U. Abresch, *Constant mean curvature tori in terms of elliptic functions*, J. Reine Angew. Math. **374** (1987) 169–192.

[2] A. D. Alexandrov, *Uniqueness theorems for surfaces in the large*. V, Vestnik Leningrad Univ. Mat. Mekh. Astronom. **13** (1958) no. 19, 5–8; Amer. Math. Soc. Transl. Ser. 2, **21**, 412–416.

[3] F. J. Almgren, Jr., *Q-valued functions minimizing Dirichlet's integral and the regularity of area minimizing rectifiable currents up to codimension two*, preprint, Princeton University, Princeton, N.J.

[4] T. Aubin, *The scalar curvature*, Differential Geometry and Relativity (Cahen and Flato, eds.), Reider, 1976.

[5] A. Bahri & J. M. Coron, *Une théorie des points critiques à l'infini pour l'equation de Yamabe et le problème de Kazdan-Warner*, C. R. Acad. Sci. Paris Sér. I Math. **15** (1985) 513–516.

[6] L. Barbosa & M. doCarmo, *Stability of hypersurfaces with constant mean curvature*, Math. Z. **185** (1984) 339–353.

[7] A. Besse, *Einstein manifolds*, Springer, Berlin, 1987.

[8] K. Brown, *Buildings*, Springer, New York, 1989.

[9] L. Caffarelli, B. Gidas & J. Spruck, *Asymptotic symmetry and local behavior of semilinear elliptic equations with critical Sobolev growth*, Comm. Pure Appl. Math. **42** (1989) 271–297.

[10] H. D. Cao, *Deformation of Kähler metrics to Kähler-Einstein metrics on compact Kähler manifolds*, Invent. Math. **81** (1985) 359–372.

[11] J. Carlson & D. Toledo, *Harmonic mappings of Kähler manifolds into locally symmetric spaces*, Inst. Hautes Études Sci. Publ. Math. **69** (1989) 173–201.

[12] A. Chang & P. Yang, *Conformal deformation of metrics on* S^2, J. Differential Geometry **27** (1988) 259–296.

[13] B. Chow, *The Ricci flow on the 2-sphere*, J. Differential Geometry **33** (1991) 325–334.

[14] B. Chow & L. F. Wu, *The Ricci flow on compact 2-orbifolds with curvature negative somewhere*, preprint.

[15] K. Corlette, *Archimedian superrigidity and hyperbolic geometry*, preprint.

[16] D. DeTurck, *Deforming metrics in the direction of their Ricci tensors*, J. Differential Geometry **18** (1983) 157–162.

[17] J. Eells & J. Sampson, *Harmonic mappings of Riemannian manifolds*, Amer. J. Math. **86** (1964) 109–160.

[18] N. Ercolani & G. Forest, *The geometry of real sine-Gordon wavetrains*, Comm. Math. Phys. **99** (1985) 1–49.

[19] J. Escobar, *Conformal deformation of a Riemannian metric to a scalar flat metric with constant mean curvature on the boundary*, preprint.

[20] ——, *The Yamabe problem on manifolds with boundary*, J. Differential Geometry (to appear).

[21] J. Escobar & R. Schoen, *Conformal metrics with prescribed scalar curvature*, Invent. Math. **86** (1986) 243–254.

[22] H. Federer, *The singular sets of area minimizing rectifiable currents with codimension one and of area minimizing flat chains modulo two with arbitrary codimension*, Bull. Amer. Math. Soc. **79** (1970) 761–771.

[23] B. Gidas, W. M. Ni & L. Nirenberg, *Symmetry and related properties via the maximum principle*, Comm. Math. Phys. **68** (1979) 209–243.

[24] M. Grüter, *Regularity of weak H-surfaces*, J. Reine Angew. Math. **329** (1981) 1–15.

[25] R. Hamilton, *Three-manifolds with positive Ricci curvature*, J. Differential Geometry **17** (1982) 255–306.

[26] ——, *Four-manifolds with positive constant operator*, J. Differential Geometry **24** (1986) 153–179.

[27] ——, *The Ricci flow on surfaces, mathematics and general relativity*, Contemp. Math. **71**, Amer. Math. Soc., Providence, RI, 1986.

[28] Z. C. Han, *Prescribing Gaussian curvature on* S^2, Duke Math. J. (to appear).

[29] R. Hardt & F. H. Lin, *The singular set of an energy minimizing map from* B^4 *to* S^2, preprint.

[30] F. Helein, *Regularite des applications faiblement harmoniques entre une surface une sphere*, preprint.

[31] H. Hopf, *Lectures on differential geometry in the large*, Notes by J. Gray, Stanford University, 1956.

[32] J. H. Jellet, *Sur la surface dont la coubure moyenne est constante*, J. Math. Pures Appl. **18** (1853) 163–167.

[33] N. Kapouleas, *Complete constant mean curvature surfaces in Euclidean three-space*, Ann. of Math. (2) **131** (1990) 239–330.

[34] ——, *Compact constant mean curvature surfaces in Euclidean three-space*, J. Differential Geometry **33** (1991) 683–715.

[35] J. Kazdan & F. Warner, *Curvature functions for compact 2-manifolds*, Ann. of Math. (2) **99** (1974) 14–47.

[36] N. Korevaar, R. Kusner & B. Solomon, *The structure of complete embedded surfaces with constant mean curvature*, J. Differential Geometry **30** (1989) 465–504.

[37] F. H. Lin, *Nonlinear theory of defects in nematic liquid crystals; phase transition and flow phenomena*, Comm. Pure Appl. Math. **42** (1989) 789–814.

[38] R. Mazzeo & N. Smale, *Conformally flat metrics of constant positive scalar curvature on subdomains of the sphere*, preprint.

[39] W. Meeks, *The topology and geometry of embedded surfaces of constant mean curvature*, J. Differential Geometry **27** (1988) 539–552.

[40] M. Micallef & J. D. Moore, *Minimal two-spheres and the topology of manifolds with positive curvature on totally isotropic two-planes*, Ann. of Math. (2) **127** (1988) 199–227.

[41] M. Min-Oo, *Almost Einstein manifolds of negative Ricci curvature*, J. Differential Geometry **32** (1990) 457–490.

[42] C. B. Morrey, *The problem of Plateau on a Riemannian manifold*, Ann. of Math. (2) **49** (1948) 807–851.

[43] J. Moser, *On a non-linear problem in differential geometry*, Dynamical Systems, Academic Press, New York, 1973.

[44] G. D. Mostow, *Strong rigidity of locally symmetric spaces*, Ann. of Math. Stud., Vol. 78, Princeton University Press, Princeton, NJ, 1973.

[45] M. Obata, *The conjectures on conformal transformations of Riemannian manifolds*, J. Differential Geometry **6** (1972) 247–258.

[46] U. Pinkall & I. Sterling, *On the classification of constant mean curvature tori*, Ann. of Math. (2) **130** (1989) 407–451.

[47] C. Poon, *Some new harmonic maps from B^3 to S^2*, J. Differential Geometry **34** (1991).
[48] J. Sacks & K. Uhlenbeck, *The existence of minimal immersions of 2-spheres*, Ann. of Math. (2) **113** (1981) 1–24.
[49] J. Sampson, *Applications of harmonic maps to Kähler geometry*, Contemp. Math. **49** (1986) 125–133.
[50] R. Schoen, *Analytic aspects of the harmonic map problem*, Math. Sci. Res. Inst. Publ. Vol. 2, Springer, Berlin, 1984, 321–358.
[51] ——, *Conformal deformation of a Riemannian metric to constant scalar curvature*, J. Differential Geometry **20** (1984) 479–495.
[52] ——, *Variational theory for the total scalar curvature functional for Riemannian metrics and related topics*, Topics in Calculus of Variations (M. Giaquinta, ed.), Lecture Notes in Math., Vol. 1365, Springer, Berlin, 1987.
[53] ——, *The existence of weak solutions with prescribed singular behavior for a conformally invariant scalar equation*, Comm. Pure Appl. Math. **41** (1988) 317–392.
[54] ——, *On the number of constant scalar curvature metrics in a conformal class*, preprint.
[55] P. Schoen & K. Uhlenbeck, *A regularity theory for harmonic maps*, J. Differential Geometry **17** (1982) 307–335.
[56] R. Schoen & S. T. Yau, *Harmonic maps and the topology of stable hypersurface and manifolds of nonnegative Ricci curvature*, Comment. Math. Helv. **39** (1976) 339–341.
[57] ——, *Conformally flat manifolds, Kleinian groups, and scalar curvature*, Invent. Math. **92** (1988) 47–71.
[58] W. X. Shi, *Deforming the metric on complete Riemannian manifolds*, J. Differential Geometry **30** (1989) 223–301.
[59] ——, *Complete noncompact Kähler manifolds with positive holomorphic bisectional curvature*, Bull. Amer. Math. Soc. (N.S.) **23** (1990) 437–440.
[60] L. Simon, *On the singularities of harmonic maps*, preprint.
[61] Y. T. Siu, *The complex-analyticity of harmonic maps and the strong rigidity of compact Kähler manifolds*, Ann. of Math. (2) **112** (1980) 73–112.
[62] N. Trudinger, *Remarks concerning the conformal deformation of Riemannian structures on compact manifolds*, Ann. Scuola Norm. Sup. Pisa Cl. Sci. (4) **22** (1968) 265–274.
[63] R. Walter, *Explicit examples to the H-problem of Heinz Hopf*, Geometriae Dedicata **23** (1987) 187–213.
[64] H. Wente, *An existence theorem for surfaces of constant mean curvature*, J. Math. Anal. Appl. **26** (1969) 318–344.
[65] ——, *A counterexample to a conjecture of H. Hopf*, Pacific J. Math. **121** (1984) 193–243.
[66] ——, *Twisted tori of constant mean curvature in* $\mathbf{R}^3$, Seminar on New Results in Nonlinear Partial Differential Equations, 1987.
[67] H. Yamabe, *On a deformation of Riemannian structures on compact manifolds*, Osaka J. Math. **12** (1960) 21–37.
[68] D. Yang, *Convergence of Riemannian manifolds with integral bounds on curvature*. I, preprint.
[69] ——, *Riemannian manifolds with small integral norm of curvature*, preprint.
[70] R. Ye, *Ricci flow, Einstein metrics and space forms*, preprint.
[71] ——, *Ricci flow and manifolds of negatively pinched curvature*, preprint.
[72] D. Zhang, *New results on geometric variational problems*, Stanford Dissertation, 1990.
[73] R. Zimmer, *Ergodic theory and semisimple groups*, Birkhäuser, Boston, 1964.

STANFORD UNIVERSITY

SURVEYS IN DIFFERENTIAL GEOMETRY
1 (1991) 243–310

TWO-DIMENSIONAL GRAVITY AND INTERSECTION THEORY ON MODULI SPACE

EDWARD WITTEN

Abstract

These are notes based on two lectures given at the Conference on Geometry and Topology (Harvard University, April 1990). The first was mainly devoted to explaining a conjecture according to which stable intersection theory on moduli space of Riemann surfaces is governed by the KdV hierarchy of integrable equations. The second lecture was primarily an introduction to the "hermitian matrix model" of two-dimensional gravity, which is a crucial part of the background for the conjecture. Analogous but more general theories also exist and are sketched in these notes. The generalization in the first lecture involves considering intersection theory on the moduli space of pairs consisting of a Riemann surface Σ and a holomorphic map of Σ to a fixed Kähler manifold K. The simplest analogous generalization in the second lecture involves a chain of hermitian matrices.

1. Introduction

At first sight, two-dimensional general relativity appears "trivial," at least as a physical theory, since for instance the Einstein-Hilbert action

$$I = \frac{1}{2\pi}\int \sqrt{g}R \tag{1.1}$$

is a topological invariant, so that the Einstein field equations are automatically obeyed.

Yet actually, on further investigation, two-dimensional quantum general relativity proves to be a strikingly rich theory. What is loosely called "critical" two-dimensional gravity is an essential ingredient in string theory. "Noncritical" two-dimensional gravity is a much more difficult subject which has been intensively studied with various motivations including possible applications to string theory and to the large N limit of quantum gauge theories with gauge group $SU(N)$.

In the earliest approach to the subject, introduced by Polyakov [54], noncritical two-dimensional gravity is related to a quantum field theory

Received June 14, 1990. Research supported in part by National Science Foundation Grant 86-20266 and National Science Foundation Waterman Grant 88-17521.

with Liouville action. This has been intensively studied [32], [11] by seeking to exploit at the quantum level the integrability of the classical Liouville equation $-\Box\phi+e^{\phi}=0$. A variant of this involves a different gauge choice that is claimed to lead to a sort of $SL(2, \mathbb{R})$ structure [42]. In a very different approach, two-dimensional gravity has been studied by counting triangulations of surfaces, which can be related to "matrix models" [14], [1], [41]. This approach has been developed with spectacular success, most recently with complete solutions in the "double scaling limit" by Brezin and Kazakov, Douglas and Shenker, and Gross and Migdal [13], [25], [35]. Another approach [52], [46] uses ideas of "topological quantum field theory" and can be reduced to a description in terms of the cohomology of the moduli space of Riemann surfaces. (Yet another approach was proposed at this meeting by I. Singer.)

A variety of arguments indicate that the theories constructed by these different approaches are equivalent. In addition to heuristic arguments, Liouville theory can be compared to the matrix models by comparison of critical exponents (which in Liouville theory can be computed by a scaling argument [21], [14]). Topological gravity is related to Liouville theory by an elegant argument due to Distler [21] that involves a variant [28] of the usual bose-fermi equivalence on Riemann surfaces. The topological field theory approach is related to the matrix models by explicit comparison in genus ≤ 3, by the "string equation" and another similar equation that can be derived in both frameworks, and by formal analogies.

Purely in mathematical terms, the proposal that topological gravity is equivalent to the one matrix model leads to a striking conjecture. Since topological gravity amounts to the study of stable intersection theory on the moduli space of Riemann surfaces, while the one matrix model is a soluble problem related to the KdV hierarchy, the conjecture that these are equivalent amounts to a conjecture that the KdV hierarchy governs the stable intersection theory on moduli space. §2 of this paper is devoted to a precise and self-contained formulation of this conjecture, and a description of the evidence for it. §3 is devoted to a generalization in which one considers a Riemann surface Σ together with a holomorphic map of Σ to a fixed complex manifold M. §4 is an introduction to the one matrix model and its relation to the KdV hiearchy. This section can be read independently of §§2 and 3. At the end of §4 we also briefly consider a matrix model analog of the generalization of the topological theory to include M.

§§2 and 3 are primarily an exposition of ideas that have appeared elsewhere [59], [19], with a few details added. §4 is an exposition of work of many authors, including the recent work of Brezin and Kazakov, Douglas and Shenker, and Gross and Migdal [13], [25], [35].

We are grateful to P. Deligne for help in understanding the compactification of moduli space and the string equation. Key points were also raised by M. F. Atiyah, H. Neuberger, D. Kazhdan, G. Segal, and I. M. Singer.

2. Stable intersection theory on moduli space

Let $\mathcal{M}_{g,n}$ be the moduli space of Riemann surfaces of genus g with n ordered punctures, and let $\overline{\mathcal{M}}_{g,n}$ be its compactification obtained by adjoining curves with double points [16]. This is the (compactified) "moduli space of stable curves," which arises naturally in string theory. $\overline{\mathcal{M}}_{g,n}$ is not a manifold but an orbifold (locally the quotient of a manifold by a finite group), so intersection theory is well defined on $\overline{\mathcal{M}}_{g,n}$, but intersection numbers are in general rational numbers rather than integers.

Such moduli spaces are endowed with natural cohomology classes, as described by Atiyah and Bott in the gauge theory case [4] and by Mumford, Morita, and Miller in the case we will be considering [50], [48], [45]. Let Σ be a stable curve with marked points $x_1, x_2, \cdots, x_n$. It is essential that, though Σ may have singularities (double points), the moduli space of stable curves is defined in such a way that the x_i never coincide with these singularities. Thus, each x_i has its complex cotangent space $T^*\Sigma|_{x_i}$, and these vary holomorphically with x_i, giving n holomorphic line bundles $\mathcal{L}_{(i)}$ over $\overline{\mathcal{M}}_{g,n}$. One can think of the x_i as sections of the universal curve $\mathcal{C}\overline{\mathcal{M}}_{g,n}$ over $\overline{\mathcal{M}}_{g,n}$. If $K_{\mathcal{C}|\mathcal{M}}$ is the cotangent bundle to the fibers of $\mathcal{C}\overline{\mathcal{M}}_{g,n} \to \overline{\mathcal{M}}_{g,n}$, then $\mathcal{L}_{(i)} = x_i^*(K_{\mathcal{C}|\mathcal{M}})$.

Let $d_1, d_2, \cdots, d_n$ be nonnegative integers such that

$$\sum_{i=1}^{n} d_i = 3g - 3 + n. \tag{2.1}$$

This is the dimensional condition under which the intersection number

$$\langle \bigwedge_{i=1}^{n} c_1(\mathcal{L}_{(i)})^{d_i}, \overline{\mathcal{M}}_{g,n} \rangle \tag{2.2}$$

may be nonzero. We will denote this number as

$$\langle \tau_{d_1} \tau_{d_2} \cdots \tau_{d_n} \rangle. \tag{2.3}$$

These quantities, which will be our main interest, are closely analogous to the intersection numbers on instanton moduli space that Donaldson introduced [22] in studying smooth four-manifolds.[1] The ordering of factors in (2.2) and (2.3) is of course immaterial, since the cohomology classes in question are even dimensional. If r_0 of the d_i are equal to 0, r_1 of them are equal to 1, r_2 of them are equal to 2, etc., then it is sometimes convenient to write

$$\langle \tau_{d_1}\tau_{d_2}\cdots\tau_{d_n}\rangle = \langle \tau_0^{r_0}\tau_1^{r_1}\tau_2^{r_2}\cdots\rangle. \tag{2.4}$$

The notation in (2.3) reflects the fact that (like their analogs in Donaldson theory), these numbers have a quantum field theory interpretation [52], [46], [9]. Indeed, $\langle \tau_{d_1}\tau_{d_2}\cdots\tau_{d_n}\rangle$ is the (unnormalized) expectation value of a product of "local operators" τ_{d_i} with respect to a certain Feynman path integral measure. Though we will not explain this path integral interpretation here, its existence is one of the things that makes plausible the conjecture that these objects are related to the hermitian one matrix model, which is also defined by a kind of path integral.

The $\langle \tau_{d_1}\tau_{d_2}\cdots\tau_{d_n}\rangle$ are closely related to the intersection numbers of the stable cohomology classes on moduli space studied by Mumford, Morita, and Miller [50], [48], [45]. In that formulation, one considers the projection $\pi:\overline{\mathcal{M}}_{g,1}\to\overline{\mathcal{M}}_{g,0}=\overline{\mathcal{M}}_g$, and defines $2n$-dimensional cohomology classes κ_n on $\overline{\mathcal{M}}_{g,0}$ by

$$\kappa_n = \pi_*(c_1(\mathcal{L})^{n+1}). \tag{2.5}$$

($\mathcal{L}$ is again the line bundle whose fiber is the cotangent space to the one marked point of $\overline{\mathcal{M}}_{g,1}$. We define $\kappa_{-1}=0$.) It is known that the κ's obey no stable relations [45], and it is conjectured that the stable cohomology of moduli space (in a sense explained in [45], [36]) is a polynomial algebra generated by the κ's. It is natural to consider intersection numbers of the κ's, which we will denote as

$$\langle \kappa_{r_1}\kappa_{r_2}\cdots\kappa_{r_n}\rangle = \langle \kappa_{r_1}\wedge\kappa_{r_2}\wedge\cdots\wedge\kappa_{r_n}, \overline{\mathcal{M}}_g\rangle. \tag{2.6}$$

As a special case of the comparison between the τ's and the κ's, consider first the expectation value of a single τ,

$$\langle \tau_d\rangle = \int_{\overline{\mathcal{M}}_{g,1}} c_1(\mathcal{L})^d. \tag{2.7}$$

---

[1] In [59], we worked with $\sigma_d = d!\tau_d$, in order to agree with the conventions of the literature on matrix models.

(Of course, this is nonzero only if $d = 3g - 2$.) By performing first the integral over the fiber of $\pi : \overline{\mathcal{M}}_{g,1} \to \overline{\mathcal{M}}_g$, we get immediately that

$$\langle \tau_d \rangle = \int_{\overline{\mathcal{M}}_g} \kappa_{d-1} = \langle \kappa_{d-1} \rangle. \tag{2.8}$$

In general, we have by definition

$$\langle \tau_{d_1} \tau_{d_2} \cdots \tau_{d_n} \rangle = \int_{\overline{\mathcal{M}}_{g,n}} c_1(\mathcal{L}_{(1)})^{d_1} \wedge \cdots \wedge c_1(\mathcal{L}_{(n)})^{d_n}. \tag{2.9}$$

To get a similar expression for $\langle \kappa_{r_1} \cdots \kappa_{r_n} \rangle$, let $\mathcal{C}_n(\overline{\mathcal{M}}_g) = \mathcal{C}_{(1)}\overline{\mathcal{M}}_g \times_{\overline{\mathcal{M}}_g} \mathcal{C}_{(2)}\overline{\mathcal{M}}_g \times_{\overline{\mathcal{M}}_g} \cdots \times_{\overline{\mathcal{M}}_g} \mathcal{C}_{(n)}\overline{\mathcal{M}}_g$ be the n-fold fiber product of n copies $\mathcal{C}_{(1)}\overline{\mathcal{M}}_g, \cdots, \mathcal{C}_{(n)}\overline{\mathcal{M}}_g$ of the universal curve over $\overline{\mathcal{M}}_g$. Let π_i be the projection of $\mathcal{C}_n(\overline{\mathcal{M}}_g)$ onto the ith factor $\mathcal{C}_{(i)}\overline{\mathcal{M}}_g$, let $K_{\mathcal{C}_{(i)}/\mathcal{M}}$ be the cotangent bundle to the fibers of $\mathcal{C}_{(i)}\overline{\mathcal{M}}_g$, and let $\widehat{\mathcal{L}}_{(i)} = \pi_i^*(K_{\mathcal{C}_{(i)}/\mathcal{M}})$. Then by the definition of the κ's, we have

$$\langle \kappa_{d_1 - 1} \kappa_{d_2 - 1} \cdots \kappa_{d_n - 1} \rangle = \int_{\mathcal{C}_n(\overline{\mathcal{M}}_g)} c_1(\widehat{\mathcal{L}}_{(1)})^{d_1} \wedge \cdots \wedge c_1(\widehat{\mathcal{L}}_{(n)})^{d_n}. \tag{2.10}$$

The key observation is now the following. A point in $\mathcal{C}_n(\overline{\mathcal{M}}_g)$ labels a stable curve Σ and n ordered marked points in Σ which are arbitrary so that in particular two or more of them are permitted to coincide. As long as we keep away from the locus in $\mathcal{C}_n(\overline{\mathcal{M}}_g)$ on which two or more of these points coincide, there is a natural 1-1 map $\mathcal{C}_n(\overline{\mathcal{M}}_g) \to \overline{\mathcal{M}}_{g,n}$. These varieties are thus birationally equivalent. The equivalence is only birational since $\overline{\mathcal{M}}_{g,n}$ parametrizes a family of genus g curves with n marked points which are never permitted to coincide. (Compactification is achieved by permitting Σ to degenerate to a curve with a larger number of components when naively two or more points are becoming coincident.) Thus, though $\mathcal{C}_n(\overline{\mathcal{M}}_g)$ and $\overline{\mathcal{M}}_{g,n}$ are birationally equivalent, they (and the curves they parametrize) differ on a certain divisor at infinity.

On the Zariski open set on which $\mathcal{C}_n(\overline{\mathcal{M}}_g)$ and $\overline{\mathcal{M}}_{g,n}$ (and the curves they parametrize) have a natural identification, the line bundles $\mathcal{L}_{(j)}$ and $\widehat{\mathcal{L}}_{(j)}$ also have a natural identification, as is immediate from their definitions. Thus, (2.9) and (2.10) differ only from the contribution of the divisor at infinity. The analysis of the effects of this divisor in comparing (2.9) and (2.10) is a universal local problem which naturally leads to additional terms involving the conjectured generators of the stable cohomology.

One finds

$$\langle \tau_{d_1}\tau_{d_2}\rangle = \langle \kappa_{d_1-1}\kappa_{d_2-1}\rangle + \langle \kappa_{d_1+d_2-2}\rangle,$$

$$(2.11)\quad \langle \tau_{d_1}\tau_{d_2}\tau_{d_3}\rangle = \langle \kappa_{d_1-1}\kappa_{d_2-1}\kappa_{d_3-1}\rangle + \langle \kappa_{d_1+d_2-2}\kappa_{d_3-1}\rangle + \langle \kappa_{d_1+d_3-2}\kappa_{d_2-1}\rangle + \langle \kappa_{d_2+d_3-2}\kappa_{d_1-1}\rangle + 2\langle \kappa_{d_1+d_2+d_3-3}\rangle,$$

and so on, as we will sketch after explaining the "string equation." These relations are invertible (they are given by a triangular matrix with 1's on the diagonal) and show that the information contained in the intersection theory of the τ's is the same as the information contained in the intersection theory of the κ's.

Let us now heuristically explain the motivation for the way that we will organize the data. Given a quantum field theory Lagrangian $\mathcal{L}_0$ and operators τ_i, it is natural to consider a more general Lagrangian

$$(2.12)\qquad \mathcal{L} = \mathcal{L}_0 - \sum_i t_i \int_\Sigma \tau_i,$$

where the t_i are known physically as "coupling constants." Thus the Feynman integral, in genus g, becomes

$$(2.13)\qquad F_g(t_i) = \int (\text{FIELDS}) e^{-\mathcal{L}_0} e^{\sum_i t_i \int_\Sigma \tau_i}.$$

We can expand the exponential

$$(2.14)\qquad e^{\sum_i t_i \int_\Sigma \tau_i} = \sum_{\{n_i\}} \prod_{i=0}^{\infty} \frac{t_i^{n_i}}{n_i!}\left(\int_\Sigma \tau_i\right)^{n_i},$$

with the sum running over all sequences $\{n_i\}$ of nonnegative integers, almost all of which are zero. So one sees that the path integral in (2.13) is the generating function of the intersection numbers $\langle \tau_0^{n_0}\tau_1^{n_1}\cdots\rangle$. Summing over genus, as is natural in string theory, we would need to consider the "total free energy"

$$(2.15)\qquad F(t_i) = \sum_{g=0}^{\infty} F_g(t_i).$$

With this motivation, the natural object that we wish to consider is the generating function of the stable intersection theory on moduli space, defined by

$$(2.16)\qquad F(t_0, t_1, \cdots) = \sum_{\{n_i\}} \prod_{i=0}^{\infty} \frac{t_i^{n_i}}{n_i!} \langle \tau_0^{n_0}\tau_1^{n_1}\tau_2^{n-2}\cdots\rangle.$$

Here it is understood that for every sequence $\{n_i\}$, $\langle \tau_0^{n_0}\tau_1^{n_1}\tau_2^{n_2}\cdots\rangle$ is to be computed in genus g, where

$$3g-3=\sum_i n_i(i-1). \tag{2.17}$$

If the g determined by this formula is not a nonnegative integer, one defines $\langle \tau_0^{n_0}\tau_1^{n_1}\tau_2^{n_2}\cdots\rangle$ to be zero. For $g=0,1,2,\cdots$, the genus g contribution $F_g(t_i)$ is defined by restricting the sum in (2.16) to sequences $\{n_i\}$ obeying (2.17). For our purposes, the sum in (2.16) can be regarded as a formal series, but the conjecture that will be stated presently would mean that (2.16) is an expansion of a natural function, defined in an open set in the space of the t_i.

There is actually a slight imprecision in the definition (2.16), since we have not given a meaning to the symbol $\langle 1\rangle$, which is the contribution from the zero sequence $n_0=n_1=\cdots=0$. This sequence corresponds according to (2.17) to curves of genus 1 with no marked points. This is a degenerate case, since the virtual dimension of $\overline{\mathcal{M}}_{1,0}$ (predicted from the Riemann-Roch formula for the moduli problem) is 0 but the actual dimension is 1. In the present paper, the object $\langle 1\rangle$ will play no role, and we could simply set it to zero, but the natural value is the Euler characteristic of $\overline{\mathcal{M}}_{1,0}$ as an orbifold, which is

$$\langle 1\rangle=-\tfrac{1}{12}. \tag{2.18}$$

We will now introduce a convenient notation for the derivatives of F. We define

$$\langle\langle \tau_{d_1}\tau_{d_2}\cdots\tau_{d_n}\rangle\rangle=\frac{\partial}{\partial t_{d_1}}\frac{\partial}{\partial t_{d_2}}\cdots\frac{\partial}{\partial t_{d_n}}F(t_0,t_1,\cdots). \tag{2.19}$$

It is evident that $\langle\langle \tau_{d_1}\cdots\tau_{d_n}\rangle\rangle$ reduces to $\langle \tau_{d_1}\cdots\tau_{d_n}\rangle$ if one sets $t_0=t_1=\cdots=0$. As a special case of (2.19), one occasionally uses the symbol $\langle\langle 1\rangle\rangle$ to represent the functional $F(t_0,t_1,\cdots)$. We also write

$$\langle\langle \tau_{d_1}\cdots\tau_{d_n}\rangle\rangle_g=\frac{\partial}{\partial t_{d_1}}\cdots\frac{\partial}{\partial t_{d_n}}F_g. \tag{2.20}$$

2a. The conjecture. Our basic conjecture is that $F(t_0,t_1,\cdots)$ is determined by the following two conditions:

(1) The object $U=\partial^2F/\partial t_0^2$ obeys the KdV equations,

$$\frac{\partial U}{\partial t_n}=\frac{\partial}{\partial t_0}R_{n+1}(U,\dot U,\ddot U,\cdots), \tag{2.21}$$

where $\dot{U} = \partial U/\partial t_0$, $\ddot{U} = \partial^2 U/\partial t_0^2$, etc., are the derivatives of U with respect to t_0, and $R_{n+1}(U, \dot{U}, \ddot{U}, \cdots)$ are certain polynomials in U and its t_0 derivatives that are well known in the theory of the KdV equations (and can be defined by a recursion relation that is given below).

(2) In addition, F obeys the "string equation,"

$$\frac{\partial F}{\partial t_0} = \frac{t_0^2}{2} + \sum_{i=0}^{\infty} t_{i+1} \frac{\partial F}{\partial t_i}. \tag{2.22}$$

The two statements can be summarized by saying that stable intersection theory on moduli space is equivalent to the "hermitian matrix model" of two-dimensional gravity. That formulation was the original context for the conjecture.

Consequences of the conjecture. We will first verify that the conjecture, if true, does uniquely determine F.

The first part of the conjecture obviously determines $U(t_0, t_1, t_2, t_3, \cdots)$ in terms of the "initial data" $U(t_0, 0, 0, 0, \cdots)$. The second part of the conjecture implies upon setting $t_i = 0$, for $i > 0$, that the initial data are

$$U(t_0, 0, 0, \cdots) = t_0. \tag{2.23}$$

Thus the conjecture suffices to determine U.

It is easy to see from the point of view of intersection theory on moduli space why the initial conditions for U must be those given in (2.23). In fact, the dimensional condition (2.17) implies right away that the numbers $\langle \tau_0^n \rangle$ for $n = 1, 2, 3, \cdots$ vanish unless $n = 3$, while $\langle \tau_0^3 \rangle$ receives a contribution only in genus zero. The moduli space of genus zero curves with three labeled marked points consists of a single point without symmetries (since the three points can be uniquely mapped to $0, 1, \infty$ by $SL(2, \mathbb{C})$), so $\langle \tau_0^n \rangle = \delta_{n,3}$. Hence (using (2.18) for the $n = 0$ contribution),

$$F(t_0, 0, 0, 0, \cdots) = \sum_{n=0}^{\infty} \frac{t_0^n}{n!} \langle \tau_0^n \rangle = \frac{t_0^3}{6} - \frac{1}{12}. \tag{2.24}$$

This indeed corresponds to the initial conditions $U = t_0$.

We now want to show that the conjecture suffices to determine F and not only $U = \ddot{F}$. Note first that the string equation is equivalent to an explicit relation among the intersection numbers

$$\left\langle \tau_0 \prod_{i=1}^{n} \tau_{d_i} \right\rangle = \sum_{j=1}^{n} \left\langle \prod_{i=1}^{n} \tau_{d_i - \delta_{ij}} \right\rangle + \delta_{n,2}\delta_{d_1,0}\delta_{d_2,0}. \tag{2.25}$$

(It is in this form that the string equation can naturally be deduced from algebraic geometry, as we will see soon.) This can be used to determine all intersection numbers, and thus F, once U is known. As a simple example, it follows from (2.25) that

$$\langle\tau_n\rangle = \langle\tau_{n+2}\tau_0\tau_0\rangle = \left.\frac{\partial U}{\partial t_{n+2}}\right|_{t_i=0}. \tag{2.26}$$

(In fact, one can easily show inductively that the KdV prediction for (2.26) is $\langle\tau_{3g-2}\rangle = 1/((24)^g \cdot g!)$.) This can be generalized in the following elementary but clumsy way to give an algorithm to compute an arbitrary genus g intersection number:

$$W = \langle\tau_{d_1}\cdots\tau_{d_k}\rangle. \tag{2.27}$$

We can suppose that none of the d_i are 0, since factors of τ_0 can be eliminated using (2.25). This being so, the dimensional equation $\Sigma_i(d_i - 1) = 3g - 3$ gives an upper bound $d_i \leq 3g - 2$ for (2.27) to be nonzero. Suppose, inductively, that for some integer r it is known that the KdV equations plus the string equation determine all intersection numbers in which all $d_i \geq 1$ and one of the d_i, say d_1, is $\geq r$. We can start the induction with $r = 3g - 1$. We want to improve the bound from r to $r - 1$. The quantity

$$W' = \langle\tau_{d_1+2}\tau_{d_2}\cdots\tau_{d_k}\tau_0\tau_0\rangle \tag{2.28}$$

can be determined form the KdV equations plus the string equation, since it is

$$W' = \left[\frac{\partial}{\partial t_{d_1+2}}\frac{\partial}{\partial t_{d_2}}\cdots\frac{\partial}{\partial t_{d_k}}U\right]_{t_i=0}, \tag{2.29}$$

and we know that the KdV equations plus the string equation suffice to determine U. Now, two uses of the string equation (2.25) to eliminate the two factors of τ_0 in (2.28) may leave us with an expression still containing τ_0's, since there may be factors of τ_1 or τ_2 in (2.28), and these may become τ_0's upon using the string equation. If so, use the string equation again until after finitely many steps all τ_0's are eliminated. After doing so, one obtains an expression exhibiting W' as a positive multiple of W plus genus g intersection numbers containing τ_{d_1+2} or τ_{d_1+1} that are already known from the induction hypothesis plus genus g correlation functions containing τ_{d_1} and a smaller number of τ_1's than are present in (2.27). Repetition of this procedure to eliminate all τ_1's eventually expresses W in terms of objects that are already known by the induction hypothesis. This completes the demonstration that the conjecture suffices to determine the generating function F.

Alternative ways of writing the equations. The polynomials R_n that appear in the first part of the conjecture can be defined inductively by the formulas

$$(2.30)\quad R_1 = U, \qquad \frac{\partial R_{n+1}}{\partial t_0} = \frac{1}{2n+1}\left(\frac{\partial U}{\partial t_0}R_n + 2U\frac{\partial R_n}{\partial t_0} + \frac{1}{4}\frac{\partial^3}{\partial t_0^3}R_n\right).$$

(This recursion relation among the generalized KdV potentials R_n was obtained in [31] from the theory of the resolvent of the Schrödinger operator. The proof establishes the not obvious fact that the right-hand side of the second equation in (2.30) is indeed the t_0 derivative of a polynomial R_{n+1}. The Schrödinger operator enters via the inverse scattering method [30] which is the basis for the integrability of the KdV equation.) This recursion relation can be interpreted as stating that the KdV flows are Hamiltonian flows for two different symplectic structures. Since, in view of the definition of U, the left-hand side of (2.21) is the same as

$$(2.31)\qquad \frac{\partial}{\partial t_0}\langle\langle\tau_n\tau_0\rangle\rangle,$$

it is clear that upon integrating once in t_0 (a step that can be justified using the string equation), (2.21) amounts to the statement that

$$(2.32)\qquad \langle\langle\tau_n\tau_0\rangle\rangle = R_{n+1}(U, \dot{U}, \ddot{U}, \cdots).$$

With the aid of the recursion relation, we get the alternative version

$$(2.33)\qquad \begin{aligned}&\langle\langle\tau_n\tau_0\tau_0\rangle\rangle\\ &= \frac{1}{2n+1}\left(\langle\langle\tau_{n-1}\tau_0\rangle\rangle\langle\langle\tau_0^3\rangle\rangle + 2\langle\langle\tau_{n-1}\tau_0^2\rangle\rangle\langle\langle\tau_0^2\rangle\rangle + \frac{1}{4}\langle\langle\tau_{n-1}\tau_0^4\rangle\rangle\right),\end{aligned}$$

which captures the full content of the KdV equations. A still more explicit version is

$$(2.34)\qquad \begin{aligned}\langle\langle\tau_n\tau_0\tau_0\rangle\rangle_g = \frac{1}{2n+1}\Bigg(&\sum_{g'=0}^{g}\langle\langle\tau_{n-1}\tau_0\rangle\rangle_{g'}\langle\langle\tau_0^3\rangle\rangle_{g-g'}\\ &+2\sum_{g'=0}^{g}\langle\langle\tau_{n-1}\tau_0^2\rangle\rangle_{g'}\langle\langle\tau_0^2\rangle\rangle_{g-g'} + \frac{1}{4}\langle\langle\tau_{n-1}\tau_0^4\rangle\rangle_{g-1}\Bigg).\end{aligned}$$

2b. Evidence for the Conjecture. "Matrix models" and "topological gravity" were both proposed originally as candidates for simple approaches to two-dimensional gravity. They are based on similar apparently "trivial" Lagrangians, and this suggested the conjecture advanced in §2a that they might be equivalent. This thought was encouraged by formal analogies between the two that will be apparent in §4. It was also encouraged by the fact that the methods used [37], [51] to compute the Euler characteristic of moduli space (which is the partition function of an appropriate topological field theory) are close cousins of the methods used in the matrix models.

Apart from such heuristic considerations, the evidence for the conjecture consists mainly of the following:

(a) The "string equation" (2.30) can be verified directly, in the intersection theory.

(b) One can verify directly that the KdV equations (2.21) hold in the intersection theory for genus $g \leq 3$. (We will here only consider $g \leq 2$, and refer to work of Horne [37], comparing to results of Faber [26], for $g = 3$.)

(c) One can also, by direct methods, prove another equation analogous to the string equation which follows from the string equation together with the KdV equations.

This is the evidence for the conjecture that can be stated without any reference to physics. Physicists consider Distler's relation of topological gravity to Liouville theory [20] to be an important indication that the conjecture is true. Also, E. and H. Verlinde have proposed [57] physical arguments that may eventually lead to a proof of the conjecture.

The string equation and its cousins. To obtain the string equation, one considers the moduli space $\overline{\mathcal{M}}_{g,n+1}$ of stable curves Σ with $n+1$ marked points $x_0, \cdots, x_n$, which we regard as sections of the universal curve $\mathcal{C}\overline{\mathcal{M}}_{g,n+1} \to \overline{\mathcal{M}}_{g,n+1}$. Except for certain low values of g and n, which will have to be treated separately, there is a projection $\pi: \overline{\mathcal{M}}_{g,n+1} \to \overline{\mathcal{M}}_{g,n}$ that forgets the first point x_0. (This projection does not exist for $g = 0$, $n = 2$, and $g = 1, n = 0$, because then forgetting x_0 will render Σ unstable.) The line bundles of interest on $\overline{\mathcal{M}}_{g,n+1}$ and $\overline{\mathcal{M}}_{g,n}$ are $\mathcal{L}_{(j)} = x_j^*(K_{\mathcal{C}|\mathcal{M}})$ and $\mathcal{L}'_{(j)} = x_j^*(K'_{\mathcal{C}|\mathcal{M}})$, respectively, where $K_{\mathcal{C}|\mathcal{M}}$ and $K'_{\mathcal{C}|\mathcal{M}}$ are the cotangent bundles along the fibers of $\mathcal{C}\overline{\mathcal{M}}_{g,n+1} \to \overline{\mathcal{M}}_{g,n+1}$ and $\mathcal{C}\overline{\mathcal{M}}_{g,n} \to \overline{\mathcal{M}}_{g,n}$.

There is a subtlety here that plays a crucial role in understanding the string equation and the analogous formulas (2.11) relating the τ_n's to the Mumford-Morita-Miller classes κ_{n-1}. A point in $\overline{\mathcal{M}}_{g,n+1}$ corresponds to a connected, stable curve Σ perhaps with more than one component. Stability means in particular that any genus zero component of Σ has at least three marked points, including possible double points. It may happen that "forgetting x_0" causes a particular genus zero component to contain only two marked points. If so, that component must be contracted to a point.

This occurs on the divisors D_j, indicated in Figure 1, in which a genus zero component contains x_0, x_j, and precisely one double point. (It also occurs if a genus zero component contains x_0 and two double points, but that happens in complex codimension two and will not affect comparisons of line bundles.) Because of this, though there is, analogous to the map $\pi:\overline{\mathcal{M}}_{g,n+1} \to \overline{\mathcal{M}}_{g,n}$ that "forgets x_0", a corresponding natural map of universal curves $\pi_{\mathscr{C}}:\mathscr{C}\overline{\mathcal{M}}_{g,n+1} \to \mathscr{C}\overline{\mathcal{M}}_{g,n}$, this map is not a fibering. $\pi_{\mathscr{C}}$ does not just "forget x_0"; it may be described by the instruction "forget x_0 and contract any genus zero components that become unstable as a result."

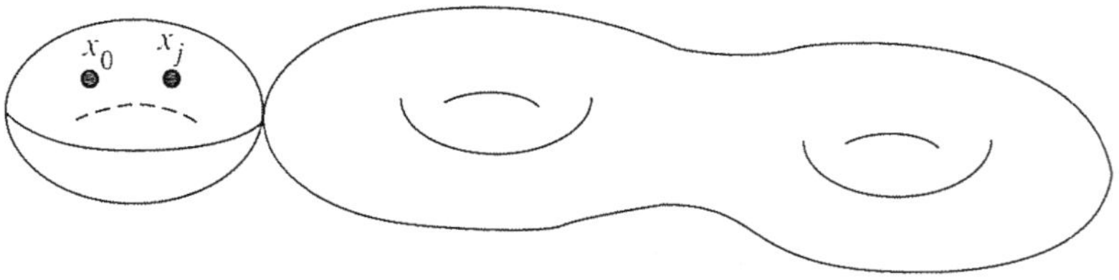

FIGURE 1. A CURVE THAT WILL BECOME UNSTABLE IF x_0 IS "FORGOTTEN."

Because of this, it is not the case, as one might have naively expected, that $\mathscr{L}_{(j)} = \pi^*(\mathscr{L}'_{(j)})$. To work out the discrepancy, note that a nonzero local section s of $K'_{\mathscr{C}|\mathcal{M}}$ determines a nonzero local section $x_j^*(s)$ of $\mathscr{L}'_{(j)}$ which intuitively is obtained by "evaluating s at x_j." Via the forgetful map $\pi:\overline{\mathcal{M}}_{g,n+1} \to \overline{\mathcal{M}}_{g,n}$, $x_j^*(s)$ pulls back to a section $\pi^*x_j^*(s)$ that vanishes precisely on the divisors D_j. In fact, if $\pi_{\mathscr{C}}:\Sigma' \to \Sigma$ is any map between curves, and s is any local holomorphic differential on Σ, then $\pi_{\mathscr{C}}^*(s)$ vanishes on any component of Σ' that is mapped to a point in Σ. Applying this to the fibers of $\pi_{\mathscr{C}}:\mathscr{C}\overline{\mathcal{M}}_{g,n+1} \to \mathscr{C}\overline{\mathcal{M}}_{g,n}$, we see that $\pi_{\mathscr{C}}^*(s) = 0$ on the locus $x_j(D_j) \subset \mathscr{C}\overline{\mathcal{M}}_{g,n+1}$. Hence, $\pi^*x_j^*(s) = x_j^*\pi_{\mathscr{C}}^*(s)$ vanishes on the divisor D_j, with a simple zero as one sees on more careful examination. The fact that a local nonzero section $x_j^*(s)$ of $\mathscr{L}'_{(j)}$ pulls back to a section $\pi^*x_j^*(s)$ with a simple zero on D_j corresponds to a formula

$$\mathscr{L}_{(j)} = \mathscr{L}'_{(j)} \bigotimes_{j=1}^{n} \mathscr{O}(D_j). \tag{2.35}$$

At the level of first Chern classes, this means that

$$c_1(\mathcal{L}_{(j)}) = \pi^*(c_1(\mathcal{L}'_{(j)})) + [D_j]. \tag{2.36}$$

We are now ready to consider the intersection number

$$\left\langle \tau_0 \prod_{i=1}^{n} \tau_{d_i} \right\rangle = \int_{\overline{\mathcal{M}}_{g,n+1}} 1 \cdot \bigwedge_{j=1}^{n} c_1(\mathcal{L}_{(j)})^{d_j}, \tag{2.37}$$

temporarily avoiding the special cases $g = 0$, $n = 2$ and $g = 1$, $n = 0$. Note the factor of $1 = c_1(\mathcal{L}_0)^0$ in (2.37). Now, obviously,

$$0 = \int_{\overline{\mathcal{M}}_{g,n+1}} \pi^* \left(\bigwedge_{j=1}^{n} c_1(\mathcal{L}'_{(j)})^{d_j} \right), \tag{2.38}$$

since the pullback of a cohomology class from $\overline{\mathcal{M}}_{g,n}$ could not be a top-dimensional class on $\overline{\mathcal{M}}_{g,n+1}$. If it were so that $\mathcal{L}_{(j)} = \pi^*(\mathcal{L}'_{(j)})$, for $j = 1, \cdots, n$, then (2.37) would vanish. We must use instead the correct formula (2.35), which implies

$$c_1(\mathcal{L}_{(j)})^n = (\pi^* c_1(\mathcal{L}'_{(j)}))^n + [D_j] \cdot \sum_{m=0}^{n-1} c_1(\mathcal{L}_{(j)})^m \cdot (\pi^*(c_1(\mathcal{L}'_{(j)})))^{n-1-m}. \tag{2.39}$$

Now, the line bundle $\mathcal{L}_{(j)}$ is trivial when restricted to the divisor D_j, since on the universal curve over D_j, the point x_j is on a rigid object, a genus zero component with three marked points. So we can discard terms in (2.39) with $m > 0$. Since it follows directly from the definition that $D_i \cap D_j = 0$ for $i \neq j$, in evaluating (2.37), we can drop terms proportional to $[D_i] \cdot [D_j]$, so (2.37) becomes

$$\left\langle \tau_0 \prod_{i=1}^{n} \tau_{d_i} \right\rangle = \sum_{j=1}^{n} \int_{\overline{\mathcal{M}}_{g,n+1}} [D_j] \cdot \bigwedge_{i=1}^{n} c_1(\mathcal{L}_{(i)})^{d_i - \delta_{ij}}. \tag{2.40}$$

(In case one of the exponents is negative, we set $c_1(\mathcal{L}_{(j)})^{-1} = 0$.) Integrating over the fibers of $\pi: \overline{\mathcal{M}}_{g,n+1} \to \overline{\mathcal{M}}_{g,n}$, we get the final result

$$\left\langle \tau_0 \prod_{i=1}^{n} \tau_{d_i} \right\rangle = \sum_{j=1}^{n} \left\langle \prod_{i=1}^{n} \tau_{d_i - \delta_{ij}} \right\rangle. \tag{2.41}$$

We still must consider the special cases $g = 0$, $n = 2$ and $g = 1$, $n = 0$. The only nonvanishing intersection number of this type is

$$\langle \tau_0 \tau_0 \tau_0 \rangle = 1 \tag{2.42}$$

for $g = 0$.

We will leave it to the reader to verify that (2.41) and (2.42) are precisely equivalent to the string equation (2.22). (2.41) alone, without the exceptional contribution for $g = 0$, $n = 2$, would give (2.22) without the $t_0^2/2$ term.

This completes the explanation of part (a) of the evidence for the conjecture. As for (c), another equation of a similar nature can be obtained by looking at

$$\left\langle \tau_1 \cdot \prod_{i=1}^{n} \tau_{d_i} \right\rangle = \int_{\overline{\mathcal{M}}_{g,n+1}} c_1(\mathcal{L}_{(0)}) \wedge_i (c_1(\mathcal{L}_{(i)}))^{d_i}. \tag{2.43}$$

Now, in evaluating (2.43) we may actually replace the $\mathcal{L}_{(i)}$ by $\pi^*(\mathcal{L}'_{(i)})$ for $i = 1, \cdots, n$. The second term in (2.35) does not contribute, since $c_1(\mathcal{L}_{(0)}) \cdot [D_j] = 0$. In evaluating (2.43) by integrating over the fiber of $\pi: \overline{\mathcal{M}}_{g,n+1} \to \overline{\mathcal{M}}_{g,n}$, one has a natural identification $\alpha: \overline{\mathcal{M}}_{g,n+1} \cong \mathcal{C}\overline{\mathcal{M}}_{g,n}$. The relative canonical bundle $K'_{\mathcal{C}|\mathcal{M}}$ has degree $2g-2$ along the fibers of $\mathcal{C}\overline{\mathcal{M}}_{g,n} \to \overline{\mathcal{M}}_{g,n}$. It is not true, as one might think naively, that $\mathcal{L}_{(0)} \equiv \alpha^*(K'_{\mathcal{C}|\mathcal{M}})$. The correct relation, by reasoning just as that which led to (2.35), is

$$\mathcal{L}_{(0)} \cong \alpha^*(K'_{\mathcal{C}|\mathcal{M}}) \otimes_{j=1}^{n} \mathcal{O}(D_j). \tag{2.44}$$

(Intuitively, a differential form on a curve Σ with n marked points is permitted to have poles at the marked points.) Thus, $\mathcal{L}_{(0)}$ is a line bundle of degree $2g-2+n$ along the fibers of $\overline{\mathcal{M}}_{g,n+1} \to \overline{\mathcal{M}}_{g,n}$. So integrating along the fibers in (2.43) we get

$$\left\langle \tau_1 \cdot \prod_{i=1}^{n} \tau_{d_i} \right\rangle = (2g-2+n) \left\langle \prod_{i=1}^{n} \tau_{d_i} \right\rangle, \tag{2.45}$$

with $2g-2+n$ being the degree of the canonical line bundle of a genus g curve with n marked points.

As in the discussion of the string equation, there is an exceptional case here, which arises for $g = 1$, $n = 0$, where there is no projection map $\overline{\mathcal{M}}_{1,1} \to \overline{\mathcal{M}}_{1,0}$. In this case, the exceptional contribution is

$$\langle \tau_1 \rangle = \tfrac{1}{24}. \tag{2.46}$$

This comes from a factor of $\frac{1}{12}$ which can be understood, for instance, for the existence of an elliptic modular form of weight 12 with a simple zero at the cusp, and a factor of $\frac{1}{2}$ because the generic elliptic curve has two symmetries. Now, (2.45) and (2.46) are equivalent to a differential equation

$$\frac{\partial F_g}{\partial t_1} = \left(2g - 2 + \sum_{i=0}^{\infty} t_i \frac{\partial}{\partial t_i}\right) F_g + \frac{1}{24}\delta_{g,1}. \tag{2.47}$$

This can be rewritten in a way that does not refer to g if one recalls that the right-hand side of (2.45) is nonzero only if $3g-3+n = \sum_i d_i$, which amounts to the statement that genus g correlation functions obey

$$\left(\sum_i (i-1)t_i \frac{\partial}{\partial t_i} - (3g-3)\right) F_g = 0. \tag{2.48}$$

The above equations combine to give

$$\frac{\partial F}{\partial t_1} = \frac{1}{3}\sum_{i=0}^{\infty}(2i+1)t_i \frac{\partial F}{\partial t_i} + \frac{1}{24}. \tag{2.49}$$

This result is part (c) of the evidence for our conjecture, because in fact it can alternatively be deduced if one assumes that $U = \ddot{F}$ obeys the KdV equations as well as the string equation.

Comparison of τ*'s and* κ*'s.* Let us now sketch how one similarly obtains the formulas (2.11) relating the τ_n's to the κ_{n-1}'s. To explain the ideas, it should suffice to indicate the origin of the first equation in (2.11),

$$\langle \tau_{d_1}\tau_{d_2}\rangle = \langle \kappa_{d_1-1}\kappa_{d_2-1}\rangle + \langle \kappa_{d_1+d_2-2}\rangle. \tag{2.50}$$

(We may assume that $d_1, d_2 > 0$, since otherwise, with $\kappa_{-1} = 0$, (2.50) is a consequence of the string equation.) To analyze this, we consider the moduli space $\overline{\mathcal{M}}_{g,2}$ of curves Σ with two marked points x_1 and x_2. It has two projections

$$\pi_i : \overline{\mathcal{M}}_{g,2} \to \overline{\mathcal{M}}^{(i)}_{g,1}, \tag{2.51}$$

where π_1 "forgets" x_2 and π_2 "forgets" x_1. (The inverted naming of the π's will make later formulas less painful.) Here $\overline{\mathcal{M}}^{(i)}_{g,1}$, $i = 1, 2$, are the two copies of $\overline{\mathcal{M}}_{g,1}$ obtained by "forgetting" one of the points. We also have the usual two line bundles over $\overline{\mathcal{M}}_{g,2}$ defined by $\mathcal{L}_{(i)} = x_i^*(K_{\mathcal{C}/\mathcal{M}})$; $K_{\mathcal{C}/\mathcal{M}}$ is the relative cotangent bundle of the universal curve. Similarly on $\pi_i(\overline{\mathcal{M}}_{g,2})$, we define $\mathcal{L}'_{(i)} = x_i^*(K'_{\mathcal{C}/\mathcal{M}})$, where $K'_{\mathcal{C}/\mathcal{M}}$ is the relative cotangent bundle to the universal curve over $\overline{\mathcal{M}}^{(i)}_{g,1}$.

According to (2.35),

$$c_1(\mathcal{L}_{(2)}) = \pi_2^*(c_1(\mathcal{L}'_{(2)})) + [D], \tag{2.52}$$

where D is the divisor in $\overline{\mathcal{M}}_{g,2}$ parametrizing curves which have a genus zero component containing x_1, x_2, and precisely one double point. Since $d_1 > 0$ and $\mathcal{L}_{(1)}$ is trivial when restricted to D, we can rewrite the definition of $\langle \tau_{d_1} \tau_{d_2} \rangle$ in the form

$$\langle \tau_{d_1} \wedge \tau_{d_2} \rangle = \left(c_1(\mathcal{L}_{(1)})^{d_1} \wedge \pi_2^*(c_1(\mathcal{L}'_{(2)}))^{d_2}, \overline{\mathcal{M}}_{g,2} \right), \tag{2.53}$$

dropping the second term in (2.52). Similarly,

$$c_1(\mathcal{L}_{(1)}) = \pi_1^*(c_1(\mathcal{L}'_{(1)})) + [D]. \tag{2.54}$$

Writing out the analog of (2.39) and discarding the terms with $m \neq 0$ for the same reason as before, we get

$$c_1(\mathcal{L}_{(1)})^{d_1} = \pi_1^*(c_1(\mathcal{L}'_{(1)}))^{d_1} + [D] \cdot \pi_1^*(c_1(\mathcal{L}'_{(1)}))^{d_1 - 1}. \tag{2.55}$$

So

$$\begin{aligned} \langle \tau_{d_1} \tau_{d_2} \rangle = & \left(\pi_1^*(c_1(\mathcal{L}'_{(1)}))^{d_1} \wedge \pi_2^*(c_1(\mathcal{L}'_{(2)}))^{d_2}, \overline{\mathcal{M}}_{g,2} \right) \\ & + (\pi_1^*(c_1(\mathcal{L}'_{(1)}))^{d_1 - 1} \wedge \pi_2^*(c_1(\mathcal{L}'_{(2)}))^{d_2}, D). \end{aligned} \tag{2.56}$$

Now, D is a copy of $\overline{\mathcal{M}}_{g,1}$. When restricted to D, $\pi_1^*(\mathcal{L}'_{(1)}) \cong \pi_2^*(\mathcal{L}'_{(2)}) \cong K_{D/\mathcal{M}}$, where $K_{D/\mathcal{M}}$ is the canonical bundle to the fibers of $D \cong \overline{\mathcal{M}}_{g,1} \to \overline{\mathcal{M}}_g$. (Indeed, if one restricts to D and then forgets x_1 or x_2, then the genus zero component containing x_1 and x_2 in the curve parametrized by D "collapses" and x_2 or x_1 is identified with the one marked point of $D \cong \overline{\mathcal{M}}_{g,1}$.) Hence, by definition of the Mumford-Morita-Miller classes,

$$\left(\pi_1^*(c_1(\mathcal{L}'_{(1)}))^{d_1 - 1} \wedge \pi_2^*(c_1(\mathcal{L}'_{(2)}))^{d_2}, D \right) = (K_{D/\mathcal{M}}^{d_1 + d_2 - 1}, D) = \langle \kappa_{d_1 + d_2 - 2} \rangle. \tag{2.57}$$

Now, practically from the definition in (2.10), we have

$$\langle \kappa_{d_1 - 1} \kappa_{d_2 - 1} \rangle = \left(\pi_1^*(c_1(\mathcal{L}'_{(1)}))^{d_1} \wedge \pi_2^*(c_1(\mathcal{L}'_{(2)}))^{d_2}, \overline{\mathcal{M}}_{g,2} \right). \tag{2.58}$$

In fact, we can identify $\overline{\mathcal{M}}_{g,2}$ with $\mathcal{C}_2\overline{\mathcal{M}}_g = \mathcal{C}_{(1)}\overline{\mathcal{M}}_g \times_{\overline{\mathcal{M}}_g} \mathcal{C}_{(2)}\overline{\mathcal{M}}_g$.[2] The two factors of $\mathcal{C}\overline{\mathcal{M}}_g$ are $\overline{\mathcal{M}}_{g,1}^{(i)}$, and the two projections defined in (2.51) are the projections of the fiber product onto the factors. With this interpretation, the $\pi_i^*(\mathcal{L}'_{(i)})$ indeed coincide with $\widehat{\mathcal{L}}_{(i)}$ of (2.10), so (2.10) is equivalent to (2.58).

[2] That is, these two varieties are naturally isomorphic. The families of curves that they parametrize are not the same. Computing the effect of the difference is the point of the present computation.

Combining (2.56), (2.57), and (2.58), we have obtained (2.50). The analogous equations expressing $\langle\tau_{d_1}\cdots\tau_{d_n}\rangle$ for $n>2$ in terms of intersection products of κ's (and vice-versa) can be obtained similarly.

Analogy with homotopy theory. We will now make a brief digression. Equations (2.22) and (2.49) can be formulated as the statement that the functional $Z=e^F$, which physicists call the "partition function," is annihilated by the linear operators

$$(2.59)\qquad \begin{aligned} L_{-1} &= -\frac{\partial}{\partial t_0}+\frac{1}{2}t_0^2+\sum_{i=0}^{\infty} t_{i+1}\frac{\partial}{\partial t_i},\\ L_0 &= -\frac{3}{2}\frac{\partial}{\partial t_1}+\sum_{i=0}^{\infty}\frac{2i+1}{2}t_i\frac{\partial}{\partial t_i}+\frac{1}{16}. \end{aligned}$$

These operators generate a Lie algebra with $[L_0, L_{-1}]=L_{-1}$. This algebra is a subalgebra of the Virasoro algebra (the universal central extension of $\mathrm{Diff}(\mathbf{S}^1)$), and (2.59) has an obvious similarity to standard realizations of the Virasoro algebra. This fact has suggested a flight of fancy. The subject we are investigating in these notes has some notable analogies to the generalized K-theory investigated in [47]. The parameters t_i are analogous to the parameters $\tau_n=\tau(\mathbf{CP}^n)$ which in [47] determine a ring homomorphism $\tau:\mathscr{U}^*\to\mathbb{Z}$, with $\mathscr{U}^*$ being the complex cobordism ring. The critical hypersurfaces (corresponding to formal group laws of height n, for various n) of that theory have an analog in the present theory which will be apparent in §4. The invariants of almost complex manifolds that we will consider in §3, which depend on the parameters t_n, are somewhat similar to the complex cobordism invariant determined by τ which is essential in [47] (but the invariants considered in §3 are not cobordism invariants, so something is wrong with the analogy in its present form). Now, in K-theory, a sort of Virasoro algebra enters in the form of the Landweber-Novikov operations, and this motivated the guess (made in different forms by the author, G. Segal, and J. Morava) that (2.59) is in fact part of a Virasoro algebra that is relevant to the intersection theory problem. Recently, it has been shown to follow from the KdV equations that the partition function Z is indeed a highest weight vector for a Virasoro algebra of which (2.59) is part [18], [29].

Verification of the conjecture for low genus. In our sketch of the evidence for the conjecture, what remains is to explain statement (b)—that, at any rate, the genus ≤ 3 contributions to U obey the relations that follow from the KdV equations. We will first describe a shortcut for verifying this for $g\leq 2$, referring to [37] for a similar discussion in genus three, and then we will reconsider the genus zero and one cases in a more leisurely and perhaps more informative way.

In genus zero and one, the dimensional condition $3g-3=\sum_i(d_i-1)$ makes it impossible to have all $d_i>1$ in a nonzero intersection number $\langle\tau_{d_1}\tau_{d_2}\cdots\tau_{d_n}\rangle$. A nonzero result requires factors of τ_0 or τ_1. On the other hand, (2.22) and (2.49) can be used to eliminate factors of τ_0 and τ_1, reducing everything in genus zero to the special case $\langle\tau_0^3\rangle=1$ and reducing everything in genus one to the special case $\langle\tau_1\rangle=\frac{1}{24}$. Since (2.22) and (2.49) as well as the values for these special cases follow either from algebraic geometry or from the KdV conjecture that we are testing, the conjecture is valid for arbitrary correlation functions in genus zero and one.

In genus two, (2.22) and (2.49) can be used to reduce everything to a knowledge of $\langle\tau_4\rangle$, $\langle\tau_2\tau_3\rangle$, and $\langle\tau_2^3\rangle$. Using the KdV equations plus the string equation, one determines these to be

$$\langle\tau_4\rangle=\frac{1}{1152},\qquad\langle\tau_2\tau_3\rangle=\frac{29}{5760},\qquad\langle\tau_2^3\rangle=\frac{7}{240}.\tag{2.60}$$

(An algorithm for computing these numbers was explained in (2.26) and the following discussion. An algorithm that is longer to prove but much quicker to use follows from the Virasoro equations of [18], [29].) Using (2.8) and (2.11), one has

$$\begin{aligned}\langle\tau_4\rangle&=\langle\kappa_3\rangle,\\ \langle\tau_2\tau_3\rangle&=\langle\kappa_1\kappa_2\rangle+\langle\kappa_3\rangle,\\ \langle\tau_2^3\rangle&=\langle\kappa_1^3\rangle+3\langle\kappa_1\kappa_2\rangle+2\langle\kappa_3\rangle.\end{aligned}\tag{2.61}$$

Mumford's formulas in genus two give

$$\langle\kappa_3\rangle=\frac{1}{1152},\qquad\langle\kappa_1\kappa_2\rangle=\frac{1}{240},\qquad\langle\kappa_1^3\rangle=\frac{43}{2880}.\tag{2.62}$$

From these one can verify the genus two KdV formulas (2.60), completing the verification that the KdV and intersection theory results coincide in genus ≤ 2.

This completes the promised shortcut. To give a clearer picture of what is going on, we will now reconsider the genus zero and one situation in somewhat more detail.

2c. Leisurely approach to genus zero and one. Perhaps it is time to explain what is surprising about our conjecture and what is the difficulty in proving it. In genus zero, one, and two, the *uncompactified* moduli spaces $\mathcal{M}_{g,n}$ are affine varieties, and the cohomology classes $\kappa_n\in H^*(\overline{\mathcal{M}}_{g,n})$ vanish when restricted to $\mathcal{M}_{g,n}$. The κ_n's may thus be taken to have their support on the compactification divisor at infinity in $\overline{\mathcal{M}}_{g,n}$, and this makes computations relatively easy. In higher genus, $\mathcal{M}_{g,n}$ is far from being an affine variety, and [45] the κ_n's are definitely not zero when restricted to the "finite" part of moduli space.

Nevertheless, the KdV relations are surprisingly close to the sort of results that would hold if the cohomology classes of interest *were* trivial when restricted to $\mathcal{M}_{g,n}$. Let us reconsider (2.34):

$$\langle\langle\tau_n\tau_0\tau_0\rangle\rangle_g = \frac{1}{2n+1}\left(\sum_{g'=0}^{g}\langle\langle\tau_{n-1}\tau_0\rangle\rangle_{g'}\langle\langle\tau_0^3\rangle\rangle_{g-g'} + 2\sum_{g'=0}^{g}\langle\langle\tau_{n-1}\tau_0^2\rangle\rangle_{g'}\langle\langle\tau_0^2\rangle\rangle_{g-g'} + \frac{1}{4}\langle\langle\tau_{n-1}\tau_0^4\rangle\rangle_{g-1}\right). \tag{2.63}$$

The right-hand side of (2.63) looks very much like a sum over possible degenerations of a stable curve, to two branches of genus g' and $g-g'$, for $0 \le g' \le g$, or to a single branch of genus $g-1$. The possibilities are sketched in Figure 2. In each degeneration, a double point appears, which leads to two additional factors of τ_0 (one on each branch in the case of a separating degeneration).

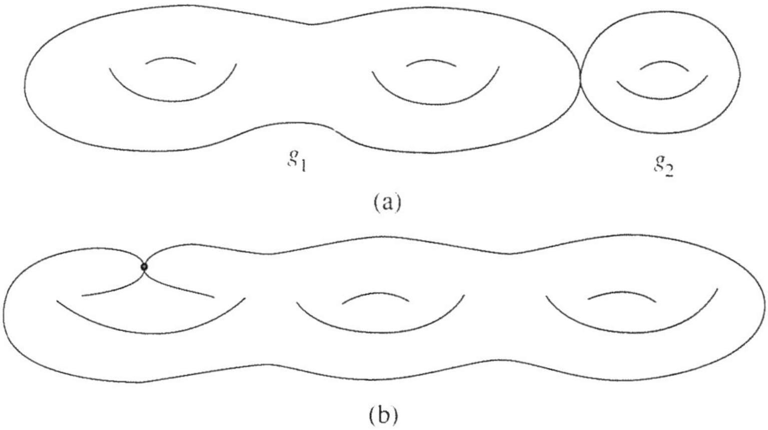

FIGURE 2. THE POSSIBLE DEGENERATIONS OF A STABLE CURVE OF GENUS g TO (A) TWO COMPONENTS WHOSE GENERA ADD TO g; OR (B) ONE COMPONENT OF GENUS $g-1$.

Now, as we will see, (2.63) is rather similar to the type of formula that would hold if the line bundles of interest were trivial on the finite part of moduli space for all g, n—yet strikingly different, because of the innocent looking factor of $1/(2n+1)$. To understand these assertions, we will consider genus zero (and one) in more detail. In genus zero, (2.63) reduces to

$$(2.64)\qquad \langle\langle\tau_n\tau_0\tau_0\rangle\rangle_0 = \frac{1}{2n+1}(\langle\langle\langle\tau_{n-1}\tau_0\rangle\rangle_0\langle\langle\tau_0^3\rangle\rangle_0 + 2\langle\langle\tau_{n-1}\tau_0^2\rangle\rangle_0\langle\langle\tau_0^2\rangle\rangle_0).$$

We will see how a similar but not identical formula arises from algebraic geometry.

Explicit treatment of genus zero. We will study the general n point function in genus zero:

$$(2.65)\qquad \langle\tau_{d_1}\tau_{d_2}\cdots\tau_{d_n}\rangle.$$

We recall that this is to be computed by intersection theory on $\overline{\mathscr{M}}_{0,n}$, which is the moduli space of stable genus zero curves Σ with n marked points $x_1, x_2, \cdots, x_n$. In particular, τ_{d_1} represents $c_1(\mathscr{L}_{(1)})^{d_1}$, where $\mathscr{L}_{(1)}$ is the cotangent bundle to Σ at x_1. Assuming $d_1 > 0$, we write schematically $\tau_{d_1} = c_1(\mathscr{L}_{(1)})\cdot\tau_{d_1-1}$, and we will evaluate $c_1(\mathscr{L}_{(1)})$ explicitly by computing the divisor of a rational section of $\mathscr{L}_{(1)}$. To write such a section explicitly, we use the fact that the finite part $\mathscr{M}_{0,n}$ of the moduli space consists of configurations of n distinct points on $\mathbf{CP}^1$ (which we represent as $\mathbf{C}\cup\infty$) modulo the action of $SL(2,\mathbb{C})$. A convenient section s of $\mathscr{L}_{(1)}$ on the finite part of moduli space can be described by the formula

$$(2.66)\qquad s = dx_1\left(\frac{1}{x_1 - x_{n-1}} - \frac{1}{x_1 - x_n}\right),$$

which has the requisite $SL(2,\mathbb{C})$ invariance. This section obviously has neither zeros nor poles on the finite part $\mathscr{M}_{0,n}$ of moduli space. But we have to consider the possible degenerations.

The differential

$$(2.67)\qquad \omega = dx\left(\frac{1}{x - x_{n-1}} - \frac{1}{x - x_n}\right)$$

on a smooth genus zero curve Σ may be characterized by saying that it has poles only at x_{n-1} and x_n, with residues 1 and -1, and no zeros. If Σ degenerates to a curve with two branches Σ_1 and Σ_2, one defines the sheaf of differentials on Σ as follows: a differential on Σ is a pair

(ω_1, ω_2), where ω_i for $i = 1, 2$ are differentials on Σ_i, and the ω_i are permitted to have simple poles, with equal and opposite residues, at the double point. With this definition, on a stable curve of genus zero, even a degenerate one, there is a unique differential ω with poles only at two given marked points x_{n-1} and x_n, of residues 1 and -1. This differential has no zeros on branches containing x_{n-1} or x_n, but, depending on the nature of the degeneration, may be identically zero on other branches. For instance, if, as in Figure 3, Σ has two branches Σ_1 and Σ_2 with x_{n-1} and x_n on the same branch, say Σ_2, then s is identically 0 on Σ_1 since otherwise its restriction to Σ_1 would be a differential with at most only one simple pole (at the node). Let D be the divisor that parametrizes such curves.

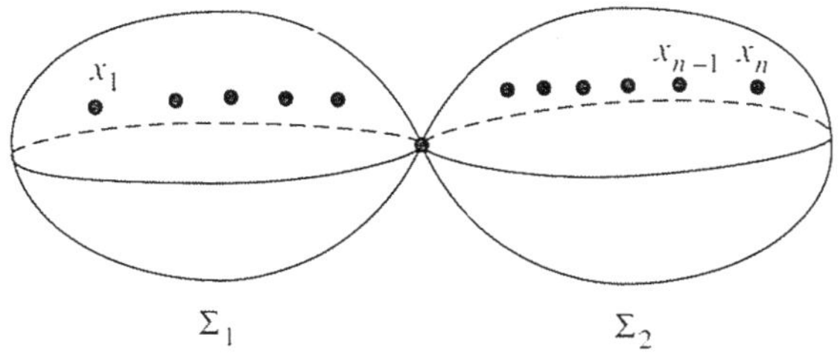

FIGURE 3. A GENUS ZERO CURVE WITH MARKED POINTS $x_1, \cdots, x_n$ DEGENERATING TO TWO COMPONENTS, ONE OF WHICH CONTAINS x_1 AND ONE OF WHICH CONTAINS x_{n-1}, x_n. THE OTHER POINTS MAY BE DISTRIBUTED IN AN ARBITRARY FASHION.

The section s in (2.66) is obtained by evaluating ω at $x = x_1$. Therefore, in view of the facts noted in the last paragraph, s has no poles, even at infinity in moduli space. But s vanishes on the divisor D. A closer examination shows that s has a simple zero on D. Let S denote the finite set $\{2, 3, \cdots, n-2\}$. The divisor D of zeros of s is a union of components $D_{X,Y}$, where $S = X \cup Y$ is a decomposition of S into disjoint subsets, and $D_{X,Y}$ is the divisor consisting of two component curves Σ, with one of the two components containing precisely x_1 and the x_j, $j \in X$, while the other component contains precisely x_{n-1}, x_n, and the x_j, $j \in Y$. We have schematically $\tau_{d_1} = [D] \cdot \tau_{d_1 - 1} = \sum_{S=X\cup Y} [D_{X,Y}] \cdot \tau_{d_1-1}$. This gives

$$\langle \tau_{d_1} \tau_{d_2} \tau_{d_3} \cdots \tau_{d_n} \rangle = \sum_{S=X\cup Y} (\tau_{d_1-1} \wedge \tau_{d_2} \wedge \tau_{d_3} \wedge \cdots \wedge \tau_{d_n}, D_{X,Y}). \tag{2.68}$$

Part of the beauty of the compactified moduli spaces $\overline{\mathcal{M}}_{g,n}$ is, however, that the divisors at infinity are themselves moduli spaces of the same type. In this case, $D_{X,Y}$ can be identified with the product $\overline{\mathcal{M}}_{0,2+n_X} \times \overline{\mathcal{M}}_{0,3+n_Y}$, where n_X and n_Y are the cardinalities of the finite sets X and Y. So (2.68) is equivalent to the much more useful expression

$$\begin{aligned}&\langle \tau_{d_1}\tau_{d_2}\tau_{d_3}\cdots\tau_{d_n}\rangle \\ &= \sum_{S=X\cup Y} \langle \tau_{d_1-1}\cdot \prod_{j\in X}\tau_{d_j}\cdot\tau_0\rangle\cdot\langle\tau_0\cdot\prod_{k\in Y}\tau_{d_k}\cdot\tau_{d_{n-1}}\tau_{d_n}\rangle.\end{aligned} \tag{2.69}$$

This is an inductive formula that determines the left-hand side in terms of a product of similar expressions with smaller values of n or of the d's. The factors of τ_0 that appear explicitly on the right-hand side of (2.69) represent the double point that appears on each branch in Figure 3.

Now, it is useful to write (2.69) in the following way. A special case of (2.69) is the case $n = 3$, in which S is empty. One gets then

$$\langle \tau_{d_1}\tau_{d_2}\tau_{d_3}\rangle = \langle\tau_{d_1-1}\tau_0\rangle\langle\tau_0\tau_{d_2}\tau_{d_3}\rangle. \tag{2.70}$$

Of course, we are working here at $t_i = 0$, as is indicated by the use of the symbol $\langle\cdots\rangle$ (rather than $\langle\langle\cdots\rangle\rangle$). Let us differentiate (2.70) with respect to t_i for some i. The resulting equation

$$\langle\tau_{d_1}\tau_i\tau_{d_2}\tau_{d_3}\rangle = \langle\tau_{d_1-1}\tau_i\tau_0\rangle\langle\tau_0\tau_{d_2}\tau_{d_3}\rangle + \langle\tau_{d_1-1}\tau_0\rangle\langle\tau_0\tau_i\tau_{d_2}\tau_{d_3}\rangle \tag{2.71}$$

is valid since it is simply the $n = 4$ case of (2.69). In a similar way, one sees that the multiple derivatives of (2.70) with respect to the t_i all vanish; indeed the vanishing of the k th derivative of (2.70) is equivalent to the validity of (2.69) for $n = k + 3$.

The fact that (2.70) vanishes *together with all of its derivatives* at $t_i = 0$ is equivalent to the single statement

$$\langle\langle\tau_{d_1}\tau_{d_2}\tau_{d_3}\rangle\rangle_0 = \langle\langle\tau_{d_1-1}\tau_0\rangle\rangle_0\langle\langle\tau_0\tau_{d_2}\tau_{d_3}\rangle\rangle_0 \tag{2.72}$$

(at least as a statement about formal power series, which is all we claim since here we are not considering analytical questions concerning the nature of these functions of the t_k). Indeed, this one equation is the generating function for the derivatives of (2.70) and thus for the totality of equations (2.69).

The special case $d_2 = d_3 = 0$ of (2.72) is clearly very similar to (2.64). From this point of view, however, the factor of $1/(2n + 1)$ in (2.64), which has no counterpart in (2.72), appears rather strange. Because of its dependence on n, it could not arise in a derivation on the above lines.

Actually, (2.64) and (2.72) are so similar without being identical that at first sight one is tempted to think that they could scarcely be consistent.

However, there are many ways to demonstrate their consistency. We will give one argument that will be useful background for §3.

To begin with, consider the objects $\langle\langle\tau_n\tau_m\rangle\rangle_0$. These are functions of $t_0, t_1, t_2, \cdots$. Let us, however, evaluate them at $t_1 = t_2 = \cdots = 0$ to get functions of t_0 only, defined by

$$G_{n,m}(t_0) = \langle\langle\tau_n\tau_m\rangle\rangle|_{t_1=t_2=\cdots=0}. \tag{2.73}$$

This process of restricting to functions of t_0 only will be important, so let us introduce some terminology. We will call the infinite-dimensional affine space $\mathbf{A}^\infty$ of the t_i the "phase space" (or "full phase space") of the theory, and we will call the line defined by $t_i = 0$, $i > 0$, the "small phase space." From (2.24), we have $t_0 = \langle\langle\tau_0\tau_0\rangle\rangle_0$ on the small phase space, so (2.73) is equivalent to the statement that on the small phase space,

$$\langle\langle\tau_n\tau_m\rangle\rangle_0 = G_{n,m}(\langle\langle\tau_0\tau_0\rangle\rangle_0). \tag{2.74}$$

We claim that (2.72) means that (2.74) is true, without modification, on the full phase space. We already know, of course, that (2.74), and therefore also its t_0 derivatives, vanish on the small phase space, so in particular

$$\langle\langle\tau_0\tau_n\tau_m\rangle\rangle_0 = G'_{n,m}(\langle\langle\tau_0\tau_0\rangle\rangle_0)\cdot\langle\langle\tau_0^3\rangle\rangle_0 \tag{2.75}$$

on the small phase space. We will use (2.72) to show that the t_k derivatives of (2.74) vanish on the small phase space also for $k > 0$.

The first derivative of the left-hand side of (2.74) with respect to t_k, on the small phase space, is

$$\begin{aligned}\langle\langle\tau_k\tau_n\tau_m\rangle\rangle_0 &= \langle\langle\tau_{k-1}\tau_0\rangle\rangle_0\langle\langle\tau_0\tau_n\tau_m\rangle\rangle_0\\ &= \langle\langle\tau_{k-1}\tau_0\rangle\rangle_0\cdot G'_{n,m}(\langle\langle\tau_0\tau_0\rangle\rangle_0)\cdot\langle\langle\tau_0^3\rangle\rangle_0,\end{aligned} \tag{2.76}$$

where (2.72) and (2.75) have been used. The first derivative of the right-hand side of (2.74) with respect to t_k is

$$G'_{n,m}(\langle\langle\tau_0\tau_0\rangle\rangle_0)\cdot\langle\langle\tau_k\tau_0\tau_0\rangle\rangle_0 = G'_{n,m}(\langle\langle\tau_0\tau_0\rangle\rangle_0)\cdot\langle\langle\tau_{k-1}\tau_0\rangle\rangle_0\cdot\langle\langle\tau_0^3\rangle\rangle_0. \tag{2.77}$$

Comparing these, one sees that the first derivative of (2.74) with respect to the t_k vanishes on the small phase space.

Inductively, if it is known that the rth derivatives of (2.74) with respect to the t_k all vanish on the small phase space, then precisely the same argument shows that the $(r+1)$th derivatives of (2.74) vanish on the small phase space. Therefore, (2.74) is valid to all orders in an expansion in powers of the t_k and therefore (2.74) is valid on the full phase space.

To make this more concrete, let us now determine the functions $G_{n,m}$. We leave it to the reader to deduce from (2.69) that

$$\langle\tau_n\tau_m\tau_0^s\rangle = \delta_{s,n+m+1}, \tag{2.78}$$

from which it follows that on the small phase space

$$\langle\langle\tau_n\tau_m\rangle\rangle_0 = \frac{t_0^{n+m+1}}{(n+m+1)!}. \tag{2.79}$$

Thus,

$$G_{n,m}(t_0) = \frac{t_0^{n+m+1}}{(n+m+1)!}. \tag{2.80}$$

In particular, setting $m = 0$, we have

$$\langle\langle\tau_n\tau_0\rangle\rangle_0 = \frac{\langle\langle\tau_0\tau_0\rangle\rangle_0^{n+1}}{(n+1)!}. \tag{2.81}$$

Differentiating with respect to t_0, we then also have

$$\langle\langle\tau_n\tau_0^2\rangle\rangle_0 = \frac{\langle\langle\tau_0^3\rangle\rangle_0\langle\langle\tau_0^2\rangle\rangle_0^n}{n!}. \tag{2.82}$$

From these facts, the reader may straightforwardly deduce the genus zero KdV relation (2.64). Alternatively, we may say the following. In view of (2.32), (2.81) for $m = 0$ implies that the function which, according to our conjecture, is the generalized KdV potential $R_{n+1}(U, \dot U, \ddot U, \cdots)$ is

$$R_{n+1}(U, \dot U, \ddot U, \cdots) = \frac{U^{n+1}}{(n+1)!} + \cdots, \tag{2.83}$$

where '$\cdots$' denote terms involving derivatives of U which will arise as contributions from genus $g \geq 1$. It can indeed be seen (from the explicit form (2.34) of the KdV recursion relations) that if we consider the kth derivative $\partial^k U/\partial t_0^k$ to be of degree k, then the genus g contribution to R_{n+1} is homogeneous of degree $2g$. Of course, (2.83) agrees with the KdV theory for the contribution of degree 0.

In addition, (2.81) makes it possible to rewrite the genus zero approximation to the string equation in an interesting way. The t_0 derivative of the string equation (2.22) is

$$U = t_0 + \sum_{i=0}^{\infty} t_{i+1}\langle\langle\tau_i\tau_0\rangle\rangle, \tag{2.84}$$

and—in a genus zero approximation—we can now write this as

$$U = \sum_{i=0}^{\infty} t_i \frac{U^i}{i!}. \tag{2.85}$$

This looks like an equation for a fixed point of a general formal transformation $U \to \sum t_i U^i/i!$ of the affine line. This again suggests the analogy with homotopy theory that has been mentioned earlier.

Genus one. We will now, more briefly, indicate a similar treatment in genus one. We consider a genus one curve Σ with n marked points $x_1, \cdots, x_n$ and the general correlation function

$$\langle \tau_{d_1} \tau_{d_2} \cdots \tau_{d_n} \rangle_1. \tag{2.86}$$

Again the basic idea is to find a suitable section s of $\mathscr{L}_{(1)}$ and write $\tau_{d_1} = [s] \cdot \tau_{d_1 - 1}$. Actually, it is more convenient to take s to be a section of $\mathscr{L}_{(1)}^{\otimes 12}$, and then write

$$\tau_{d_1} = \frac{[s]}{12} \cdot \tau_{d-1}. \tag{2.87}$$

Indeed, if (on the Zariski open set in $\overline{\mathscr{M}}_{1,n}$ in which Σ has only one component) we regard Σ as an elliptic curve with x_1 as the origin, then the elliptic modular form Δ of weight 12 with a simple zero at the cusp can be interpreted as a section of $\mathscr{L}_{(1)}^{\otimes 12}$. It has no poles, and vanishes only when Σ degenerates. In the theory of modular forms, the only relevant degeneration is the degeneration of Σ to a rational curve with double point, where Δ and hence also s has a simple zero. Let us call this divisor D_0. In the present context there is an additional possibility: s vanishes when Σ degenerates to a union of two components, of genus zero and one, respectively, as sketched in Figure 4, provided that x_1 is on the genus zero component. (The reasoning showing that s vanishes in this situation, with a 12th order zero since it is a section of $\mathscr{L}_{(1)}^{\otimes 12}$, is similar to the reasoning that we used at an analogous point in the genus zero discussion. Heuristically, an elliptic curve, even if it degenerates to two components of genus zero and one, has a nonzero holomorphic differential, but this vanishes identically on the genus zero component as it would otherwise be a nonzero differential on that component with at most only a single pole at the node.) We will refer to this divisor as $D_{0,1}$. So we have

$$\frac{1}{12}[s] = \frac{[D_0]}{12} + [D_{0,1}]. \tag{2.88}$$

As in the genus zero discussion, it is essential that the divisors on which s vanishes are themselves moduli spaces of stable curves with marked points. Thus, the divisor D_0 on which Σ degenerates to a rational curve with double point is in the orbifold sense

$$D_0 = \tfrac{1}{2}\overline{\mathscr{M}}_{0,n+2}. \tag{2.89}$$

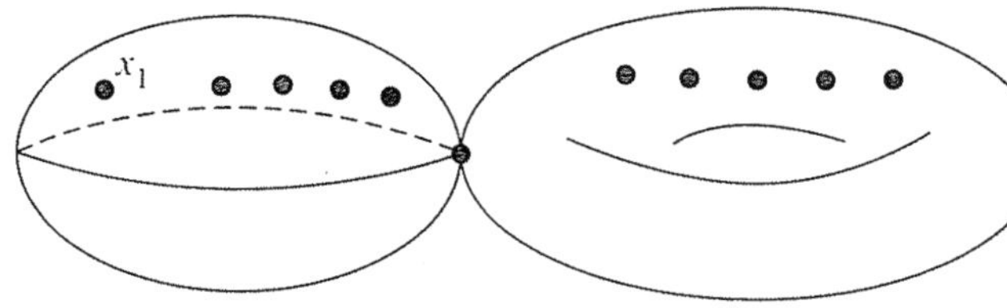

FIGURE 4. A GENUS ONE CURVE Σ WITH MARKED POINTS $x_1, x_2, \cdots, x_n$ DEGENERATING TO TWO COMPONENTS OF GENUS ZERO AND GENUS ONE RESPECTIVELY, WITH x_1 ON THE GENUS ZERO COMPONENT AND AN ARBITRARY DISTRIBUTION OF THE OTHER POINTS.

The $n+2$ marked points here are the n that were present originally and the two copies of the double point on the normalization of Σ; and the factor of $\frac{1}{2}$ arises because the two copies of the double point have no preferred ordering, but we have defined $\overline{\mathcal{M}}_{0,n}$ in terms of configurations of *ordered* points. The divisor $D_{0,1}$ is reducible, corresponding to the fact that the points $x_2, x_3, \cdots, x_n$ may be distributed on the two branches in an arbitrary fashion. For every decomposition $S = X \cup Y$ of the set $S = \{2, 3, \cdots, n\}$ as a union of disjoint subsets X and Y, $D_{0,1}$ has a component $D_{0,1;X,Y}$ in which precisely the x_j, $j \in Y$, are on the genus one branch. $D_{0,1;X,Y}$ is a copy of $\overline{\mathcal{M}}_{2+n_X} \times \overline{\mathcal{M}}_{1+n_Y}$, where n_X and n_Y are the cardinalities of X and Y, respectively. Writing $\langle \tau_{d_1}\tau_{d_2}\cdots\tau_{d_n}\rangle = (\tau_{d_1-1}\tau_{d_2}\cdots\tau_{d_n}, [s])/12$ and using (2.88) and the facts just indicated, we get

$$\begin{aligned} \langle \tau_{d_1}\tau_{d_2}\cdots\tau_{d_n}\rangle_1 = &\sum_{S=X\cup Y} \langle \tau_{d_1-1}\prod_{j\in X}\tau_{d_j}\cdot\tau_0\rangle_0\cdot\langle\tau_0\cdot\prod_{j\in Y}\tau_{d_j}\rangle_1 \\ &+ \frac{1}{24}\langle\tau_{d_1-1}\tau_{d_2}\cdots\tau_{d_n}\cdot\tau_0^2\rangle_0. \end{aligned} \tag{2.90}$$

By reasoning exactly analogous to that which led from (2.69) to (2.72), the totality of equations (2.90) is equivalent to a single statement about generating functionals,

$$\langle\langle\tau_n\rangle\rangle_1 = \langle\langle\tau_{n-1}\tau_0\rangle\rangle_0\langle\langle\tau_0\rangle\rangle_1 + \tfrac{1}{24}\langle\langle\tau_{n-1}\tau_0\tau_0\rangle\rangle_0. \tag{2.91}$$

One can also deduce a sort of genus one analog of (2.74), namely

$$F_1 = \tfrac{1}{24}(\ln(\langle\langle\tau_0^3\rangle\rangle_0) - 2). \tag{2.92}$$

Indeed, this equation holds on the small phase space (where $\langle\tau_0^3\rangle_0 = 1$; the -2 in (2.92) comes from (2.18)). Its derivatives with respect to the t_n

can, inductively, be shown to vanish, using (2.91) and the genus zero recursion relations; this establishes that (2.92) holds on the full phase space. By differentiating (2.92) with respect to the t_n's, one can get formulas expressing $\langle\langle\tau_n\rangle\rangle_1$, $\langle\langle\tau_n\tau_m\rangle\rangle_1$, etc., in terms of genus zero quantities (which, using (2.74), can all be expressed in terms of the $\langle\langle\tau_0^n\rangle\rangle_0$).

In §3, we will generalize this story by introducing a rather general "target space" M, and see that a considerable amount of the structure that we have described generalizes.

3. Coupling to sigma models

The analogy of the problem that we have been studying to ordinary string theory is greatly strengthened if one couples the purely "gravitational" problem to "topological sigma models." (These correspond to a mathematical problem studied by Floer, Gromov, and others [27], [34]. The relation of this problem to quantum field theory was suggested in [2]. A Lagrangian realization was found in [58], and was developed further in [8].)

Fix a compact Kähler manifold M and a Riemann surface Σ with a fixed complex structure, to begin with. (The Kähler condition on M can be weakened, as noted below, but this is not central for our purposes.) Let $\mathcal{S}_\lambda$ be the moduli space of holomorphic maps of Σ to M, of a fixed homotopy type λ, and let $\mathcal{S} = \bigcup_\lambda \mathcal{S}_\lambda$.

Let us begin with a few comments on different methods and goals in the mathematical and physical work in this area. In the mathematical literature, the goal has been to relax the Kähler condition. Intersection theory on the $\mathcal{S}_\lambda$ is usually regarded as the definition of the problem, and analytical problems involving these spaces are the crux of the matter. From this point of view, the theory has been developed for general compact symplectic manifolds; a symplectic structure determines an almost complex structure uniquely up to homotopy by requiring that the symplectic form is positive and of type $(1, 1)$, and a notion of almost holomorphic maps from Riemann surfaces exists for general almost complex manifolds. The main success of the mathematical theory has been to obtain exotic invariants of symplectic manifolds and to use them to prove theorems about such manifolds.

Physically, the starting point is not intersection theory on moduli space but the existence of an appropriate topological sigma model Lagrangian, which one attempts to quantize. The Lagrangian exists for arbitrary almost complex manifolds M, and one aims to develop the theory in this generality. From this point of view, a priori one is studying Feynman path integrals, not intersection theory on moduli space. However, there are formal arguments [58] which, when the $\mathcal{S}_\lambda$ are smooth and compact, give a reduction of the Feynman path integral to classical intersection theory. Those arguments break down when the behavior of the moduli spaces $\mathcal{S}_\lambda$ is bad, but one would expect the Feynman integral itself to provide a more general definition of the desired "intersection numbers" even in such a case. The goal of the physical discussion is to construct quantum field theories, to explore a possible "unbroken phase" of string theory, and, in the present context, to explore how much of the discussion of the last section has an analog.

Letting K denote the canonical line bundle of M, the content of the theory that we will be discussing depends very much on the sign of $c_1(K)$. If $c_1(K) > 0$ and the dimension of M is not very small, the theory will be rather dull since the formal dimensions of the moduli spaces will almost all be negative. A much more interesting situation arises when $c_1(K) < 0$ (a condition that singles out a much smaller class of Kähler manifolds, including $\mathbb{C}P^n$), since then the moduli spaces generally have positive formal dimension.

For $c_1(K) < 0$, one has nice finiteness conditions; any correlation function (as introduced in the next subsection) receives contributions only from finitely many components of moduli space, and it seems likely that all of the unknown functions that will be introduced later can be determined from computation of finitely many special cases. Also, for $c_1(K) < 0$, the topological sigma model if studied by conventional physical methods is "asymptotically free," and has much better properties.

The discussion in this section will be particularly informal. We will not attempt to determine the appropriate class of target spaces M, and we will assume that the moduli spaces $\mathcal{S}_\lambda$ behave favorably. We should also note that, although we will concentrate on models derived from Kahler manifolds, there are other classes of models that obey the same general conditions (such as the models derived from matrix chains, which we will consider in §4d, and much less well understood models associated with the two-dimensional analogs of Donaldson theory.)

We will limit ourselves to the case $\pi_1(M) = 0$ to avoid a number of questions that have not yet been elucidated.

3a. Correlation functions. Let $\mathcal{S}$ be the moduli space of holomorphic maps of $\varphi: \Sigma \to M$, and let $\Phi: \mathcal{S} \times \Sigma \to M$ be the corresponding "universal instanton." For x a point in Σ, let Φ_x be the restriction of Φ to $\mathcal{S} \times x$. Every cohomology class $\alpha \in H^*(M)$ determines a cohomology class $W_\alpha = \Phi_x^*(\alpha) \in H^*(\mathcal{S})$, which is obviously independent of x since any two points on Σ are cohomologous. (One can consider integral cohomology at this point, but we will eventually be thinking in terms of real cohomology.) If α is the Poincaré dual to a submanifold $H \subset M$, then W_α is Poincaré dual to $W_H = \{\varphi | \varphi(x) \in H\}$. One denotes the intersection numbers of the W_α as

$$(3.1) \qquad \langle \mathcal{O}_{\alpha_1}(x_1)\mathcal{O}_{\alpha_2}(x_2)\cdots\mathcal{O}_{\alpha_n}(x_n)\rangle = (W_{\alpha_1} \wedge W_{\alpha_2} \wedge \cdots \wedge W_{\alpha_n}, \mathcal{S}).$$

As in (2.3), the motivation for the notation is that these intersection numbers can be represented as the expectation value of a product of local operators $\mathcal{O}_\alpha$, with respect to some Feynman path integral measure. In (3.1), the x_i are arbitrary distinct points in Σ, and $W_{\alpha_i} = \Phi_{x_i}^*(\alpha_i)$. To avoid cluttering this section with minus signs, we will assume that M has only even-dimensional cohomology so that the ordering of terms in (3.1) is immaterial.

Of course, (3.1) vanishes except for contributions from homotopy classes λ such that

$$(3.2) \qquad \sum_i \dim \alpha_i = \dim \mathcal{S}_\lambda.$$

According to the Riemann-Roch formula, the virtual dimension of $\mathcal{S}_\lambda$ is

$$(3.3) \qquad \dim_{\mathbb{C}} \mathcal{S}_\lambda = (1-g)\cdot \dim_{\mathbb{C}} M - (\varphi^*(c_1(K)), \Sigma),$$

where φ is any map of the homotopy type λ. Thus, for $c_1(K) < 0$, only finitely many $\mathcal{S}_\lambda$'s have the right dimension to contribute to (3.1), and there are no problems of infinite sums. In general, we would have to modify the discussion to keep track of the homotopy type. (The right modification is actually part of what we will do anyway later in forming the generating functional F, but we do not wish to introduce it now in an ad hoc fashion.)

Actually (for an appropriate class of target spaces M), the intersection numbers (3.1) are independent of the complex structure on Σ. Mathematically, this is a corollary of the existence of an appropriate moduli space $\mathcal{N}_{g,n}$ of pairs of objects (Σ, φ), where Σ is a stable curve of genus g with n marked points $x_1, \cdots, x_n$, and $\varphi: \Sigma \to M$ is a holomorphic map. Physically, one would deduce the independence of complex structure by using the BRST invariance of the Lagrangian.

3b. Topological field theory. In fact, these intersection numbers obey the axioms of a "topological quantum field theory," in a sense spelled out in detail in [3], adapted from Segal's axiomatization of conformal field theory [55]. These axioms require one to associate to every circle C a vector space V_C, and to every Riemann surface Σ bounding a collection of "incoming" circles C_i, $i \in X$, and "outgoing" circles C_j, $j \in Y$, a linear transformation $\Phi_\Sigma: \bigotimes_{i\in X} V_{C_i} \to \bigotimes_{j\in Y} V_{C_j}$. The main requirement (sketched in Figure 5) is that if Σ is obtained by joining the outgoing boundary of Σ_1 to the incoming boundary of Σ_2, then one wants

$$\Phi_\Sigma = \Phi_{\Sigma_2} \circ \Phi_{\Sigma_1}. \tag{3.4}$$

Physicists describe this by saying that "one can calculate the transition amplitude by summing over physical intermediate states." In addition to (3.4), one imposes a similar condition relating Φ_Σ and Φ'_Σ, where Σ' is obtained from Σ by cutting on a nonseparating cycle.

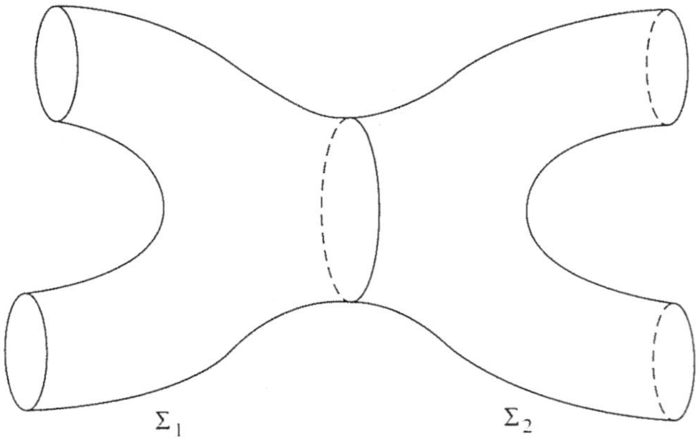

FIGURE 5. A KEY TOPOLOGICAL FIELD THEORY AXIOM IS A COMPOSITION LAW $\Phi_\Sigma = \Phi_{\Sigma_2} \circ \Phi_{\Sigma_1}$ THAT MUST HOLD IN THIS SITUATION.

To realize these axioms in the case at hand, one takes V_C for every circle C to be a copy of $H^*(M, \mathbb{R})$. This vector space has a natural metric given by Poincaré duality, so one need to distinguish incoming and outgoing circles. The metric will play an important role in what follows. If H_σ, $\sigma \in L$, is a basis for the real cohomology of M, then the metric is $\eta_{\sigma\tau} = (H_\sigma \wedge H_\tau, M)$. This is an invertible matrix whose inverse will be denoted by $\eta^{\sigma\tau}$.

If Σ is a surface bounded by circles $C_1, \cdots, C_n$, then the linear transformation Φ_Σ which is part of the topological field theory data is simply a vector in $\bigotimes_{i=1}^n V_{C_i} = \bigotimes_{i=1}^n H^*(M, \mathbb{R})$. This vector is given by the correlation function $\langle \mathcal{O}_{\alpha_1} \mathcal{O}_{\alpha_2} \cdots \mathcal{O}_{\alpha_n} \rangle$. The only property that needs to be verified is the composition law (3.4) (and its analog for nonseparating cuts). We will investigate this in the next paragraph. To simplify notation, we will specialize the following discussion to the case of genus 0 with $n = 4$, which proves to play a special role. Other cases (and the rest of the topological field theory axioms) can be discussed similarly.

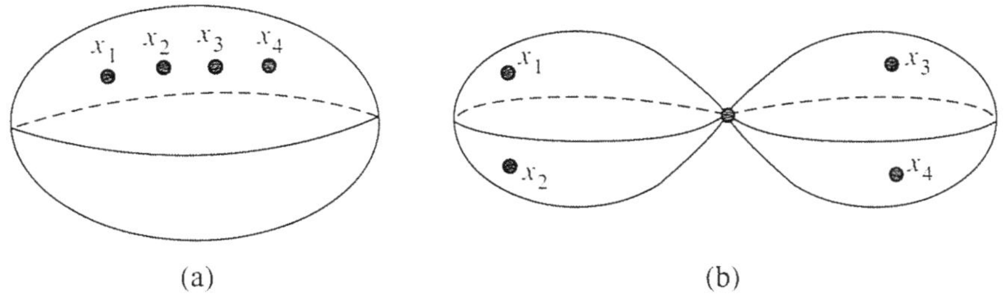

FIGURE 6. A CONFIGURATION OF FOUR POINTS $x_1, \cdots, x_4$ IN GENUS ZERO, AND ONE OF ITS THREE DEGENERATIONS.

The four point function $\langle \mathcal{O}_\alpha(x_1)\mathcal{O}_\beta(x_2)\mathcal{O}_\gamma(x_3)\mathcal{O}_\delta(x_4)\rangle$ is computed, as in Figure 6, by fixing a configuration of four points $x_1, \cdots, x_4$ on a curve Σ of genus zero, and evaluating an appropriate intersection on $\mathcal{S}_{0,4}$. Suppose that one degenerates the configuration of four points to the boundary of moduli space, where Σ decomposes into two components Σ_1 and Σ_2 which share a double point P. A holomorphic map $\varphi : \Sigma \to M$ is by definition a pair (φ_1, φ_2), where for $i = 1, 2$, φ_i is a holomorphic map $\Sigma_i \to M$, obeying $\varphi_1(P) = \varphi_2(P)$. If $\mathcal{S}$, $\mathcal{S}_1$, and $\mathcal{S}_2$ are the moduli spaces of holomorphic maps of Σ, Σ_1, and Σ_2 to M, then the condition $\varphi_1(P) = \varphi_2(P)$ defines a cycle X in $\mathcal{S}_1 \times \mathcal{S}_2$, and we can identify $\mathcal{S}$ with X. Since $X \subset \mathcal{S}_1 \times \mathcal{S}_2$, we can compute the four point function by counting intersections on $\mathcal{S}_1 \times \mathcal{S}_2$. Moreover, if x_1, x_2 lie on Σ_1 and x_3, x_4 lie on Σ_2, then we can think of $\mathcal{O}_\alpha, \mathcal{O}_\beta$ as representing classes $W_\alpha^{(1)}, W_\beta^{(1)}$ in $H^*(\mathcal{S}_1)$, and $\mathcal{O}_\gamma, \mathcal{O}_\delta$ as representing classes $W_\gamma^{(2)}, W_\delta^{(2)}$ in $H^*(\mathcal{S}_2)$, which are then pulled back to $H^*(\mathcal{S}_1 \times \mathcal{S}_2)$. We then have

$$\langle \mathcal{O}_\alpha \mathcal{O}_\beta \mathcal{O}_\gamma \mathcal{O}_\delta \rangle = (W_\alpha^{(1)} \wedge W_\beta^{(1)} \wedge W_\gamma^{(2)} \wedge W_\delta^{(2)} \wedge [X], \mathcal{S}_1 \times \mathcal{S}_2). \tag{3.5}$$

In the last step we are thinking of $[X]$, the Poincaré dual to X, as a class in $H^*(\mathcal{S}_1 \times \mathcal{S}_2)$. To put this in a more useful form, we reexpress $[X]$ by using the Kunneth decomposition of the diagonal $\Delta \subset M \times M$, which in real cohomology reads

$$[\Delta] = \sum_{\sigma, \tau \in L} \eta^{\sigma\tau} H_\sigma \times H_\tau. \tag{3.6}$$

Here $[\Delta]$ is the Poincaré dual of the diagonal in $M\times M$. Correspondingly, one has $[X]=\sum_{\sigma,\tau}\eta^{\sigma\tau}W_\sigma^{(1)}\times W_\tau^{(2)}$, so (3.5) becomes

$$\langle\mathscr{O}_\alpha\mathscr{O}_\beta\mathscr{O}_\gamma\mathscr{O}_\delta\rangle=\sum_{\sigma\tau}\langle\mathscr{O}_\alpha\mathscr{O}_\beta\mathscr{O}_\sigma\rangle\eta^{\sigma\tau}\langle\mathscr{O}_\tau\mathscr{O}_\gamma\mathscr{O}_\delta\rangle. \tag{3.7}$$

This is precisely the topological field theory axiom which we had aimed to explain, and though we have focused on the four point function in genus zero, the general case is no different.

The four point function in genus zero has, however, a special significance. Since the left-hand side is symmetric in α, β, γ, and δ, but the right-hand side does not obviously possess this symmetry, we deduce that

$$\sum_{\sigma,\tau}\eta^{\sigma\tau}\langle\mathscr{O}_\alpha\mathscr{O}_\beta\mathscr{O}_\sigma\rangle\langle\mathscr{O}_\tau\mathscr{O}_\gamma\mathscr{O}_\delta\rangle=\sum_{\sigma,\tau}\eta^{\sigma\tau}\langle\mathscr{O}_\alpha\mathscr{O}_\delta\mathscr{O}_\sigma\rangle\langle\mathscr{O}_\tau\mathscr{O}_\gamma\mathscr{O}_\beta\rangle. \tag{3.8}$$

This equation means that if we define

$$f_{\alpha\beta\gamma}=\langle\mathscr{O}_\alpha\mathscr{O}_\beta\mathscr{O}_\gamma\rangle \tag{3.9}$$

and

$$f_{\alpha\beta}{}^{\delta}=\eta^{\gamma\delta}f_{\alpha\beta\gamma}, \tag{3.10}$$

then the formula

$$\mathscr{O}_\alpha\mathscr{O}_\beta=\sum_\gamma f_{\alpha\beta}{}^{\gamma}\mathscr{O}_\gamma \tag{3.11}$$

defines a commutative, associative multiplication law on $H^*(M,\mathbb{R})$ (which in addition is compatible with the metric η in the sense that $\eta(A,BC)=\eta(BA,C)$ for any A, B, C; this amounts to the statement, clear from the definition, that $f_{\alpha\beta\gamma}$ is completely symmetric). This ring has an identity, namely $\mathscr{O}_1$, where $1\in H^0(M,\mathbb{R})$ is the identity in the ring $H^*(M,\mathbb{R})$. (Conversely, it can be shown in an elementary fashion [17], [59] that from such a ring structure one can reconstruct a two-dimensional topological field theory. In higher dimensions, topological field theories are not classified so easily.)

By our definitions, (3.9) is to be computed by summing over all homotopy classes of holomorphic maps $\Sigma\to M$, with Σ a curve of genus zero. A particularly simple role is played by the null-homotopic maps. This component of $\mathscr{S}$ can be identified with M itself, since a holomorphic map that is also homotopic to zero is constant. Unwinding the definitions, one finds that if one considers only the null-homotopic maps in computing $f_{\alpha\beta\gamma}$, then one recovers the classical ring structure on $H^*(M,\mathbb{R})$. This is a graded ring in which all elements of positive degree are nilpotent. The higher homotopy classes of holomorphic maps contribute in such a way as to deform $H^*(M, \mathbb{R})$ to a structure which tends to be less degenerate, especially if $c_1(K) > 0$, and in general is no longer graded. For example, for $M = \mathbf{CP}^n$, the classical cohomology ring is $\mathbb{R}[x]/(x^{n+1})$, where x is a generator in degree 2, but it is straightforward to compute that the "quantum corrections," that is, the contributions of the higher homotopy classes, deform this to a "quantum cohomology ring" which is isomorphic to $\mathbb{R}[x]/(x^{n+1}-1)$. If $c_1(M)$, as an integral cohomology class, is divisible by r, then the quantum cohomology ring is graded by $\mathbb{Z}/2r\mathbb{Z}$, as one can see from the Riemann-Roch formula for the dimensions of the $\mathcal{N}_\alpha$'s.

We will later generalize this in the following perhaps surprising way. We will find a function F on the vector space $H^*(M, \mathbb{R})$, with the property that at any point in $H^*(M, \mathbb{R})$, the third derivatives of F are the structure constants of a commutative associative algebra. Thus, if y^σ, $\sigma \in L$, are affine coordinates for $H^*(M, \mathbb{R})$, and

$$f_{\alpha\beta\gamma} = \frac{\partial^3 F}{\partial y^\alpha \partial y^\beta \partial y^\gamma} \tag{3.12}$$

(F will not be a cubic function, in general, so the f's are not constant) and $f_{\alpha\beta}{}^{\delta} = \eta^{\gamma\delta} f_{\alpha\beta\gamma}$, then at *any* point in $H^*(M, \mathbb{R})$, the formula

$$\mathcal{O}_\alpha \mathcal{O}_\beta = f_{\alpha\beta}{}^{\gamma} \mathcal{O}_\gamma \tag{3.13}$$

defines a commutative, associative (and of course metric compatible) algebra. This is equivalent to saying that F obeys the overdetermined system of equations

$$\eta^{\sigma\tau} \frac{\partial^3 F}{\partial y^\alpha \partial y^\beta \partial y^\sigma} \cdot \frac{\partial^3 F}{\partial y^\tau \partial y^\gamma \partial y^\delta} = \eta^{\sigma\tau} \frac{\partial^3 F}{\partial y^\alpha \partial y^\gamma \partial y^\sigma} \cdot \frac{\partial^3 F}{\partial y^\tau \partial y^\beta \partial y^\delta}, \tag{3.14}$$

for which we will find for each M a canonical solution.

3c. Coupling to gravity. This structure becomes considerably more interesting if we let the complex structure on Σ vary. Thus, the basic object of study will henceforth be the (compactified) moduli space $\mathcal{N}_{g,n}$ of stable pairs (Σ, φ), where Σ is a curve of genus g with n distinct, ordered marked points $x_1, \cdots, x_n$ and $\varphi: \Sigma \to M$ is a holomorphic map. It is now possible to combine the two constructions that we have considered in this paper. On the one hand, each marked point x_i has a complex cotangent bundle $T^*\Sigma|_{x_i}$, and these vary over $\mathcal{N}_{g,n}$ to give n line bundles $\mathcal{L}_{(i)}$. On the other hand, let $\mathcal{C}\mathcal{N}_{g,n}$ be the "universal curve" over $\mathcal{N}_{g,n}$. We can regard the marked point $x_i: \mathcal{N}_{g,n} \to \mathcal{C}\mathcal{N}_{g,n}$ as a section of the universal curve. If $\Phi: \mathscr{C}\mathscr{N}_{g,n} \to M$ is the "universal instanton," then for $i = 1, \cdots, n$ one has maps $\Phi \circ x_i : \mathscr{N}_{g,n} \to M$, and this gives for each i a natural map from $\alpha \in H^*(M)$ to $W_{i,\alpha} = (\Phi \circ x_i)^*(\alpha) \in H^*(\mathscr{N}_{g,n})$. Corresponding to each of the marked points x_i, we can thus define natural cohomology classes in $H^*(\mathscr{N}_{g,n})$ of the form

$$c_1(\mathscr{L}_{(i)})^d \cdot W_{i,\alpha}. \tag{3.15}$$

Here d is a nonnegative integer, and we may as well consider α to run over a finite set corresponding to a basis L of $H^*(M, \mathbb{R})$. Symbolically, as in §2, we represent these classes by "quantum field theory operators" $\tau_d(\mathscr{O}_\alpha)$,[3] or, for brevity, simply $\tau_{d,\alpha}$, and we write

$$\begin{aligned} &\langle \tau_{d_1,\alpha_1} \tau_{d_2,\alpha_2} \cdots \tau_{d_n,\alpha_n} \rangle \\ &= (c_1(\mathscr{L}_{(1)})^{d_1} \wedge W_{1,\alpha_1} \cdot \cdots \cdot c_1(\mathscr{L}_{(n)})^{d_n} \wedge W_{n,\alpha_n}, \mathscr{N}_{g,n}). \end{aligned} \tag{3.16}$$

Again as in §2, we introduce formal variables ("coupling constants") t_r^α, $r = 0, 1, 2, \cdots$, $\alpha \in L$, and we define the "generating functional"

$$F(t_r^\alpha) = \langle e^{\Sigma_{r,\alpha} t_r^\alpha \tau_{r,\alpha}} \rangle. \tag{3.17}$$

More concretely, this is to be

$$F(t_r^\alpha) = \sum_{\{n_{r,\alpha}\}} \prod_{r,\alpha} \frac{(t_r^\alpha)^{n_{r,\alpha}}}{n_{r,\alpha}!} \cdot \left\langle \prod_{r,\alpha} \tau_{r,\alpha}^{n_{r,\alpha}} \right\rangle, \tag{3.18}$$

where the $n_{r,\alpha}$ are arbitrary collections of nonnegative integers, almost all zero, labeled by r, α. The correlation function on the right-hand side is to be summed over all values of the genus g of Σ, and all homotopy classes of holomorphic maps $\varphi: \Sigma \to M$. However, a nonzero contribution arises only if the genus and the homotopy class obey an appropriate dimensional condition. If $d(\alpha)$ is the dimension of the cohomology class $\alpha \in H^*(M, \mathbb{R})$, then the dimensional condition is now, from the Riemann-Roch formula,

$$6g - 6 + (2g-2)\dim_{\mathbb{C}} M - 2 \int_\Sigma \varphi^*(c_1(K)) = \Sigma_{r,\alpha} n_{r,\alpha}(2r - 2 + d(\alpha)). \tag{3.19}$$

For $c_1(K) < 0$, this condition ensures that the coefficient of a given monomial $\prod (t_r^\alpha)^{n_{r,\alpha}}$ receives a contribution only from finitely many homotopy classes of maps; in this case F can be regarded as a formal power series,

[3] By analogy with standard terminology in conformal field theory, one sometimes refers to the $\mathscr{O}_\alpha = \tau_0(\mathscr{O}_\alpha)$ as "primaries" and the $\tau_d(\mathscr{O}_\alpha)$, $d > 0$, as "descendants." If $m = \dim H^*(M, \mathbb{R})$, the model we are discussing then "has m primaries."

as in §2. (Otherwise, F must be expanded as a series in $e^{\sum_i n_i y_i}$, where the n_i are integers and $y_i = t_0^{\alpha_i}$; here the α_i run over a basis of the *two-dimensional* cohomology of M.)

As in §2, it is convenient to write

$$\langle\langle \tau_{d_1}(\alpha_1)\cdots\tau_{d_n}(\alpha_n)\rangle\rangle = \frac{\partial}{\partial t_{d_1}^{\alpha_1}}\cdot\cdots\cdot\frac{\partial}{\partial t_{d_n}^{\alpha_n}}\cdot F(t_r^{\alpha}). \tag{3.20}$$

Thus, if one sets all $t_r^{\alpha} = 0$, the symbol $\langle\langle\ \rangle\rangle$ reduces to $\langle\ \rangle$. As a special case of (3.20), we write $\langle\langle 1\rangle\rangle = F$.

Our goal in the rest of this section is to show that at least part of the discussion of §2 generalizes in this more elaborate situation.

3d. Analogs of the string and KdV equations. First of all, one important part of the discussion of §2, namely the string equation (2.22), generalizes straightforwardly. We recall that the origin of that equation was that one of the operators (namely τ_0) correspond to the identity, a zero-dimensional cohomology class. The analogous object in the present context is $\tau_0(\mathscr{O}_1)$ (or simply $\tau_{0,1}$), where $\mathscr{O}_1$ corresponds to $1 \in H^0(M, \mathbb{R})$. We consider a general genus g correlation function

$$\left\langle \tau_{0,1}\cdot\prod_{i=1}^{n}\tau_{d_i,\alpha_i}\right\rangle \tag{3.21}$$

of this operator with n other operators. As in §2, this is to be evaluated by integrating a certain cohomology class over $\mathscr{N}_{g,n+1}$; if the cohomology class in question were a pullback from $\mathscr{N}_{g,n}$, then (3.21) would vanish. A nonzero result comes, again, only from the second term in (2.35). Treating this in a similar way, we arrive at the generalized string equation:

$$\frac{\partial F}{\partial t_0^1} = \frac{1}{2}\eta_{\sigma\tau}t_0^{\sigma}t_0^{\tau} + \sum_{i=0}^{\infty}\sum_{\alpha} t_{i+1}^{\alpha}\frac{\partial}{\partial t_i^{\alpha}}F. \tag{3.22}$$

It is also possible to find a general analog of (2.49), but we will not enter into this here.

In particular, it follows from the string equation that if all $t_r^{\alpha} = 0$ for $r > 0$, then

$$\langle\langle \tau_{0,1}\tau_{0,\alpha}\rangle\rangle = \eta_{\alpha\beta}t_0^{\beta}, \tag{3.23}$$

which will be useful later.

In the rest of this section, we will see that at least in genus zero and genus one, one can find analogs of the KdV flows. To begin with, we consider a genus zero curve Σ with marked points $x_1, \cdots, x_n$, and a general correlation function

$$\langle \tau_{d_1,\alpha_1}\tau_{d_2,\alpha_2}\cdots\tau_{d_{n-1},\alpha_{n-1}}\tau_{d_n,\alpha_n}\rangle. \tag{3.24}$$

This can be treated precisely as we treated (2.65) in §2. Symbolically, $\tau_{d_1,\alpha_1} = [s]\cdot\tau_{d_1-1,\alpha_1}$, where $[s]$ is the divisor of a section s of $\mathscr{L}_{(1)}$. For s we pick the same section that we used in §2, namely

$$s = dx_1\cdot\left(\frac{1}{x_1 - x_{n-1}} - \frac{1}{x_1 - x_n}\right). \tag{3.25}$$

Its divisor again consists of certain degenerate configurations in which Σ has two branches Σ_1 and Σ_2. As a holomorphic map $\varphi:\Sigma \to M$ is a pair $\varphi_i:\Sigma_i \to M$ obeying a condition $\varphi_1(P) = \varphi_2(P)$ at the node P, we must again, as in the derivation of (3.7), carry out the Kunneth decomposition of the diagonal in $M\times M$ to express this condition in terms of the cohomology classes we are using. Upon thus modifying the derivation of (2.69), we arrive at the generalization of that equation, namely

$$\begin{aligned}\langle \tau_{d_1,\alpha_1}\tau_{d_2,\alpha_2}&\cdots\tau_{d_{n-1},\alpha_{n-1}}\tau_{d_n,\alpha_n}\rangle\\ &=\sum_{\sigma,\tau}\sum_{S=X\cup Y}\left\langle \tau_{d_1-1,\alpha_1}\prod_{j\in X}\tau_{d_j,\alpha_j}\cdot\tau_{0,\sigma}\right\rangle\\ &\quad\cdot\eta^{\sigma\tau}\cdot\left\langle\tau_{0,\tau}\cdot\prod_{j\in Y}\tau_{d_j,\alpha_j}\cdot\tau_{d_{n-1},\alpha_{n-1}}\tau_{d_n,\alpha_n}\right\rangle.\end{aligned}\tag{3.26}$$

Just as in the derivation of (2.72), we may now assert that the totality of equations (3.26) for $n \geq 3$ is actually equivalent to a single relation among the genus zero generating functions, namely

(3.27)

$$\langle\langle\tau_{d_1,\alpha_1}\tau_{d_2,\alpha_2}\tau_{d_3,\alpha_3}\rangle\rangle_0 = \sum_{\sigma,\tau}\langle\langle\tau_{d_1-1,\alpha_1}\tau_{0,\sigma}\rangle\rangle_0\cdot\eta^{\sigma\tau}\langle\langle\tau_{0,\tau}\tau_{d_2,\alpha_2}\tau_{d_3,\alpha_3}\rangle\rangle_0.$$

Now, (3.27) has the following consequence. The derivative of (3.27) with respect to $t_{d_4}^{\alpha_4}$ is the equation

$$\begin{aligned}\langle\langle\tau_{d_1,\alpha_1}&\tau_{d_2,\alpha_2}\tau_{d_3,\alpha_3}\tau_{d_4,\alpha_4}\rangle\rangle_0\\ &=\sum_{\sigma\tau}\langle\langle\tau_{d_1-1,\alpha_1}\tau_{d_4,\alpha_4}\tau_{0,\sigma}\rangle\rangle_0\cdot\eta^{\sigma\tau}\langle\langle\tau_{0,\tau}\tau_{d_2,\alpha_2}\tau_{d_3,\alpha_3}\rangle\rangle_0\\ &\quad+\sum_{\sigma\tau}\langle\langle\tau_{d_1-1,\alpha_1}\tau_{0,\sigma}\rangle\rangle_0\cdot\eta^{\sigma\tau}\langle\langle\tau_{0,\tau}\tau_{d_2,\alpha_2}\tau_{d_3,\alpha_3}\tau_{d_4,\alpha_4}\rangle\rangle_0.\end{aligned}\tag{3.28}$$

The left-hand side of (3.28) is symmetric under permutations of (d_2, α_2), (d_3, α_3), and (d_4, α_4), but the right-hand side is not. Therefore we can infer that

$$\begin{aligned}\sum_{\sigma,\tau}\langle\langle\tau_{d_1,\alpha_1}\tau_{d_2,\alpha_2}\tau_{0,\sigma}\rangle\rangle_0\eta^{\sigma\tau}\langle\langle\tau_{0,\tau}\tau_{d_3,\alpha_3}\tau_{d_4,\alpha_4}\rangle\rangle_0 \\ = \sum_{\sigma,\tau}\langle\langle\tau_{d_1,\alpha_1}\tau_{d_4,\alpha_4}\tau_{0,\sigma}\rangle\rangle_0\cdot\eta^{\sigma\tau}\langle\langle\tau_{0,\tau}\tau_{d_2,\alpha_2}\tau_{d_3,\alpha_3}\rangle\rangle_0.\end{aligned} \tag{3.29}$$

This amounts to the statement that

$$\eta^{\sigma\tau}\frac{\partial^3 F}{\partial t_{d_1}^{\alpha_1}\partial t_{d_2}^{\alpha_2}\partial t_0^{\sigma}}\frac{\partial^3 F}{\partial t_0^{\tau}\partial t_{d_3}^{\alpha_3}\partial t_{d_4}^{\alpha_4}} = \eta^{\sigma\tau}\frac{\partial^3 F}{\partial t_{d_1}^{\alpha_1}\partial t_{d_3}^{\alpha_3}\partial t_0^{\sigma}}\frac{\partial^3 F}{\partial t_0^{\tau}\partial t_{d_2}^{\alpha_2}\partial t_{d_4}^{\alpha_4}}. \tag{3.30}$$

If all d_i are set to zero, this reduces to (3.14), so the free energy F is the promised function whose third derivatives with respect to the $t_{0,\alpha}$ at any point define a commutative, associative algebra.

Actually, (3.30) is more than was promised. Thus, (3.14) was formulated as an equation for a function defined on the finite-dimensional vector space $H^*(M, \mathbb{R})$, but in (3.30) we have a function F of infinitely many variables t_r^α. To reduce to (3.14), in addition to setting $d_i = 0$, we restrict F to a finite-dimensional subspace characterized to $t_r^\alpha = c_r^\alpha$, $r \geq 1$ (where the c_r^α are arbitrary constants). Thus, F is really a family of solutions of (3.14) depending on the c_r^α as parameters.

To give a very simple concrete example, let $M = \mathbf{CP}^1$. $H^*(\mathbf{CP}^1, \mathbb{R})$ is two dimensional, generated by a zero-form 1 and a two-form ω. If $x = t_0^1$ and $y = t_0^\omega$, then, from §2.3 of [19], the function F on the small phase space is $F(x, y) = x^2y + e^y$.

Equation (3.27) can be given an interpretation analogous to (2.74). We introduce the infinite-dimensional affine "phase space" $\mathbf{A}^\infty$ of the t_r^α, and the "small phase space" characterized by $t_r^\alpha = 0$, for $r \geq 1$. Thus, the small phase space is a copy of $H^*(M, \mathbb{R})$, and has coordinates t_0^α. On the small phase space, the genus zero two point functions $\langle\langle\tau_{n,\alpha}\tau_{m,\beta}\rangle\rangle_0$ are functions of the t_0^γ:

$$\langle\langle\tau_{n,\alpha}\tau_{m,\beta}\rangle\rangle_0 = G_{n,\alpha;m,\beta}(t_0^\gamma). \tag{3.31}$$

According to (3.23), if we define

$$U_\alpha = \langle\langle\tau_{0,1}\tau_{0,\alpha}\rangle\rangle_0, \qquad U^\alpha = \eta^{\alpha\beta}U_\beta, \tag{3.32}$$

then we can rewrite this as

$$\langle\langle\tau_{n,\alpha}\tau_{m,\beta}\rangle\rangle_0 = G_{n,\alpha;m,\beta}(U^\gamma). \tag{3.33}$$

Precisely the argument that led to (2.74) can now be repeated to show that (3.33) is valid, without modification, on the full phase space, not just the small phase space.

As in §2, it is possible at this stage to reinterpret the string equation (3.22). Differentiating (3.22) with respect to $t_{0,\alpha}$, we get

$$U_\alpha = \eta_{\alpha\beta} t_0^\beta + \sum_{i=0}^{\infty} \sum_\beta t_{i+1}^\beta \langle\langle \tau_{i,\beta} \tau_{0,\alpha} \rangle\rangle \tag{3.34}$$

for all α. Using (3.33), we can rewrite this as

$$U_\alpha = \eta_{\alpha\beta} t_0^\beta + \sum_{i=0}^{\infty} \sum_\beta t_{i+1}^\beta G_{i,\beta;0,\alpha}(U^\gamma). \tag{3.35}$$

If $H^*(M, \mathbb{R})$ is m dimensional, this is a system of m equations for the m unknowns U_α; these equations (insofar as their solution is unique, which is actually true in an open set in phase space) determine the U_α's as functions of the parameters t_i^β.

It is interesting to note that the equations (3.35) can be given an interpretation as the equations for a critical point (with respect to the U's) of a certain generalized potential $W(U_\sigma; t_i^\beta)$.[4] To see this, note that on the small phase space there are some functions $G_{i,\alpha}(t_0^\gamma)$ such that

$$\langle \tau_{i,\alpha} \rangle = G_{i,\alpha}(t_0^\gamma). \tag{3.36}$$

(As far as we know, (3.36) does not extend in any nice way on the full phase space.) Comparing (3.36) and (3.33), we see that

$$G_{i,\alpha;0,\beta}(t_0^\gamma) = \frac{\partial G_{i,\alpha}(t_0^\gamma)}{\partial t_0^\beta}. \tag{3.37}$$

This identity of course remains valid if the arguments of the functions on the left and right are U^γ instead of t_0^γ. So (3.35) can be written

$$U_\alpha = \eta_{\alpha\beta} t_0^\beta + \sum_{i=0}^{\infty} \sum_\beta t_{i+1}^\beta \frac{\partial}{\partial U^\alpha} G_{i,\beta}(U^\sigma). \tag{3.38}$$

This is tantamount to the critical point equation

$$\frac{\partial W}{\partial U^\sigma} = 0, \tag{3.39}$$

[4] In the context of matrix chains, of which we will give a very brief sketch in §4d, Ginsparg, Goulian, Plesser, and Zinn-Justin [33] and Jevicki and Yoneya [38] have shown that the string equations are the variational equations of an appropriate Lagrangian. The argument that we are about to give shows that, at least in genus zero, this is true for arbitrary topological field theories coupled to topological gravity.

where

$$W = -\frac{1}{2}U^\alpha U_\alpha + t_0^\beta U_\beta + \sum_{i=0}^{\infty}\sum_\beta t_{i+1}^\beta G_{i,\beta}. \tag{3.40}$$

Conjectured generalization. What will become of the structure that we have found here for genus zero when one considers contributions of higher genus? The analogy with the results of matrix models and the structure of the generalized KdV equations suggests the following. Use the symbols $\dot U_\sigma$, $\ddot U_\sigma$, etc., to denote derivatives of U_σ with respect to t_0^1, so

$$\dot U_\sigma = \frac{\partial U}{\partial t_0^1} = \langle\langle \tau_{0,1}^2 \tau_{0,\sigma}\rangle\rangle, \qquad \ddot U_\sigma = \frac{\partial^2 U_\sigma}{\partial (t_0^1)^2} = \langle\langle \tau_{0,1}^3 \tau_{0,\sigma}\rangle\rangle, \tag{3.41}$$

and so on. Let us consider U, $\dot U$, $\ddot U$, $\dddot U$, etc., to be of degree 0, $1, 2, 3, \cdots$. By a differential function of degree k we mean a function $G(U, \dot U, \ddot U, \dddot U, \cdots)$ which is of degree k in that sense. (Thus, in particular, such a function has only a polynomial dependence on $\dot U$, $\ddot U$, $\dddot U, \cdots$, but its dependence on U need not be polynomial.)

Let us recall now that the free energy F_g has an expansion $F = \sum_{g=0}^\infty F_g$, where F_g is the genus g contribution. Similarly, all other generating functionals that we have considered, such as $U_\sigma = \langle\langle \tau_{0,1}\tau_{0,\sigma}\rangle\rangle$, or $\langle\langle \tau_{n,\alpha}\tau_{m,\beta}\rangle\rangle$, etc., are derivatives of F and in particular have similar expansions. More generally, we may be interested in products of generating functionals. Such a product of course also has a genus expansion, which explicitly is

$$(\langle\langle A\rangle\rangle\langle\langle B\rangle\rangle)_g = \sum_{g'=0}^{g} \langle\langle A\rangle\rangle_{g'}\langle\langle B\rangle\rangle_{g-g'}. \tag{3.42}$$

Then the following conjecture is a tempting generalization of (3.33):

Conjecture. For every $g \geq 0$, there are differential functions $G_{m,\alpha;n,\beta}(U_\sigma, \dot U_\sigma, \ddot U_\sigma, \cdots)$ of degree $2g$ such that

$$\langle\langle \tau_{n,\alpha}\tau_{m,\beta}\rangle\rangle = G_{m,\alpha;n,\beta}(U_\sigma, \dot U_\sigma, \ddot U_\sigma, \cdots) \tag{3.43}$$

up to and including terms of genus g.

To explain the rationale for the conjecture, let me point out that for $M =$ a point, it is a consequence of the main conjecture of §2, since the KdV hierarchy has the stated property. Indeed, the KdV hierarchy has a stronger property— $G_{m,\alpha;n,\beta}$ is a differential function of degree at most $2(m+n)$. This means that for any given correlation function of fixed m and n, the $G_{m,\alpha;n,\beta}$ are differential functions of finite degree even for

$g \to \infty$. To put it more forcefully, this means that in the KdV case, there are differential functions $G_{m,\alpha;n,\beta}$ of finite degree (depending on m and n) such that (3.43) is true *exactly*, not just up to some genus g. But the conjecture stated above permits the possibility that for general M, in going to higher and higher genus, one will have to add to the $G_{m,\alpha;n,\beta}$ terms of higher and higher degree.

The conjecture is an attempt to interpret (3.33), which hold for a very large class of target manifolds M, as the genus zero approximation to a systematic picture that would hold in arbitrary genus g, without proposing that there is an integrable hierarchy of differential equations associated with every compact Kähler manifold M or even every such manifold in a large class. Apart from the case $M =$ a point, $g \leq 3$, the only situation in which we know the conjecture is true is the following. If the dimension of $H^*(M, \mathbb{R})$ is 2, then the conjecture can be verified in genus one (in a tedious and unilluminating way, which we will not present here) using the formulas of the next subsection. (In practice, $\dim H^*(M, \mathbb{R}) = 2$ only for $M = \mathrm{CP}^1$, but the reasoning applies also to an arbitrary model that obeys the general properties assumed here and has "two primaries" in a sense described in a previous footnote.)

To make the conjecture sound a little more plausible, let me point out the following reinterpretation of the above genus zero equations. We have

$$
\begin{aligned}
\frac{\partial}{\partial t_n^\alpha} U_\sigma &= \langle\langle \tau_{0,1} \tau_{n,\alpha} \tau_{0,\sigma} \rangle\rangle = \frac{\partial}{\partial t_0^1} \langle\langle \tau_{n,\alpha} \tau_{0,\sigma} \rangle\rangle \\
&= \frac{\partial}{\partial t_0^1} G_{n,\alpha;0,\sigma}(U_\gamma) = \frac{\partial}{\partial t_0^1} \frac{\partial}{\partial U^\sigma} G_{n,\alpha}(U_\gamma).
\end{aligned}
\tag{3.44}
$$

In the last two steps, (3.33) and (3.37) have been used.

Now, (3.44) has the following interpretation. Think of the U_σ as functions of $x = t_0^1$, and introduce Poisson brackets, with

$$
\{U_\sigma(x), U_\tau(x')\} = \eta_{\sigma\tau} \frac{d}{dx} \delta(x - x'). \tag{3.45}
$$

These Poisson brackets correspond to one of the two symplectic structures of the KdV equations. Introduce the "Hamiltonians"

$$
H_{n,\alpha} = \int dx\, G_{n,\alpha}(U_\gamma). \tag{3.46}
$$

Then (3.44) can be regarded as the Hamiltonian equation of motion:

$$
\frac{\partial U_\sigma}{\partial t_n^\alpha} = \{U_\sigma, H_{n,\alpha}\}. \tag{3.47}
$$

Thus, the genus zero correlation functions for any M (of an appropriate type to justify the above considerations) can be described by a family of commuting Hamiltonian flows! In genus zero, the Hamiltonian densities $G_{n,\alpha}$ are simply functions of the U's. A somewhat sharpened version of the above conjecture would assert that in a genus g approximation, the correlation functions are generated by a system of Hamiltonian flows with the Hamiltonian densities being differential functions of degree $2g$ (which Poisson commute up to terms of degree $2g+2$). If the conjecture is true, then one would expect, upon taking the limit as $g \to \infty$ in a suitable sense, to obtain commuting Hamiltonians that would no longer be differential functions of finite order, so that the commuting Hamiltonian flows would be governed by integral equations rather than differential equations.

The KdV flows have the much stronger property of being bi-Hamiltonian; that is, they preserve two different symplectic structures. This is closely related to the fact that the differential functions in equation (3.43) have a degree that is bounded by $2(m+n)$, independent of g, and thus, one really gets commuting differential operators of finite order. (The other symplectic structure is also related to a kind of Virasoro algebra.) We do not know of any evidence for a second symplectic structure playing a role for general M.

Genus one structure. We will now much more briefly discuss how the genus one equations of §2 generalize in the present situation. The generalization of (2.91) to include a target space M can be obtained by reasoning that should by now be familiar, giving

(3.48)
$$\langle\langle\tau_{n,\alpha}\rangle\rangle_1 = \sum_{\sigma,\tau}\langle\langle\tau_{n-1,\alpha}\tau_{0,\sigma}\rangle\rangle_0\eta^{\sigma\tau}\langle\langle\tau_{0,\tau}\rangle\rangle_1 + \frac{\eta^{\sigma\tau}}{24}\langle\langle\tau_{n-1,\alpha}\tau_{0,\sigma}\tau_{0,\tau}\rangle\rangle_0.$$

It is also possible to obtain an analog of (2.92). To this aim, introduce the matrix

(3.49)
$$M_{\sigma\tau} = \langle\langle\tau_{0,1}\tau_{0,\sigma}\tau_{0,\tau}\rangle\rangle_0.$$

Define a function $E(t_0^\alpha)$ by requiring that the genus one part of the free energy, on the small phase space, is

(3.50)
$$F_1 = \tfrac{1}{24}\ln\det M + E(t_0^\alpha).$$

Then we claim that the genus one free energy, on the full phase space, is

(3.51)
$$F_1 = \tfrac{1}{24}\ln\det M + E(U^\alpha).$$

Since (3.51) is valid on the small phase space (by definition of E), it suffices to prove that the repeated derivatives of (3.51) with respect to the $t_{n,\alpha}$ all vanish. This can be proved inductively using (3.48).[5]

As we have already mentioned, with the above equations and some patience, one can verify the conjecture of the last subsection for genus one in the case $\dim H^*(M) = 2$. We do not know if this restriction is necessary.

4. Introduction to matrix models

Our goal in this section is to give a relatively self-contained but far from complete introduction to the matrix model approach to two-dimensional gravity and some of the remarkable results obtained recently by Brezin and Kazakov, Douglas and Shenker, and Gross and Migdal [3], [25], [35]. In §4a, we explain the physical problem and the strategy for discretizing it; §4b is an explanation of how the discretized problem can be interpreted in terms of matrix integrals, and in §4c, the matrix integrals are described in terms of (discrete analogs of) the KdV flows. The reader who is willing to take it on faith that the problem of interest is to compute matrix integrals $\int (dM)\exp(-\mathrm{tr}(V(M)))$ can read §4c without understanding all the previous details.

4a. The physical problem. Let Σ be a smooth two-dimensional surface of genus g (no complex structure given), and let h be a metric on Σ. The curvature scalar of this metric will be denoted as R. The space MET_g of metrics is itself an infinite-dimensional Riemannian manifold. Indeed, let h_t be a one-parameter family of metrics. Then $\delta h = (dh_t/dt)_{t=0}$ is a tangent vector to MET_g at $h = h_{t=0}$, and one defines its norm to be

$$|\delta h|^2 = \int_\Sigma \sqrt{h}(\delta h, \delta h) = \int_\Sigma \sqrt{h} h^{\alpha\gamma} h^{\beta\delta} \delta h_{\alpha\beta} \delta h_{\gamma\beta}. \tag{4.1}$$

This determines a metric on MET_g and thus, formally, a Riemannian measure, which we will denote as (Dh). The physical problem is to learn how to integrate over MET_g. Naively speaking, one would like to compute

---

[5] In models based on matrix chains, there is strong evidence that all general relations that hold for arbitrary M are valid [19], and therefore, by the reasoning just indicated, one expects that (3.51) holds with some E. However, in the matrix chains, the "primary fields" (the $\tau_{0,\sigma}$) have negative dimension, and the virtual dimension of moduli space in genus one is zero. These facts force $E = \text{constant}$. This explains results that were noted in [19].

the integral

$$F(g) = \int_{\mathrm{MET}_g} (Dh) \exp\left(-\lambda_1 \int_\Sigma \sqrt{h} - \lambda_2 \int_\Sigma \sqrt{h} \frac{R}{2\pi}\right), \tag{4.2}$$

with arbitrary real numbers λ_1 and λ_2. Of course, the term multiplying λ_2 is a topological invariant, the Euler characteristic $\chi(\Sigma) = 2 - 2g$, and plays a trivial role as long as g is fixed; but we will be interested in the dependence on g.

Let $\mathrm{MET}_{A,g}$ be the space of metrics of total area A on a genus g surface. It too has an induced Riemannian structure, and therefore it should have a volume $\mathrm{Vol}(g, A)$. Computing (4.2) is equivalent to knowing $\mathrm{Vol}(g, A)$, since from

$$\begin{aligned} F(g, A) &= \int_{\mathrm{Met}_{A,g}} (Dh) \exp\left(-\lambda_1 A - \lambda_2 \int_\Sigma \sqrt{h} \frac{R}{2\pi}\right) \\ &= \mathrm{Vol}(g, A) \cdot \exp(-\lambda_1 A - \lambda_2 \chi(\Sigma)) \end{aligned} \tag{4.3}$$

we can recover (4.2) by integrating over A.

Of course, a priori one does not quite know what integration theory on these infinite-dimensional spaces is supposed to mean. Usually, in quantum field theory one introduces some sort of "cut-off," which one might imagine to be an approximation to MET_g of some finite-dimension Λ, such that the desired integrals become well defined. Then one tries to "remove the cutoff," that is, one considers a sequence of better and better approximations to MET_g with increasing Λ, and one tries to determine the limit of the integrals for $\Lambda \to \infty$. It then will typically occur even in good cases that such a limit does not exist unless one adjusts ("renormalizes") the "coupling constants" λ_1 and λ_2 in a suitable fashion. So we come to the basic problem of renormalization theory:

Problem. Adjust λ_1 and λ_2 as $\Lambda \to \infty$ so that $F(g, A)$ and $F(g)$ converge to well-defined functions of g and A.

Now, notice that if this problem has a solution, the solution cannot quite be unique. For one could add to λ_2 a finite constant, that is, a constant c independent of the cutoff, and the F's would change by

$$F(g, A) \to F(g, A) \cdot e^{c\chi(\Sigma)}. \tag{4.4}$$

Similarly, there is a potential ambiguity from the ability to add a constant to λ_1, but it turns out that this ambiguity can be canonically removed by requiring that $F(g, A)$ varies only as a power of A (adding a constant to λ_1 would introduce an exponential factor).

The problem of renormalization posed above has analogs in many other quantum field theories; and, usually, it is very difficult to get full control over this problem (including possible strong coupling fixed points)—or to explain the partial results in an introductory lecture. But the particular problem we are discussing here can be treated in a remarkably effective way using a discrete cutoff that was proposed in [14], [41], [1]. In this approach, one considers not metrics on Σ, but triangulations of Σ, or certain generalizations that we will consider later. Every triangulation of Σ determines a metric; for instance, one can consider the triangles to be equilateral triangles of area ε. Of course, the metrics determined this way, except in a flat case where all the coordination numbers are six, cannot be smooth; the curvature consists of delta functions with coefficients that are integral multiples of $2\pi/6$. Nevertheless, one can hope that if a surface is covered with a very large number of triangles, and one averages over the local irregularities, then on a large scale one can effectively see a general metric.

Let $V(g, n)$ be the number of isomorphism classes of triangulations of a genus g surface with n triangles.[6] Now, for small n, $V(g, n)$ is determined by "accidents." But for large n, we can hope that the metrics determined by triangulations approximate any given point in MET_g with equal probability, and that counting triangulations becomes an approximation to computing integrals on MET_g. In fact, one proves (a rather precise account is given in [10]) that the large n behavior of V is

$$V(g, n) \sim e^{cn} \cdot n^{\gamma(2-2g)-1} \cdot b_g \cdot (1 + O(1/n)). \tag{4.5}$$

If we consider every triangle to have area ε, then the total area is $A = n\varepsilon$, so

$$n = A/\varepsilon. \tag{4.6}$$

We regard ε as a cutoff, and we regard $\sum_n V(g, n)$, which is the sum over metrics of arbitrary area, as an approximation to $\int dA\, \mathrm{Vol}(g, A)$. With $\sum_n \sim \int dA/\varepsilon$, we interpret $V(g, n)/\varepsilon$ as an approximation $\mathrm{Vol}_\varepsilon(g, A)$ to the volume of $\mathrm{MET}_{A,g}$. We want to take $\varepsilon \to 0$ while keeping A fixed. (4.5) means that

$$\mathrm{Vol}_\varepsilon(g, A) \sim \frac{1}{\varepsilon} \cdot e^{cA/\varepsilon} \cdot \left(\frac{A}{\varepsilon}\right)^{\gamma(2-2g)-1} \cdot b_g. \tag{4.7}$$

---

[6] To get a precise relation to the matrix model formulation later, one should define $V = \sum_T 1/q(T)$, where the T's are isomorphism classes of triangulations, and $q(T)$ is the order of the automorphism group of the triangulation T. We may ignore this at present, since $q(T)$ is 1 for almost all T, and the factors of $q(T)$ do not affect the large n behavior.

So from (4.3), the corresponding cutoff version of $F(g, A)$ is

$$\begin{aligned} F_\varepsilon(g, A) &= \mathrm{Vol}_\varepsilon(g, A)\exp(-\lambda_1 A - \lambda_2(2-2g)) \\ &= \frac{1}{\varepsilon}\cdot e^{cA/\varepsilon}\cdot\left(\frac{A}{\varepsilon}\right)^{\gamma(2-2g)-1}\cdot b_g\cdot\exp(-\lambda_1 A - \lambda_2(2-2g)). \end{aligned} \tag{4.8}$$

Now we must "renormalize," that is, take $\varepsilon \to 0$ while adjusting λ_1 and λ_2 as functions of ε so that $F_\varepsilon(g, A)$ converges to a well-defined function $F(g, A)$. Obviously, what we need is

$$\lambda_1 = c/\varepsilon, \qquad \lambda_2 = \gamma\ln(A_0/\varepsilon), \tag{4.9}$$

where A_0 is a constant. The limiting or "renormalized" function $F(g, A)$ is then

$$F(g, A) = \frac{1}{A}\cdot\left(\frac{A}{A_0}\right)^{\gamma(2-2g)}\cdot b_g. \tag{4.10}$$

A_0 is the arbitrary constant discussed already in (4.4). In the above derivation, one might feel that it is natural to set $A_0 = 1$. Intuitively, this is in fact unnatural, because A, ε, and A_0 all have dimensions of area, and there is no natural unit of area. In including the arbitrary A_0 in (4.10), we are simply bringing out into the open the need to choose such an arbitrary unit. (Shortly we will replace triangles with squares. If one sets $A_0 = 1$ using triangles, one gets a different answer from what one would get if one sets $A_0 = 1$ using squares, so neither choice is truly natural.)

Now, apart from the fundamental fact that renormalization works and the theory exists, the moral of the above discussion is that the dominant looking term e^{cn} in (4.5) did not matter and disappeared after renormalization. On the other hand, the subleading power $n^{\gamma(2-2g)-1}$ does matter, as does b_g. However, the b_g are well defined only up to

$$b_g \to b_g\cdot t^{1-g}, \tag{4.11}$$

for constant t.

The above computation is interesting, but it is not so fundamental if we are just exploring quirks of triangles. We want to see that we would obtain the same theory if we make different arbitrary choices of regularization. For instance, we could construct the theory with squares instead of triangles. Let $W(g, n)$ be the number of ways to cover a genus g surface with n squares. Then[7] one finds that the large n behavior of $W(g, n)$ is just like the asymptotic formula in (4.5), but with a different value of c, the

[7] Apart from a delicate factor of two that will be pointed out at the end of §4c.

same value of γ, and the b_g's differing only by a transformation of the type (4.11). Thus, up to the one inevitable ambiguity, the same theory is obtained if one uses squares instead of triangles. Similarly, one obtains the same results if one uses the number of ways to cover a surface with pentagons, hexagons, etc., as the way to regularize $\mathrm{MET}_{A,g}$.

There is a fascinating variant of this, originally proposed in [40]. Instead of covering a surface with, say, only squares or only hexagons, we could permit both squares and hexagons. Let $W(g; n_4, n_6)$ be the number of ways to cover a genus g surface with n_4 squares and n_6 hexagons. We pick a real number x and let

$$W_x(g, n) = \sum_{n_4+n_6=n} W_x(g; n_4, n_6)x^{n_6}. \tag{4.12}$$

For generic x, the large n behavior of $W_x(g, n)$ is the same as (4.5), except for the usual modifications—an x dependent value of c, and an x dependent transformation of the sort in (4.4). At a critical value of x, however, one finds a new theory, with a different value of γ and very different b_g's.

This in turn can be generalized. We can consider coverings of a surface with s-gons of various s. It turns out that nothing essential is lost if one considers the s-gons of even s only. So we let $W(g; n_2, n_4, n_6, \cdots)$ be the number of ways to cover a genus g surface with n_2 2-gons, n_4 4-gons, n_6 6-gons, and so on. Picking real numbers $x_2, x_4, x_6, \cdots$, we let

$$W_{\{x\}}(g, n) = \sum_{n_2+n_4+\cdots=n} W(g; n_2, n_4, \cdots)x_2^{n_2} \cdot x_4^{n_4} \cdot x_6^{n_6} \cdots. \tag{4.13}$$

To avoid analytical questions, we can restrict this to a theory with s_0 parameters, for arbitrary s_0, by supposing that the x_{2s} are zero for $s > s_0$. Then, the generic large n behavior of $W_{\{x\}}$, for fixed $\{x\}$, is that of (4.5) (up to the usual irrelevant modifications). On a codimension one subvariety, one finds the exceptional behavior that we already mentioned in the theory with only squares and hexagons. Generically, on this subvariety, the large n behavior of $W_{\{x\}}(g, n)$ is independent of the x's. But on the codimension two subvariety, one finds again a new theory, with a new value of γ and essentially new b_g's. This process continues indefinitely; in every codimension there is a new critical subvariety. The kth theory arises on a codimension $k-1$ subvariety, for $k = 1, 2, 3, \cdots$. This nested hierarchy of critical subvarieties is, as we have already noted in §2, reminiscent of the situation considered in [47].

Now, let us return to the theory in which Σ is covered by squares only, but let us enrich the theory by permitting a few impurities. We consider as impurities u_2 2-gons, u_4 4-gons, u_6 6-gons, etc. (Since we are covering Σ by 4-gons anyway, it is necessary to specify that by a 4-gon impurity we mean a marked 4-gon, in a sea of 4-gons which are generically unmarked.) Let $W(g, n; u_2, u_4, u_6, \cdots)$ be the number of ways to cover a surface of genus g with n unmarked 4-gons and u_{2r} $2r$-gon impurities, for $r = 1, 2, \cdots$. We consider the $2r$-gon impurities of each r to be ordered (otherwise, one must simply divide by $\prod_i u_{2i}!$).[8] Then the methods that give (4.5) yield

(4.14)

$$W(g, n; u_0, u_1, \cdots) \sim e^{cn} n^{\gamma(2-2g)-1+\sum_{r=1}^{\infty} u_{2r}\gamma_r} \cdot f_g(u_2, u_4, u_6, \cdots),$$

with certain constants γ_r and $f_g(u_2, u_4, u_6, \cdots)$. (The γ_r are "universal," that is, they are unchanged if one considers impurities in a sea of hexagons, etc., instead of squares. The $f_g(u_2, u_4, \cdots)$ similarly are universal up to a transformation, analogous to (4.11), that can be absorbed in a rescaling of the variables t_i that we will introduce in a moment.) The general methods for computing the $f_g(u_2, u_4, u_6, \cdots)$ can be found in [10] (and an introduction is given below).

The dramatic development of the last year is that it has been found [12], [23], [34], [7] that the generating function defined as

(4.15)

$$f(t_0, t_1, t_2, \cdots) = \sum_g \sum_{\{u_{2i}\}} \prod_{i=1}^{\infty} \frac{t_{i-1}^{u_{2i}}}{u_{2i}!} f_g(u_2, u_4, u_6, \cdots) \cdot \text{trivial constants}$$

obeys the KdV equations as well as the string equation, described in §2.[9] The main conjecture of §2 is equivalent to the statement that the function $F(t_0, t_i, \cdots)$ defined there coincides, after some slight shifts in the variables, with the function f. That conjecture indeed was an attempt to propose for $f(t_0, t_1, \cdots)$ a geometrical interpretation more direct than the one by which it is defined.

4b. Random matrices. A powerful tool for obtaining the results just sketched comes from the interpretation, given long ago by 't Hooft [56], of Feynman diagrams with matrix-valued fields in terms of triangulations of

---

[8] To make the ideas clear, we will in this and the next paragraph overlook a few important details including the need to take linear combinations of the u's corresponding to "scaling variables."

[9] The "trivial constants" arise because of the integral over are needed to go from (4.3) to (4.2). The f_g's were defined in terms of the behavior with a fixed large number of squares, corresponding to fixed area; but the generating function f that one really wants should be defined with an integral over the area. This gives some trivial constants, as we will discuss later.

Riemann surfaces. 'T Hooft's motivation was to understand the behavior of quantum gauge theories with gauge group $SU(N)$ in the large N limit. This problem, which is outstandingly interesting from a physical point of view, has so far been intractable except in two space time dimensions. A few years after the original suggestion, it was realized [12] that drastically simplified models of this program could be understood by methods of random matrix theory, which had been developed in the 1950s and 1960s by Wigner, Dyson, Mehta, and others with the aim of understanding the statistics of nuclear energy levels. A classic reference is Mehta's book [43]. The paper [10] gives a highly readable account, not assuming any prior familiarity with Feynman diagrams, of the application of random matrix methods to count triangulations of surfaces. The reader not acquainted with these matters is strongly urged to consult §§2 and 3 of that paper, as we will only offer a few indications here.

Suppose that we wish to compute $W(g, n)$, the number of ways to cover a genus g surface with n squares. The dual to a covering by squares is a four-valent graph, as indicated in Figure 7. So we can interpret $W(g, n)$ as the number of connected four-valent graphs that can be drawn on a surface of genus g.

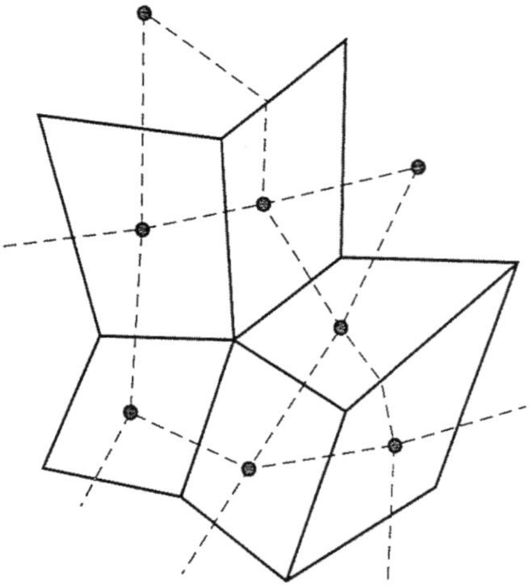

FIGURE 7. A PORTION OF A COVERING OF A SURFACE BY SQUARES AND ITS DUAL FOUR-VALENT GRAPH.

Consider first the slightly easier problem of counting abstract four-valent graphs with n vertices (without reference to any Riemann surface). Let $u(n)$ be the number of graphs which are *connected* and let $y(n)$ be the number of such graphs that are not necessarily connected.[10] The corresponding generating functions are

$$U(-\lambda)=\sum_{n=0}^{\infty}(-\lambda)^n u(n), \qquad Y(-\lambda)=\sum_{n=0}^{\infty}(-\lambda)^n y(n). \tag{4.16}$$

It is easy to see that these are simply related,

$$Y=e^U. \tag{4.17}$$

Now, Y has a convenient integral representation

$$Y(-\lambda)=\int_{-\infty}^{\infty}\frac{d\phi}{\sqrt{2\pi}}\exp\left(-\frac{\phi^2}{2}-\lambda\frac{\phi^4}{4!}\right). \tag{4.18}$$

(In other words, the function of λ that is well defined for $\operatorname{Re}\lambda>0$ by (4.18) has an asymptotic, not convergent, expansion near $\lambda=0$ with coefficients $y(n)$.) We will explain the origin of (4.18) momentarily, but first let us note that (4.18) leads to a quick determination of the large n behavior of $y(n)$ and $u(n)$. By taking the nth derivative of (4.18) we have

$$\begin{aligned} y(n)&=\frac{1}{4!^n n!}\int_{-\infty}^{\infty}\frac{d\phi}{\sqrt{2\pi}}e^{-\phi^2/2}(\phi^4)^n \\ &=\frac{1}{4!^n n!}\int_{-\infty}^{\infty}\frac{d\phi}{\sqrt{2\pi}}\exp\left(-\frac{\phi^2}{2}+4n\ln\phi\right). \end{aligned} \tag{4.19}$$

The integral in (4.19) can be estimated for large n by noting that the main contribution comes from the neighborhood of the maxima of the integrand at $\phi=\pm\sqrt{4n}$, and this gives $y(n)\sim(4n)^{2n}e^{-2n}/(4!^n n!)$. It is easy to see that the growth with n of $y(n)$ is so fast that for large n almost every four-valent graph with n vertices is connected, and thus asymptotically $u(n)\sim y(n)$. In particular, $u(n)$ and $y(n)$ grow faster than exponentially with n, so that the series in (4.16) has zero radius of convergence.

To understand (4.18), we first note the elementary integral

$$\int_{-\infty}^{\infty}\frac{d\phi}{\sqrt{2\pi}}e^{-\phi^2/2}=1, \tag{4.20}$$

and as a result

$$\int_{-\infty}^{\infty}\frac{d\phi}{\sqrt{2\pi}}e^{-\phi^2/2+J\phi}=\int_{-\infty}^{\infty}\frac{d\phi}{\sqrt{2\pi}}e^{-(\phi-J)^2/2+J^2/2}=e^{J^2/2}. \tag{4.21}$$

[10] We will consider the vertices to be unordered; otherwise, the numbers $u(n)$ and $y(n)$ are larger by a factor of $n!$, and a factor of $1/n!$ would be included in the following definition. The minus signs in the definitions of the generating functions are for later convenience.

So

$$\int_{-\infty}^{\infty} \frac{d\phi}{\sqrt{2\pi}} e^{-\phi^2/2} \phi^{2k} = \left[\frac{d^{2k}}{dJ^{2k}} \int_{-\infty}^{\infty} \frac{d\phi}{\sqrt{2\pi}} e^{-\phi^2/2 + J\phi} \right]_{J=0}$$
$$= \left[\frac{d^{2k}}{dJ^{2k}} e^{J^2/2} \right]_{J=0} . \tag{4.22}$$

Now since

$$\frac{d}{dJ} e^{J^2/2} = J e^{J^2/2} , \tag{4.23}$$

a derivative d/dJ, when acting on $e^{J^2/2}$, "creates" a factor of J. More generally, when we compute a repeated derivative

$$\frac{d}{dJ} \frac{d}{dJ} \cdots \frac{d}{dJ} e^{J^2/2} , \tag{4.24}$$

each derivative either "creates" a factor of J when it acts on the exponential, or "annihilates" a factor of J that has been created by a derivative further to the right. Since in (4.22), we are to set $J = 0$ at the end, every factor of J that is "created" by one derivative must be "annihilated" by another. So, finally, (4.22) is equal to the number of ways to group $2k$ objects in pairs.

What we actually want is

$$y(n) = \frac{1}{n!} \int_{-\infty}^{\infty} \frac{d\phi}{\sqrt{2\pi}} e^{-\phi^2/2} \left(\frac{\phi^4}{4!} \right)^n . \tag{4.25}$$

The factors of ϕ were "born" in groups of four by expanding the exponential in (4.18). So we are counting the possible ways of pairing $4n$ objects which come in groups of four. As shown in Figure 8, it is natural to represent a group of four by a vertex from which four lines emerge, and a pairing of two objects as a connection between the corresponding lines. (The factors of $1/4!$ in (4.25) mean that the four objects in each group are unordered, and the factor of $1/n!$ means that the vertices in the graph are unordered.) In such a way we obtain a four valent graph with n vertices, and the argument shows that $y(n)$ is indeed the number of such graphs. The graphs obtained by such perturbative expansions of integrals are known in quantum field theory as Feynman graphs or Feynman diagrams.

The faster than exponential growth of $y(n)$ and $u(n)$ should be compared with the prediction (4.5) that the number $V(g, n)$ or $W(g, n)$ of three-valent or four-valent graphs that can be drawn on a surface of genus

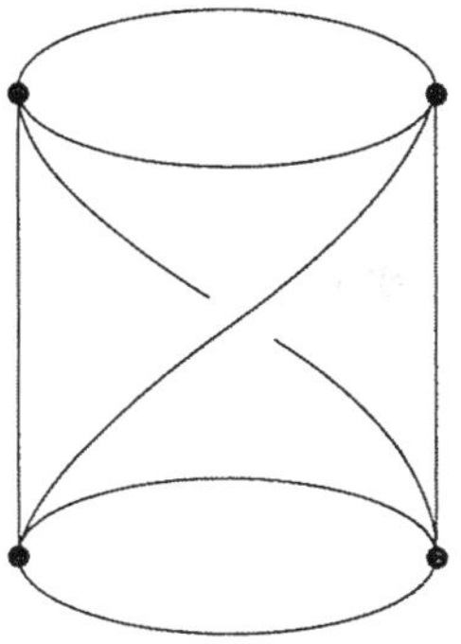

FIGURE 8. A FOUR-VALENT "FEYNMAN GRAPH" OBTAINED BY TAKING CLUSTERS OF FOUR OBJECTS (SUCH A CLUSTER IS DEPICTED AS A VERTEX FROM WHICH FOUR LINES EMERGE) AND PAIRING THEM (BY CONNECTING THE LINES IN PAIRS). IN THIS INSTANCE, THERE ARE FOUR VERTICES.

g grows only exponentially with the number of vertices. How can one modify the above graph counting to construct the generating function of the number of graphs that can be drawn on a surface of fixed genus? The simple modification that is required goes back to [56]. One simply replaces ϕ by an $N \times N$ hermitian matrix M. The space of such matrices is a Euclidean space $\mathbb{R}^{N^2}$, on which one introduces a translationally invariant measure (dM) normalized so that

$$\int (dM) \exp -\operatorname{Tr} M^2 = 1. \tag{4.26}$$

Then, the claim is that the integral

$$Z(N, -\lambda) = \int (dM) \exp -\operatorname{Tr}\left(\frac{M^2}{2} + \lambda\frac{M^4}{4N}\right) \tag{4.27}$$

is essentially the generating function that we need. Indeed, $F(N, -\lambda) = \ln Z$ has the expansion

$$F(N, -\lambda) = \sum_{g=0}^{\infty} N^{2-2g} \sum_{n=0}^{\infty} (-\lambda)^n W(g, n), \tag{4.28}$$

where $W(g, n)$ is the number of ways to cover a surface of genus g with n squares.

Derivation of (4.28). Equation (4.28), which is essentially due to 't Hooft [56], is explained in [10] and in [51]. Here is a very brief account. To begin with, by completing the square, one proves that

$$\int (dM)\exp(-\mathrm{Tr}(\tfrac{1}{2}M^2 + MJ)) = \exp(\mathrm{Tr}(J^2)/2). \tag{4.29}$$

Hence

$$\begin{aligned}
&\int (dM)\exp\left(-\mathrm{Tr}\left(\frac{M^2}{2}\right)\right) M^{i_1}_{j_1} M^{i_2}_{j_2}\cdots M^{i_n}_{j_n} \\
&= \left[\frac{\partial}{\partial J^{j_1}_{i_1}}\cdots\frac{\partial}{\partial J^{j_n}_{i_n}}\int (dM)\exp\left(-\mathrm{Tr}\left(\frac{M^2}{2} + MJ\right)\right)\right]_{J=0} \\
&= \left[\frac{\partial}{\partial J^{j_1}_{i_1}}\cdots\frac{\partial}{\partial J^{j_n}_{i_n}} e^{\mathrm{Tr}\, J^2/2}\right]_{J=0}.
\end{aligned} \tag{4.30}$$

Now, as in the previous case, each derivative $\partial/\partial J^j_i$ either "creates" or "annihilates" a factor of J^i_j. Since one is to set $J = 0$ at the end of the computation, every factor of J^i_j that is created must be annihilated, so that the evaluation of (4.30) involves a sum over pairings. Again, it is natural to represent such a pairing by a line connecting two vertices. The difference is now that there are N^2 distinct "objects" J^i_j that may be propagating in such a line. Following 't Hooft, we denote this by a "double line notation" in which each line is thickened slightly to a band, and the edges are labeled by i or j, as in Figure 9(a). The two edges of the band correspond to the two indices of the matrix J^i_j, and the N possible labels of each edge correspond to the N possible values of the corresponding index.

Now, expanding (4.27) in powers of λ, the coefficient of $(-\lambda)^n$ is

$$\left(\frac{1}{4N}\right)^n \frac{1}{n!}\int (dM)\exp\left(-\frac{1}{2}\mathrm{Tr}(M^2)\right)(\mathrm{Tr}(M^4))^n. \tag{4.31}$$

Again, we must integrate a polynomial of order $4n$ in the matrix elements of M; again, this can be done using (4.30), and will lead to a sum over four-valent graphs of an appropriate type. However, we must pay attention to just what kind of $4n$th order polynomial in matrix elements of M we have in (4.31). If we bear in mind that $\mathrm{Tr}\, M^4 = M^i_j M^j_k M^k_l M^l_i$, then in the double line notation, the four-valent vertices have the structure indicated in Figure 9(b), and the diagrams with "double lines" connecting such vertices are as in Figure 9(c). The key point is now that, though an abstract graph does not naturally determine a Riemann surface on which it can be drawn, the "double line" structure has had the effect of thickening

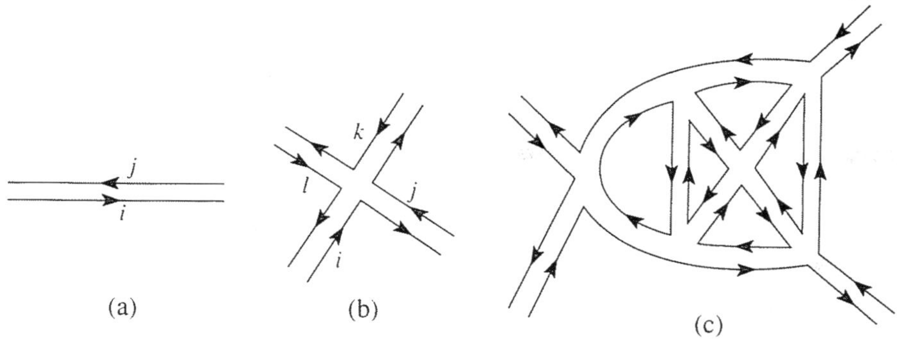

FIGURE 9. A CONVENIENT NOTATION FOR FEYNMAN DIAGRAMS OBTAINED BY PERTURBATIVE EXPANSION OF MATRIX INTEGRALS. EVERY LINE IS THICKENED AS IN (A) TO A "BAND," WHOSE TWO EDGES CORRESPOND TO THE TWO INDICES OF AN $N \times N$ MATRIX M^i_j; EACH EDGE CARRIES A LABEL THAT MAY RUN FROM $1, \cdots, N$, CORRESPONDING TO THE POSSIBLE VALUES OF THE CORRESPONDING INDEX. IN THIS NOTATION, THE FOUR-VALENT VERTEX CORRESPONDING TO A FACTOR OF $\mathrm{Tr}(M^4)$ IN AN INTEGRAL IS DEPICTED AS IN (B). COMBINING THE THICKENED LINES OF (A) WITH THE THICKENED VERTICES OF (B) ONE OBTAINS GRAPHS (C) IN WHICH THE EDGES FIT TOGETHER SMOOTHLY INTO "INDEX LOOPS." FILLING IN THE INDEX LOOPS WITH DISCS, ONE CANONICALLY CONSTRUCTS A TWO-DIMENSIONAL SURFACE.

the lines slightly, in a way which is compatible with the structure of the vertices, and this gives the extra information that is needed in order to reconstruct a Riemann surface. Indeed, with the vertices drawn as indicated, the edges of the double lines join together into circles, and upon filling in these circles with discs, we obtain a surface Σ together with a simplicial decomposition.

Let n_0, n_1, and n_2 be the number of 0, 1, and 2 simplices in this decomposition. Then n_0 is the same as the number n in equation (4.31), and it is a fact of life for graphs drawn with four-valent vertices that

$$n_1 = 2n_0. \tag{4.32}$$

On the other hand, n_2 is the same as the number of circles that were filled in to reconstruct Σ.

These circles are usually called "index loops". The terminology reflects the fact that each edge of one of the thickened lines has a labeling or "index" that takes an arbitrary value in the range $1, \cdots, N$; because of the structure of the vertices, the labelings are constant in running around the circles or index loops, and there is no correlation between the labeling of different loops.

Now, the evaluation of (4.31) proceeds by drawing all the possible four-valent thickened graphs, and then assigning to each graph a numerical factor which comes from factors explicitly present in (4.31) and from summing over the various types of "object" that can be propagating in each double line, that is, by summing over the labelings of the edges. The sum over labelings gives a factor of N for each index loop, or altogether a factor of N^{n_2}. In addition, a factor of N^{-n_0} is explicit in (4.31). The N dependence is thus

$$N^{-n_0+n_2} = N^{n_0-n_1+n_2} = N^{2(1-g)}, \tag{4.33}$$

where in the first step we use (4.32), and in the second step we use the fact that $n_0 - n_1 + n_2$ is the Euler characteristic $2-2g$. The power of N is the main result that is claimed in (4.28). The other numerical factors that arise are the trivial factors that appear explicitly in (4.31). The factor of $1/n!$ means simply that the vertices are unordered, and the factor of $(1/4)^n$ means that the four objects emanating from a vertex carry only a cyclic order.

The double scaling limit. Granting (4.28), what must we do to understand two-dimensional quantum gravity? The problem is that (4.27) generates, via (4.28), all the numbers $W(g, n)$, but this is far more than we want. According to the discussion in §4a, we are only interested in the large n behavior of $W(g, n)$, where one sees an approximation to a random metric on a surface of genus g. Therefore, we want to take a limit of (4.28) in which the extraneous information will be eliminated. This occurs in the limit in which λ approaches a critical value at which the infinite sum in (4.28) is ceasing to converge—and exhibits a singularity that is determined by the asymptotic behavior of the series. The issue has been analyzed (nonrigorously) as follows in the literature. According to (4.5), for large n, $W(g, n) \sim e^{cn} \cdot n^{\gamma(2-2g)-1} b_g$. The genus g contribution to $F(N, -\lambda)$,

$$F_g(-\lambda) = \sum_{n=0}^{\infty} (-\lambda)^n W(g, n) \tag{4.34}$$

thus has a singularity at $\lambda = \lambda_c = -e^{-c}$. The leading singular behavior of

F_g is

$$\sum_{n=0}^{\infty}\left|\frac{\lambda}{\lambda_c}\right|^n n^{\gamma(2-2g)-1} b_g \sim \left|\frac{\lambda-\lambda_c}{\lambda_c}\right|^{-\gamma(2-2g)} \cdot b_g \cdot \Gamma(\gamma(2-2g)). \tag{4.35}$$

Therefore, with

$$y = N^2\left|\frac{\lambda-\lambda_c}{\lambda_c}\right|^{-2\gamma}, \tag{4.36}$$

the sum over g of the leading singular contributions to the F_g is

$$F_{\text{sing}} = \sum_{g=0}^{\infty} y^{1-g} \cdot b_g \cdot \Gamma(\gamma(2-2g)). \tag{4.37}$$

(The sum over n in (4.35) corresponds to the integral over area to go from (4.3) to (4.2), and the resulting Γ function is the "trivial constant" in (4.15). It is the generating functional F_{sing} with these factors included that (a) corresponds to an ensemble with a random metric of any area on a surface of any genus; (b) can be represented as a matrix integral.) Thus, the prescription that has been followed in the recent literature is to extract the leading singularity of F in the limit $N \to \infty$, $\lambda \to \lambda_c$, with y fixed. It is in this limit, which is known as the double scaling limit, that the matrix integral (4.27) is governed by the interesting numbers b_g and γ.

More generally, if, as in (4.13), we wish to consider arbitrary mixtures of 2-gons, 4-gons, 6-gons, etc., one must consider a generalization of (4.27), namely

$$Z(\lambda_i) = \int (DM) \exp\left(-\operatorname{Tr}\left((1-\lambda_2)\frac{\phi^2}{2} - \lambda_4\frac{\phi^4}{4N} - \lambda_6\frac{\phi^6}{6N^2} - \cdots\right)\right). \tag{4.38}$$

In effect, in order to study coverings of genus g surfaces by n-gons of various n, we must discuss a general integral

$$Z = \int (dM) \exp(-\operatorname{Tr} V(M)), \tag{4.39}$$

for general V. It is in this form that we will discuss the problem in the next subsection.

4c. Orthogonal polynomials and discrete KdV flows. In this subsection, we will, finally, explain the origin of some of the key recent results [13], [25], [35], [7] that are important physically and motivated the conjecture about intersection theory on moduli space presented in §2. Following some preliminaries that can be found in [10], we will take a point of view that for the most part follows the exposition by Douglas [23] (with some modifications suggested in part by G. Segal).

The first step in analyzing integrals of the form (4.39) is to diagonalize the matrix M, so $M = U\Lambda U^{-1}$, where U is a unitary matrix and $\Lambda = \mathrm{diag}(s_1, s_2, \cdots, s_n)$. Then as computed in appendix (2) of [10], the measure can be written

$$(dM) = \text{constant} \cdot (dU) \cdot ds_1 ds_2 \cdots ds_n \cdot \prod_{i<j} (s_i - s_j)^2, \tag{4.40}$$

where (dU) denotes Haar measure on the unitary group, and the constant factor is not important because it does not affect the singularity in the double scaling limit. The double zero of the measure at $s_i = s_j$ reflects the fact that in the space of hermitian matrices, the matrices with two equal eigenvalues are of codimension three rather than codimension one since the stabilizer of a hermitian matrix with two equal eigenvalues has dimension two more than the stabilizer of a generic hermitian matrix.

The integral (4.39) can therefore be replaced by

$$Z = \int_{-\infty}^{\infty} ds_1 ds_2 \cdots ds_N \prod_{i<j} (s_i - s_j)^2 \cdot \prod_i e^{-V(s_i)}. \tag{4.41}$$

Let $d\mu$ be the measure $ds\, e^{-V(s)}$ on the real line. Introduce the monic orthogonal polynomials $P_r(s)$, $r = 0, 1, 2, \cdots$, for this measure, defined by

$$P_r(s) = s^r + \text{lower order terms} \tag{4.42}$$

and

$$\int d\mu\, P_k(s) P_r(s) = h_r \delta_{k,r}. \tag{4.43}$$

Let Q be the $N \times N$ matrix whose i, j matrix element is s_i^{j-1}. Then $\det Q$ is a polynomial of order $N(N-1)/2$ which vanishes whenever $s_i = s_j$ for any i, j since in that case Q has two equal rows. These facts fix the relation

$$\det Q = (-1)^{N(N-1)/2} \prod_{i<j} (s_i - s_j) \tag{4.44}$$

up to a numerical factor which can be verified by, for instance, working out the coefficient of $\prod_{i=1}^N s_i^{i-1}$. On the other hand, consider instead of Q the matrix $\tilde{Q}$ whose ij element is $P_i(s_j)$. $\det \tilde{Q} = \det Q$, since, in view of (4.42), $\tilde{Q}$ differs from Q by column rearrangements of a triangular kind. Hence, we may rewrite (4.41) in the form

$$Z = \int d\mu(s_1)\cdots d\mu(s_N)(\det \tilde{Q})^2. \tag{4.45}$$

If now we explicitly write

$$\det \tilde{Q} = \sum_{\pi}(-1)^{\pi}\prod_{i} P_{\pi(i)}(s_i), \tag{4.46}$$

where π ranges over the permutations of N objects, then the integral becomes

$$Z = \int d\mu(s_1)\cdots d\mu(s_N)\sum_{\pi,\pi'}(-1)^{\pi+\pi'}\prod_{i} P_{\pi(i)}(s_i)P_{\pi'(i)}(s_i). \tag{4.47}$$

By using the orthogonality relation (4.43), this gives

$$Z = N!\cdot h_0\cdot h_1\cdot\cdots\cdot h_{N-1}. \tag{4.48}$$

Thus, to solve the problem, it suffices to know the normalization constants h_i of the monic orthogonal polynomials. The constant $N!$ in (4.48) is, again, irrelevant in the double scaling limit.

At this point, it is convenient to switch to *orthonormal* polynomials,

$$\widehat{P}_r = \frac{P_r}{\sqrt{h_r}}. \tag{4.49}$$

Let $\mathscr{V}$ be the vector space consisting of polynomial functions $\sum_{i=0}^{n} b_i s^i$, of arbitrary degree. It is a fixed vector space, given once and for all. Since we do not want to introduce any Hilbert space structure on $\mathscr{V}$ (the only natural L^2 structure in the problem is determined by the measure $d\mu$, which depends on the potential V, but we want to consider objects that are independent of V), by a *basis* of $\mathscr{V}$ we will mean a vector space basis, that is, a set of vectors $q_r \in \mathscr{V}$ such that every element of $\mathscr{V}$ can be uniquely written as a *finite* linear combination of the q_r. In particular, every choice of a potential V determines a canonical basis, namely the basis consisting of the polynomials $\widehat{P}_i$, which may be characterized completely as the orthogonal polynomials for the measure $d\mu$ with positive leading term.

On $\mathscr{V}$, there are certain natural operators, such as the operation $\mathscr{S}$ of "multiplication by s", which maps the polynomial $P(s)$ to the polynomial $sP(s)$, and the operation $\mathscr{T}$ of "differentiation with respect to s," which maps $P(s)$ to dP/ds. If one is given a particular basis for $\mathscr{V}$, such as the basis of orthogonormal polynomials, then $\mathcal{S}$ or $\mathcal{T}$ can be written out as a concrete $(\infty \times \infty)$ matrix,

$$\mathcal{S}\widehat{P}_r = \sum_k S_{k,r}\widehat{P}_k. \tag{4.50}$$

It is obvious that $S_{k,r} = 0$ for $k - r > 1$, and noting that

$$\int d\mu(s\widehat{P}_r)\widehat{P}_k = \int d\mu\widehat{P}_r(s\widehat{P}_k), \tag{4.51}$$

we see that $S_{k,r} = S_{r,k}$, so $S_{k,r}$ vanishes for $r - k > 1$. We have learned that, in the basis of orthonormal polynomials, $\mathcal{S}$ is a "Jacobi matrix," that is, a symmetric matrix whose matrix elements $S_{k,l}$ vanish for $|k - l| > 1$. For $|k - l| = 1$, the $S_{k,l}$ are determined by the leading coefficients of the orthonormal polynomials, so concretely

$$\mathcal{S} \cdot \widehat{P}_r = \sqrt{\frac{h_{r+1}}{h_r}}\widehat{P}_{r+1} + S_r\widehat{P}_r + \sqrt{\frac{h_r}{h_{r-1}}}\widehat{P}_{r-1}, \tag{4.52}$$

with some constants S_r.

Now, for *every* choice of potential V, we get a canonical basis of orthonormal polynomials in which S can be written out as a Jacobi matrix. Considering explicitly as arbitrary polynomial V,

$$V(s) = \sum w_i s^i \tag{4.53}$$

(with all but finitely many w_i vanishing), we get a family of Jacobi matrices $S(w_1, w_2, w_3, \cdots)$. However, since the matrix $S(w_1, w_2, \cdots)$ is obtained by writing out a *fixed* operator $\mathcal{S}$ on a fixed vector space $\mathcal{V}$ in a basis that depends on the w's all that happens to it when the w_i are changed is that it is written out in terms of a new basis. If the derivative of the basis with respect to the w_i is

$$\frac{\partial \widehat{P}_k}{\partial w_i} = \sum_l (O_{(i)})_{kl}\widehat{P}_l, \tag{4.54}$$

then the derivative of the Jacobi matrix S with respect to the w_i is

$$\frac{\partial S}{\partial w_i} = [O_{(i)}, S]. \tag{4.55}$$

Since we know a priori that S is well defined as a function of the w_i, the $O_{(i)}$, which we have not yet determined, must be such that the flows defined by (4.55) in the space of Jacobi matrices are *commuting*.

Now, under appropriate conditions, a symmetric Jacobi matrix can behave as a discrete approximation to a second-order differential operator in one dimension of the form $Q = d^2/dx^2 + U(x)$. The usual KdV flows are commuting flows in the space of such operators, of the form

$$\frac{\partial Q}{\partial t_n} = [M_n, Q], \tag{4.56}$$

where the M_n are certain differential operators of order n. Now, as explained by Moser [49], the space of Jacobi matrices, like the space of second-order operators of the indicated type, has a natural symplectic structure, and moreover Moser proposed commuting flows of the type (4.55) as discrete analogs of the KdV flows. (See also P. van Moerbeke, Inv. Math. **37** (1976) 45, and references therein.)

Now actually, in the continuum case, that is, the case of differential operators, the M_n in (4.56) are almost uniquely determined by requiring that $[M_n, Q]$ is a zeroth order differential operator (which can be interpreted as $\partial U/\partial t_n$). Indeed, according to Gelfand and Dikii [31], a differential operator with this property is a linear combination of the operators

$$M_n = (Q^{n/2})_+, \tag{4.57}$$

where $Q^{n/2}$ is the $n/2$ power of Q as a pseudodifferential operator, and $(Q^{n/2})_+$ is the unique differential operator such that $Q^{n/2} - (Q^{n/2})_+$ is of negative order. The KdV flows are precisely the flows (4.56) with these M_n. Half of the flows are trivial, since if n is even, $n/2$ is an integer, and $(Q^{n/2})_+ = Q^{n/2}$ is a differential operator that commutes with Q.

Let us now consider the discrete analogs of these flows. We will call a matrix W local, of order p, if the matrix elements $W_{k,l}$ vanish for $|k-l| > p$. A local matrix is a natural candidate for approximating a differential operator. Given a Jacobi matrix S, in looking for local matrices W such that $[W, S]$ is a local matrix, the interesting W's are the antisymmetric ones. For given any W, if we write $W = W_+ + W_-$, where W_+ is symmetric and W_- is antisymmetric, the condition that $[W, S]$ is a Jacobi matrix means (since Jacobi matrices are symmetric by definition) that $[W_-, S]$ is a Jacobi matrix and $[W_+, S] = 0$. For a generic Jacobi matrix S with distinct eigenvalues, the condition on W_+ has only the trivial solutions that $W_+ = \sum_i a_i S^i$, analogous to the trivial KdV flows. So we may as well take W antisymmetric. An antisymmetric local matrix W such that $[W, S]$ is a Jacobi matrix is (as shown in the concluding pages of [49]) a linear combination of certain matrices B_p, $p = 1, 2, 3, \cdots$, with B_p being local of degree p. (If $S^p = A_+ + A_-$, where A_+ is upper triangular and A_- is the transpose of A_+, then one can take $B_p = A_+ - A_-$.) The flows

$$\frac{dS}{dt_n} = [B_n, S] \tag{4.58}$$

are discrete analogs of the KdV flows.

Let us now verify that the matrices $O_{(i)}$ defined in (4.54) are a linear combination of the B_p. By differentiating the orthonormality relation, we get

$$0 = \frac{\partial}{\partial w_i} \int d\mu \widehat{P}_k \widehat{P}_l = \int d\mu \left(\frac{d\widehat{P}_k}{dw_i} \widehat{P}_l + \widehat{P}_k \frac{d\widehat{P}_l}{dw_i} - s^i \widehat{P}_k \widehat{P}_l \right). \tag{4.59}$$

The s^i term comes from differentiating the measure $d\mu = dse^{-V}$ with respect to w_i. This gives

$$(O_{(i)})_{k,l} + (O_{(i)})_{l,k} = \int d\mu s^i \widehat{P}_k \widehat{P}_l. \tag{4.60}$$

Since the $\widehat{P}_k$ are polynomials of order k, it follows immediately from the definition in (4.54) that $(O_{(i)})_{k,l} = 0$ for $l > k$. As multiplication by s is a Jacobi matrix, it follows from (4.60) that $(O_{(i)})_{k,l}$ also vanishes for $k - l > i$, and thus $O_{(i)}$ is local, of degree i, as we wished to show. Notice that (4.60) can be written in the form

$$O_{(i)} - \frac{S^i}{2} = - \left(O_{(i)} - \frac{S^i}{2} \right)^T. \tag{4.61}$$

Thus, $\widetilde{O}_{(i)} = O_{(i)} - S^i/2$, which obviously generates the same flow as $O_{(i)}$, is antisymmetric and indeed coincides with B_i as defined above.

So far, we have determined that $S(w_1, w_2, \cdots)$ is an orbit of the discrete KdV flows. It remains to determine which orbit arises, that is, to determine the initial conditions. To this aim, we will appeal to an elegant argument by Douglas [24]. In addition to $\mathscr{S}$ = multiplication by s being local, it is also true that $\mathscr{T} = d/ds$ is local in the basis of orthonormal polynomials, provided the potential V is a polynomial (provided almost all of the w's vanish). It is indeed obvious that if we write

$$\mathscr{T} \widehat{P}_k = \sum_{r=1}^{k} T_{k,r} \widehat{P}_r, \tag{4.62}$$

then $T_{k,r} \neq 0$ only for $r < k$. By considering

$$0 = \int ds \frac{d}{ds} (e^{-V} \widehat{P}_k \widehat{P}_r) = \int d\mu \left(-\frac{dV}{ds} \widehat{P}_k \widehat{P}_r + \frac{d\widehat{P}_k}{ds} \widehat{P}_r + P_k \frac{d\widehat{P}_r}{ds} \right), \tag{4.63}$$

one sees that $T_{k,r} = 0$ unless $k - r \leq n - 1$, where n is the degree of V. Thus $\mathcal{T}$ is represented in the basis of orthonormal polynomials by a matrix T that is a local of degree $n - 1$, and so has an expansion

$$T = \sum_{j=1}^{n-1} (v_j O_{(j)} + p_j S^j), \tag{4.64}$$

for some real numbers v_j and p_j. In fact, more incisively, it follows from (4.63) that

$$\mathcal{T}' = \mathcal{T} - \tfrac{1}{2} V'(\mathcal{S}) \tag{4.65}$$

is antisymmetric, and this expression for an antisymmetric matrix as the sum of a triangular matrix and a polynomial in S determines $\mathcal{T}'$ as a linear combination of the B_p's.

The underlying relation

$$[\mathcal{T}, \mathcal{S}] = 1 = [\mathcal{T}', \mathcal{S}] \tag{4.66}$$

may then be written out in the form

$$\sum_{j=1}^{n-1} v_j [O_{(j)}, S] = 1. \tag{4.67}$$

The requirement that there exist constants v_j such that (4.67) is obeyed determines a particular orbit for the discretized KdV flows on the space of Jacobi matrices. In fact, (4.67) is a discrete analog of (2.22) which served in §2 to determine the initial conditions for the solution of the KdV equations. To see this, write (2.22) in the form

$$\langle\langle \tau_0 \rangle\rangle - \sum_{i=0}^{\infty} t_{i+1} \langle\langle \tau_i \rangle\rangle = \frac{t_0^2}{2}. \tag{4.68}$$

Differentiating twice with respect to t_0, this becomes

$$\sum_{i=0}^{\infty} y_i \langle\langle \tau_i \tau_0 \tau_0 \rangle\rangle = 1, \tag{4.69}$$

where $y_i = \delta_{i,0} - t_{i+1}$. Alternatively, this can be written

$$\sum_{i=0}^{\infty} y_i \frac{\partial}{\partial t_i} U = 1. \tag{4.70}$$

According to the main conjecture of §2, $\partial U / \partial t_i$ is the ith KdV flow.

Interpreting S as a discrete approximation to a differential operator $Q = d^2/dt_0^2 + U$, and $[O_{(i)}, S]$ as a discrete approximation to the ith KdV flow, we see that we can indeed identify the initial conditions (4.67) in the matrix model formulation as a discrete approximation to the initial conditions defined by the string equation.

We have carried out all of this discussion *without* considering the double scaling limit, discussed at the end of §4b, in which it is expected that two-dimensional gravity can be extracted from the matrix model. It is argued in the literature that in the double scaling limit, S converges to a differential operator, and the discrete KdV flows in the space of Jacobi matrices converge to the ordinary continuum KdV flows in the space of differential operators. We refer to the original papers [13], [25], [35] for these arguments.

The role of the odd polynomials. To conclude, we will attempt to explain a detail that has been left unclear in the previous literature, though somewhat similar points are raised in [24].[11] This detail is important in a careful comparison of intersection theory to matrix models.

On the space $\mathscr{V}$ of polynomials in s, let U be the operator that maps $P(s)$ to $P(-s)$. Let $\widehat{S} = -USU$. It is easy to see that like S, $\widehat{S}$ is a Jacobi matrix, or more precisely, a family of Jacobi matrices parametrized by $w_1, w_2, \cdots$. Moreover,

$$\frac{\partial \widehat{S}}{\partial w_r} = [\widehat{O}_{(r)}, \widehat{S}], \tag{4.71}$$

where $\widehat{O}_{(r)} = UO_{(r)}U$ is local of the same degree as $O_{(r)}$. So like S, $\widehat{S}$ evolves by the discrete analog of the KdV flows.

Therefore, the question arises of whether $\widehat{S}$, like S, might in the double scaling limit converge to a second order differential operator. Actually, it is really necessary to specify more precisely that the statement "S (or $\widehat{S}$) converges to a differential operator in a certain limit" will mean that in acting on vectors $\sum_i a_i \widehat{P}_i$, where the a_i are *slowly varying with* i, S (or $\widehat{S}$) approximates a differential operator. It is evident that $\widehat{S}$ converges to a differential operator in this sense if and only if in the same limit S approximates a differential operator when acting on vectors of the type $\sum_i (-1)^i b_i \widehat{P}_i$, with slowly varying b_i. Thus, the consideration of convergence of $\widehat{S}$ to a differential operator is equivalent to consideration

[11] After writing these notes, I received a paper developing the same point more extensively [6].

of a generalized possibility for how S might converge to a differential operator.[12]

Now, in general, in the double scaling limit, upon appropriate adjustment of the couplings w_i, it is possible for both S and $\widehat{S}$ to converge to differential operators, of the form $Q = d^2/dx^2 + V$ and $\widehat{Q} = d^2/dx^2 + \widehat{V}$, and there is absolutely no general relation between the two potentials V and $\widetilde{V}$[13]. Both Q and $\widehat{Q}$ evolve separately according to the KdV flows, so in fact the hermitian matrix model leads to two entirely independent, commuting copies of the KdV hierarchy! The initial conditions are of the same structure for each, since the argument that led to (4.67) could just as well be made for $\widehat{Q}$.

It is dull to study two decoupled copies of the same structure, and what is usually done in the matrix model literature is to eliminate half the variables. The usual way to do this is to take the potential V to be even, $V(s) = V(-s)$. It is evident that in this case, $\widehat{S} = S$, so one is seeing, in effect, the diagonal combination of the two theories. The free energy computed this way, which is the result usually reported in the matrix model literature, receives half its contribution from S and half its contribution from $\widetilde{S}$, and is precisely twice that of the basic system. In a generic double scaling limit with a noneven potential, S or $\widehat{S}$ would converge to a differential operator, but not both, and the free energy would be precisely half of the result for an even potential. By careful comparison of intersection theory on moduli space, discussed in §2, to the matrix model results, one can see that (at least in genus ≤ 3, where all of the conjectures of §2 have been verified) the free energy defined by intersection theory is equal to that of the matrix models for a generic, noneven potential, and is half of the matrix model result as usually quoted.

4d. Matrix chains. In §3, we generalized intersection theory on moduli space of Riemann surfaces to include maps to a Kähler manifold M. One may wonder whether the hermitian matrix model has an analogous generalization. In fact, it has a very beautiful generalization, which we will now indicate very briefly.

First of all, the physical problem is to study two-dimensional quantum gravity coupled to quantum fields. Once one agrees to describe

[12] There are yet more elaborate possibilities for how S might converge to a differential operator, but they do not arise for generic even potentials, and thus are not relevant to elucidating the existing literature, which is our goal in the present discussion.

[13] To explicitly achieve this, take a matrix model potential V which in the naive large N limit is even. Add to it odd terms, suppressed by just the right powers of N so as to give contributions of order 1 in the double scaling limit. In this way, one gets an explicit double scaling solution with Q and $\widehat{Q}$ completely independent.

quantum gravity by a sum over triangulations of a surface, it is natural to describe the quantum fields by lattice statistical mechanics on the triangulated surface. This, again, can be accommodated in the framework of matrix models—provided that one introduces more than one matrix. The most general type of example that has so far been tractable is the "matrix chain", in which one considers n hermitian matrices M_i, $i = 1, \cdots, n$, and an integral of the form

(4.72)
$$Z = \int (dM_1) \cdots (dM_n) \exp\left(-\mathrm{Tr}\left(\sum_{i=1}^{n} V_i(M_i) + \sum_{i=1}^{n-1} c_i M_i M_{i+1}\right)\right).$$

Such an integral has an interpretation as the generating function for coverings of a Riemann surface (of variable genus) by graphs with certain additional information. The additional information arises because a "vertex" in the graph may come from expanding the factor of $\exp(-\mathrm{Tr}(V_i(M_i)))$ for any value of $i = 1, \cdots, n$. In addition to summing over all isomorphism classes of graphs in evaluating (4.72), one sums over all maps of the set of vertices in the graph to the finite set $\{1, 2, 3, \cdots, n\}$. The possible maps (from a given graph) are not weighted equally; they are weighted by local factors, which one finds by further study of (4.72), and which are similar to the characteristic Boltzmann weights of statistical mechanics.

The integral in (4.72) can again be analyzed very effectively using orthogonal polynomials. One requires certain additional tricks originally introduced by Mehta and collaborators [44]. (Mehta's crucial formula for integrating over angular variables has been explained as an application of the Duistermaat-Heckman stationary phase formula [53].) The main difference in the result that eventually emerges is that the matrix analogous to S is still a local matrix but has degree > 1. As a result, S does not converge to a second-order differential operator, but in general to a differential operator of higher degree.

Let $D = d/dx$, and let S be an $(N+1)$th order differential operator of the form

$$S = D^{N+1} + \sum_{\alpha=0}^{N-1} v_\alpha D^\alpha. \tag{4.73}$$

For $n = 1, 2, 3, \cdots$, let $K_n = (S^{n/(N+1)})_+$ be the differential operator part of the pseudodifferential operator $S^{n/(N+1)}$. The flows

$$\frac{\partial S}{\partial y_n} = [K_n, S] \tag{4.74}$$

on the space of S's are the commuting flows of the Nth generalized KdV hierarchy. Arguments of Douglas [23] indicate that this hierarchy governs the double scaling limit of the N matrix chain.

In §3, we generalized intersection theory on moduli space to include a target space M, and we described general properties of the resulting models that hold for a large class of M's. It turns out that the generalized KdV hierarchies obey all of the same general properties! I refer to the second half of [19] for an explanation of this, and merely note that the Nth KdV hierarchy has a behavior similar to that of the models studied in §3 with a target space M such that the dimension of $H^*(M, \mathbb{R})$ is N (and the signature of M is 1 or 0 for odd or even N). In this correspondence, $v_{N-\alpha}$, $\alpha = 1, \cdots, N$ (or more precisely a certain differential polynomial of the form $v_{N-\alpha}+$ higher order terms), corresponds to $\langle\langle \tau_{0,1}\tau_{0,\alpha}\rangle\rangle$, where the $\tau_{0,\alpha}$ are the "primary fields" associated to a basis of $H^*(M, \mathbb{R})$. The variables $t_{n,\alpha}$ correspond to $y_{n(N+1)+\alpha}$. With this translation, the string equation of the N matrix chain has precisely the structure (3.22), and the other key conclusions of §3, such as the equations (3.27) and (3.48) that determine the genus zero and genus one correlation functions, may be deduced from standard properties of the generalized KdV hierarchies!

We do not actually believe that there is a mysterious Kähler manifold with N-dimensional cohomology that underlies the N matrix chain and on which the holomorphic curves are governed by the Nth generalized KdV hierarchy. It seems likely, though, that the model based on the N matrix chain has a geometrical interpretation in terms of an appropriate kind of intersection theory on some suitable moduli space.

Note added in proof: Recently K. Li (*Topological strings with minimal matter*, Caltech preprint CALT-68-1662) has answered the question raised in the last paragraph by showing which topological field theory coupled to topological gravity is equivalent to the N matrix model. This has been further clarified in R. Dijkgraaf and E. and H. Verlinde (*Topological strings in* $D < 1$, Institute for Advanced Study preprint, October, 1990). The interpretation of the N matrix model in algebraic geometry turns out to involve intersection theory on a cover of moduli space obtained by taking certain fractional roots of the canonical line bundle of a surface, as will be explained elsewhere (E. Witten, to appear).

Improved derivations of some of the foundational questions related to §4.3 have been given by H. Neuberger (*Regularized string and flow equations*, Rutgers preprint RU-90-50).

References

[1] J. Ambjorn, B. Durhuus & J. Forhlich, *Diseases of triangulated random surfaces models, and possible cures*, Nuclear Phys. B **257** [FS14] (1985) 433; J. Ambjorn, B. Durhuus, J. Frohlich & P. Orland, *The appearance of critical dimensions in regulated string theories*, Nuclear Phys. B **270** [FS16] (1986) 457.

[2] M. F. Atiyah, Lecture at the Schloss-Ringberg meeting, April, 1987; *New invariants of three and four-dimensional manifolds*, The Mathematical Heritage of Hermann Weyl (R. Wells, ed.), Proc. Sympos. Pure. Math. Vol. 48, Amer. Math. Soc., 1988.

[3] ——, *Topological quantum field theories*, René Thom festschrift (to appear).

[4] M. F. Atiyah & R. Bott, *The Yang-Mills equations over Riemann surfaces*, Philos. Trans. Roy. Soc. London Ser. A **308** (1982) 523.

[5] M. F. Atiyah & L. Jeffreys, *Topological Lagrangians and Cohomology*, preprint, 1990.

[6] C. Bachas & P. M. S. Petropoulos, *Doubling of equations and universality in matrix models of random surfaces*, preprint, CERN-TH5714/90.

[7] T. Banks, M. Douglas, N. Seiberg & S. Shenker, *Microscopic and macroscopic loops in nonperturbative two dimensional quantum gravity*, Rutgers University preprint RU-89-50.

[8] L. Baulieu & I. Singer, *The topological sigma model*, Comm. Math. Phys. **125** (1989) 227.

[9] ——, preprint, 1990.

[10] D. Bessis, C. Itzykson & J.-B. Zuber, *Quantum field theory techniques in graphical enumeration*, Advances in Appl. Math. **1** (1980) 109.

[11] E. Braaten, T. Curtwright & C. Thorn, *An exact operator solution of the quantum Liouville field theory*, Ann. Physics (N.Y.) **147** (1983) 365.

[12] E. Brezin, C. Itzykson, G. Parisi & J.-B. Zuber, *Planar diagrams*, Comm. Math. Phys. **59** (1978) 35.

[13] E. Brezin & V. A. Kazakov, *Exactly solvable field theories of closed strings*, Phys. Lett. B **236** (1990) 144.

[14] F. David, *Planar diagrams, two dimensional lattice gravity, and surface models*, Nuclear Phys. B **257** [FS14] (1985) 45.

[15] ——, *Conformal field theory coupled to two dimensional gravity in the conformal gauge*, Modern Phys. Lett. A **3** (1988) 509.

[16] P. Deligne & D. Mumford, *The irreducibility of the space of curves of given genus*, Inst. Hautes Études Sci. Publ. Math. **45** (1969) 75.

[17] R. Dijkgraaf, *A geometrical approach to two dimensional conformal field theory*, Ph. D. Thesis, University of Utrecht, pp. 29–34.

[18] R. Dijkgraaf, E. Verlinde & H. Verlinde, *Loop equations and Virasoro constraints in non-perturbative 2-D quantum gravity*, Inst. for Advanced Study preprint, 1990.

[19] D. Dijkgraaf & E. Witten, *Mean field theory, topological field theory, and multimatrix models*, Nucl. Phys. B **342** (1990) 486.

[20] J. Distler, *2-D quantum gravity, topological field theory, and the multicritical matrix models*, Princeton preprint PUPT-1161, 1990.

[21] J. Distler & H. Kawai, *Conformal field theory and two dimensional quantum gravity*, Nuclear Phys. B **321** (1989) 509.

[22] S. Donaldson, *Polynomial invariants of smooth four manifolds*, Topology (to appear).

[23] M. Douglas, *Strings in less than one dimension and the generalized KdV hierarchies*, Rutgers University preprint.

[24] M. Douglas, N. Seiberg & S. Shenker, *Flow and instability in quantum gravity*, Rutgers University preprint, 1990.

[25] M. Douglas & S. Shenker, *Strings in less than one dimension*, Rutgers University preprint; Nuclear Phys. B (to appear).
[26] C. Faber, *Chow rings of moduli spaces of curves*, Ph.D. thesis, and private communication.
[27] A. Floer, *Morse theory for fixed points of symplectic diffeomorphisms*, Bull. Amer. Math. Soc. **16** (1987) 279.
[28] D. Friedan, E. Martinec & S. Shenker, *Conformal invariance, supersymmetry, and string theory*, Nuclear Phys. B **271** (1986) 93.
[29] M. Fukuma, H. Kawai & R. Nakayama, *Continuum Schwinger-Dyson equations and universal structures in two-dimensional quantum gravity*, University of Tokyo preprint UT-562, 1990.
[30] C. S. Gardner, J. M. Green, M. D. Kruskal & R. M. Miura, *Methods for solving the Korteweg de Vries equation*, Phys. Rev. Lett. **19** (1967) 1095; M. D. Kruskal, R. M. Miura, C. S. Gardner & N. J. Zabusky, *Korteweg de Vries equations and generalizations*: V. *Uniqueness and existence of polynomial conservation laws*, J. Math. Phys. **11** (1970) 952.
[31] I. M. Gel′fand & L. A. Dikii, *Asymptotic behavior of the resolvent of Sturm-Liouville equations and the algebra of the Korteweg-De Vries equations*, Russian Math. Surveys **30** (1975) No. 5, 77.
[32] J.-L. Gervais & A. Neveu, *Dual string spectrum in Polyakov quantization*, I, II, Nucl. Phys. B **199** (1982) 59, B **209** (1982) 125.
[33] P. Ginsparg, M. Goulian, M. R. Plesser & J. Zinn-Justin, (p, q) *String actions*, Harvard preprint, 1990.
[34] M. Gromov, *Pseudo-holomorphic curves in symplectic manifolds*, Invent. Math. **82** (1985) 307.
[35] D. J. Gross & A. A. Migdal, *A nonperturbative treatment of two dimensional quantum gravity*, Phys. Rev. Lett. **64** (1990) 127.
[36] J. Harer, *The cohomology of the moduli space of curves*, preprint.
[37] J. Harer & D. Zagier, *The Euler characteristic of the moduli space of curves*, Invent Math. **185** (1986) 457.
[38] J. H. Horne, Princeton University preprint (to appear).
[39] J. Jevicki & T. Yoneya, *Action principle for strings in less than one dimension*, Brown University, preprint, 1990.
[40] V. A. Kazakov, *The appearance of matter field from quantum fluctuation of 2D quantum gravity*, Modern Phys. Lett. A **4** (1989) 2125.
[41] V. A. Kazakov, I. K. Kostov & A. A. Migdal, *Critical properties of randomly triangulated planar random surfaces*, Phys. Lett. B **157** (1985) 295.
[42] V. Knizhnik, A. M. Polyakov & A. Zamolodchikov, Modern Phys. Lett. A **3** (1988) 819.
[43] M. L. Mehta, *Random matrices*, Academic Press, New York, 1967.
[44] ——, *A method of integration over matrix variables*, Comm. Math. Phys. **79** (1981), 327; S. Chadha, G. Mahoux & M. L. Mehta, *A method of integration over matrix variables*. II, J. Phys. A **14** (1981) 579.
[45] E. Miller, *The homology of the mapping class group*, J. Differential Geometry **24** (1986) 1.
[46] D. Montano & J. Sonnenschein, *Topological strings*, Nuclear Phys. B **313** (1989) 258.
[47] J. Morava, *Forms of K-theory*, Math. Z. **201** (1989) 401.
[48] S. Morita, *Characteristic classes of surface bundles*, Invent. Math. **90** (1987) 551.
[49] J. Moser, *Infinitely many mass points on the line under the influence of an exponential potential—an integrable system*, Dynamical Systems, Theory and Applications, Lecture Notes in Phys. Vol. **38**, J. Moser, ed., Springer, Berlin 1975.

[50] D. Mumford, *Towards an enumerative geometry of the moduli space of curves*, Arithmetic and Geometry, (Michael Artin & John Tate, eds.), Birkhäuser, Basel, 1983.
[51] R. Penner, *Perturbative series and moduli space*, Bull. Amer. Math. Soc. (N.S.) **15** (1986) 73; *The moduli space of punctured surfaces*, preprint.
[52] M. Pernici, J. Labastida & E. Witten, *Topological gravity in two dimensions*, Nuclear Phys. B **310** (1988) 611.
[53] R. F. Picken, *The Duistermaat-Heckman integration formula on flag manifolds*, J. Math. Phys. B **31** (1990) 616.
[54] A. M. Polyakov, *Quantum geometry of bosonic strings*, Phys. Lett. B **103** (1981) 207.
[55] G. Segal, *Two dimensional conformal field theories and modular functors*, IXth Internat. Conf. Math. Phys. (Swanswea, July, 1988) (B. Simon, A. Truman, & I. M. Davies, eds.), Adam Hilger, 1989, 22; and preprint (to appear).
[56] G. 't Hooft, Nuclear Phys. B **72** (1974) 461.
[57] E. Verlinde & H. Verlinde, *A solution of topological gravity*, Inst. for Advanced Study preprint, 1990.
[58] E. Witten, *Topological sigma models*, Comm. Math. Phys. **118** (1988) 411.
[59] ——, *On the structure of the topological phase of two dimensional gravity*, Nucl. Phys. B **340** (1990) 281.

INSTITUTE FOR ADVANCED STUDY

Made in the USA
Monee, IL
08 November 2026

68055340R00175